Photochemistry in Microheterogeneous Systems

Photochemistry
in Microheterogeneous
Systems

K. KALYANASUNDARAM

Institute of Physical Chemistry
Swiss Federal Institute of Technology (EPFL)
Lausanne, Switzerland

 1987

ACADEMIC PRESS, INC.
Harcourt Brace Jovanovich, Publishers

Orlando   San Diego   New York   Austin
Boston   London   Sydney   Tokyo   Toronto

GOVERNORS STATE UNIVERSITY
UNIVERSITY PARK
IL. 60466

ACADEMIC PRESS, INC.
Orlando, Florida 32887

United Kingdom Edition published by
ACADEMIC PRESS INC. (LONDON) LTD.
24–28 Oval Road, London NW1 7DX

Library of Congress Cataloging in Publication Data

Kalyanasundaram, K.
    Photochemistry in microheterogeneous systems.

    Includes index.
    1. Photochemistry. 2. Microchemistry. I. Title.
QD715.K35   1986   541.3'5   86-10841
ISBN 0–12–394995–5 (alk. paper)

Contents

Chapter 4   Micellar Photochemistry and Photoreactions

Chapter 5   Photoprocesses in Reversed Micelles and in Microemulsions

Chapter 6   Photoprocesses in Lipids, Surfactant Vesicles, and Liposomes

Foreword

The subjects of this monograph have a long history, longer than that of chemistry itself. They are the basis of many familiar things such as foods and soaps and are at the heart of living matter. They used to form part of the discipline called colloid chemistry, which was concerned mainly with their gross physicochemical and thermodynamic properties rather than with their structure at the molecular level, although there were notable exceptions among which is the work of Irving Langmuir on monomolecular films, which must be regarded as having pioneered the whole field.

Over the last one or two decades the study of these substances has flourished to such an extent that it is only a small exaggeration to say that a new kind of chemistry has been born. It is the chemistry of aggregates of molecules, often very large numbers of them, in which the structure of the molecules, along with the solvent or other surrounding medium, determines the structure of the aggregate. But in spite of the rather large number of molecules that it contains, the aggregate is still small, usually of colloidal size, hence the name "microheterogeneous."

Photochemistry and its associated spectroscopic and optical observations are not readily carried out in normal macroscopic heterogeneous systems where light scattering makes quantitative measurements very difficult. Microheterogeneous systems on the other hand, whose particles are usually much smaller than the wavelength of light, form clear solutions having no scattering problems. Furthermore, the study of the photophysical properties of colored molecules incorporated into a microheterogeneous system often provides valuable structural information about the aggregate, indeed this may be the only method available because the aggregates often lie in an awkward size range and cannot be removed from their environment.

There were several reasons for the recent growth of interest in the microheterogeneous systems and their photochemistry in particular. First and most important, it has turned out to be extremely interesting scientifically because of the way that aggregates can be designed, on the basis of the known structure of their individual components, to have a varied range of structures, properties, and shapes. Second, new phenomena of great potential practical importance have been discovered, such as the optical properties of liquid crystals and the catalytic properties of zeolites and polyelectrolytes. Third, the great interest in photochemical methods for the storage of solar energy opened up particulate absorbers as photoelectrochemical contenders. Last, and now the main stimulus, is the realization of the overwhelming importance of lipid membrane and vesicular structures in all biological systems. The supreme example of photochemistry in microheterogeneous systems is, of course, the photochemistry of the photosynthetic unit.

Dr. Kalyanasundaram has a wide experience of most aspects of this new field, having worked in several of the principal laboratories and made some notable contributions himself. This monograph provides an extensive review of the large amount of recent work in these diverse topics, but it is very readable and will be found interesting and useful to all who wish to learn of the rapid progress in a fascinating area of science.

SIR GEORGE PORTER P.R.S.

Preface

As new instruments become available and as newer, more sophisticated techniques are created, we witness their novel application to increasingly complex chemical and biological systems. The combined efforts of scientists with very different backgrounds involved in these novel applications drive the evolution of new disciplines. The topic of this monograph concerns one such area, namely, application of photophysical and photochemical processes and techniques to the study of various microheterogeneous systems of chemical and biological interest.

This short monograph was written to provide an introduction to the subject of photochemistry in microheterogeneous systems for the student at the graduate level and to review the recent, significant developments in the field for the practicing chemist. It should be equally useful to those who intend to broaden their research in this new and exciting field. The systems considered are of interest and utility to those in a wide spectrum of research in specialized fields from chemistry to biology: colloids, interfaces, catalysis, kinetics, polymers, biomembranes, photochemistry, and photobiology, to name a few.

There are two potential approaches that we can take in discussing the photochemistry in microheterogeneous systems (MHS): We can consider each photophysical and chemical process separately and discuss its occurrence and applications to different forms of the MHS, or we can choose a certain type of MHS and outline how the existing knowledge of the systems and photoprocesses can profitably be employed to gain a better understanding of the systems and processes. We choose this latter approach for three principal reasons: (1) each system is unique in having different static and dynamic properties; (2) systems of increasing complexity are readily handled as extensions in a logical manner (chronologically the evolution of the subject has been on these lines!); and (3) most researchers' interests lie on one or more types of the MHS.

We consider a variety of simple, organized systems that are structurally well characterized. They are "microheterogeneous" in that they are heterogeneous at the microscopic level with the presence of charged interfaces in hydrophilic or hydrophobic domains. In all of them there is some kind of self-organization and order that we want to exploit. The motivations of these studies are numerous, but we can single out two main, complementary ones: (1) to use the existing knowledge of the photophysical and photochemical processes to probe the dynamic and static properties of these organized systems and (2) to use the information available about these systems to study excited-state, molecular processes under novel microenvironments. Success in both of these areas has been phenomenal and is growing all the time.

Partially reflecting the early origin and maturity of aggregated systems composed of surfactants and lipids, a major portion of the book deals with normal and inverted micelles, vesicles and liposomes, monolayers, black lipid membranes, and liquid crystalline solvents. This is followed by overviews of newer topics of current research with organic and inorganic polymers, e.g., neutral and ionic polymers, polyelectrolytes, ion-exchange membranes, polyaluminosilicates such as zeolites and clays, polysugars such as cyclodextrins, polyethers such as crown ethers and crpytands, and oxides such as alumina and silica.

It is a pleasure to express my gratitude to the many people who helped me directly and indirectly. Sincere thanks go to the "trio" who induced research interests in the topic of the book and encouraged my participation: Professors J. K. Thomas, Sir G. Porter, and M. Grätzel. Special thanks are also due to Professors D. G. Whitten, N. J. Turro, J. H. Fendler, and Dr. R. Humphry-Baker for undertaking the onerous task of reading the manuscript and offering invaluable comments and suggestions. Professor Harry B. Gray richly deserves my gratitude and appreciation for providing a very pleasant and stimulating environment for a major portion of this book to evolve in sunny California. Finally, I would like to thank my wife Uthira for willingly foregoing several evenings and weekends, which rightfully belonged to her, and for her assistance in typing several versions of the manuscript. This book is dedicated to our parents who cared and suffered so much to give us an excellent education.

Chapter 1

Introduction

This monograph is concerned with studies of unimolecular and bimolecular reactions of electronically excited molecules in nonhomogeneous media. There are two major goals for these studies: to use the existing knowledge on the photophysics and photochemical processes to probe the static and dynamic properties of a wide variety of organized microheterogeneous systems (systems of chemical, industrial, and biological interest) and conversely to use the known information on these systems to examine the excited-state processes under novel environments. The goal of this work is to illustrate the enormous progress that has been made in recent years in both of these areas.

The complementary nature of these two goals requires some knowledge in each of these areas. In order to provide the necessary background so that the present monograph is self-supporting and to set the stage for detailed discussions on the photochemistry in microheterogeneous systems per se, we briefly overview each of these areas. Readers who desire more elaborate discussions on the fundamentals or finer details on these background areas are well advised to consult comprehensive texts in these areas (photochemistry,[1-9] microheterogeneous systems[10-56]). As systems become increasingly complex, no single technique can provide all the answers and even unambiguously resolve different processes. There are numerous physical or chemical methods that we can utilize to supplement. Techniques such as magnetic resonance (NMR, ESR), diffraction methods (X-ray, neutron), electron microscopy, electrical (EMF, conductance, capacitance) and pulsed methods such as stopped flow, temperature-jump, pressure-jump, and ultrasonic relaxation are to mention a few. Often it is imperative that we use the information derived from one technique to gain insight into the results from another.

1.1 Microheterogeneous Systems, an Overview

The microheterogeneous systems that we will consider are numerous, and they can be broadly classified into two major types: molecular aggregates composed of surfactants or lipids and organic or inorganic polymeric systems and supports. Figure 1.1 presents schematically a broad subclassification of these systems as they are discussed in subsequent chapters. Organized molecular assemblies such as micelles, vesicles, microemulsions, and others have been quite intensely studied in the past decade. The field has largely matured, and the available photochemistry literature is fairly extensive. Micellar systems are often used as simpler model systems to study and understand larger, more complex aggregates. Consequently, a good proportion of our discussion concerns these systems, especially the simpler micellar systems. Herein we broadly survey the general features of various forms of microheterogeneous systems that we shall be dealing with. Later, as a prelude to the discussion of photochemistry, we will elaborate further on various structural and dynamical properties of interest in each chapter.

A few remarks on the usage of the term *microheterogeneous* are pertinent here. The systems under consideration are "heterogeneous" at the "microscopic" level. The implications are numerous. The solute distribution can be inhomogeneous throughout the entire volume of the solution/aggregate. There may be hydrophobic or hydrophilic cavities/cages/pockets/pools/pores that can sequester (or eliminate) the solutes. There may be charged interfaces where electrostatic effects can play a dominant role in influencing the solute distribution and their reactions. The systems or the process of dissolution of solutes in their interiors can be dynamic such as the continuous dissolution and reformation of the aggregates, entry, and exit of the solutes so that the solutes can experience some time-averaged effects due to the aggregates. It is important to note, however, that a good majority of these systems provide optically transparent (nonturbid) solutions readily amenable to photochemical investigations by steady-state and pulsed photolysis methods.[10]

Surfactant and lipid molecules with one or more long alkyl chains (with at least six methylene units) and a polar headgroup are called amphiphathic molecules:

X = headgroup ($\overset{+}{N}Me_3$, OSO_3^-, etc.)

$\sim\sim\sim$ = hydrocarbon tail $-(CH_2)_n$

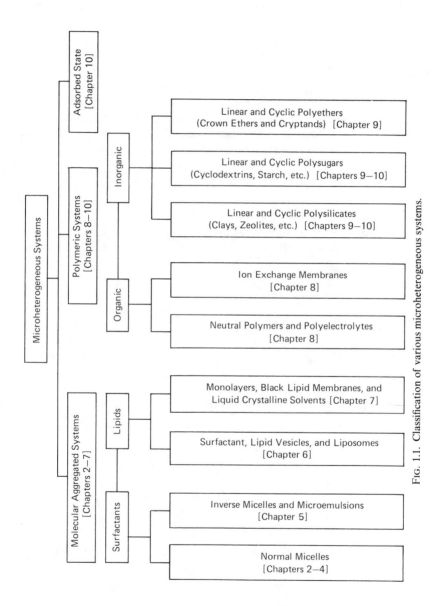

FIG. 1.1. Classification of various microheterogeneous systems.

Their unique structure confers in them both hydrophobic and hydrophilic properties. This gives them the fundamental property to form association structures of different types, some of which are shown in Fig. 1.2. Depending on the nature and length of the hydrocarbon chain, the headgroup, concentration, temperature, and other additives, the size (aggregation number) and shape of these aggregates may vary. The existence and boundaries separating different phases (forms) of the aggregates are best derived from studies of phase diagrams such as those shown in Figs. 1.3 and 1.4. Detailed discussions on the features of these phase diagrams and their roles are beyond the scope of this monograph. However, for those who envisage working with any of these organized assemblies, it is advisable that they carefully examine (or at least be aware of!) the phase diagrams for the systems of interest. During the studies of one particular system, variations in the surfactant/lipid concentration, temperature, etc. can lead to crossing of phase boundaries and drastic changes in the size and shape of the molecular aggregates under investigation.

The simplest systems to consider are those consisting of just two components, a surfactant or lipid dispersed in water. Single-chain surfactants

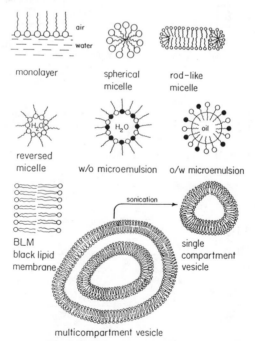

FIG. 1.2. Oversimplified representation of various forms of aggregated systems composed of surfactants and lipids. [From Fendler.[10] Copyright ©1982 John Wiley and Sons, Inc.]

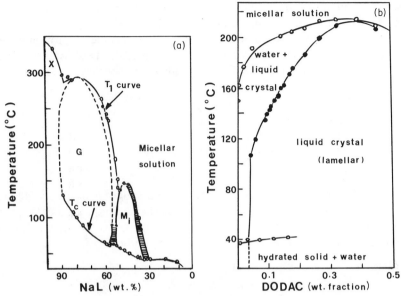

FIG. 1.3. (a) Phase diagram of sodium laurate–water system: M_1, middle soap (hexagonal phase); G, neat soap (lamellar phase); and I, isotropic (micellar solutions). (b) Phase diagram of dioctadecyldimethylammoniumchloride (DODAC)–water system. (From H. Kunieda and K. Shinoda, *J. Phys. Chem.* **82**, 1710 (1978). Copyright 1978 American Chemical Society.)

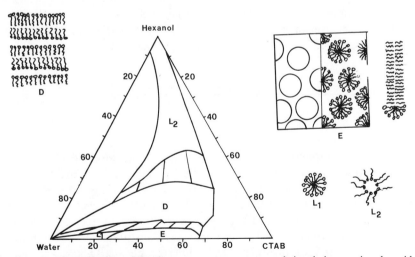

FIG. 1.4. Phase diagram of the three-component system cetyltrimethylammonium bromide (CTAB)–hexanol–water. Here L_1 and L_2 refer to regions with water-rich and hexanol-rich solutions, respectively; D and E are lamellar and hexagonal liquid crystalline phases, respectively. (From B. Lindman and H. Wennerstoron, *Top. Curr. Chem.* **87**, 26 (1980)).

such as sodium dodecylsulfate, $CH_3(CH_2)_{11}O$—$SO_3^- Na^+$ form quasi-spherical (or globular) micelles (diameter $\leqslant$ 30 Å).[11-16] Surfactants and lipids with two or more hydrocarbon chains form larger unilamellar or multilamellar vesicles, e.g., dodecyldimethylammonium chloride (DODAC) or dipalmitoylphosphatidylcholine (DPC). A unifying feature of these systems is their

$[CH_3(CH_2)_{11}]_2 N^+(CH_3)_2 Cl^-$ Dodecyldimethylammonium chloride (DODAC)

$$CH_3(CH_2)_n COOCH_2 \quad O$$
$$CH_3(CH_2)_n COOCH—P—O—CH_2CH_2N(CH_3)_2 \quad \text{Dipalmitoylphosphatidylcholine (DPC)}$$
$$O^-$$

structural design.[17-23] The hydrocarbon chains align/group themselves to form an inner hydrophobic part while the polar head groups are located at the hydrocarbon–water interface. Depending on the nature of the headgroup, these aggregates can be cationic, anionic, or nonionic:

Cationic $CH_3(CH_2)_{15}\overset{+}{N}(CH_3)_3 Br^-$ (CTAB)

Anionic $CH_3(CH_2)_{11}OSO_3^- Na^+$ (SDS)

Nonionic $CH_3(CH_2)_{12}(OCH_2CH_2)_{23}OH$ (Brij 35)

Though the formation of larger, higher-order structures such as lamellar phases at high surfactant/lipid concentrations is known, not many studies have been carried out in them. Chapters 2–4 and 6 deal with photoprocesses studied in surfactant and lipid systems, respectively.

The association of surfactants is not restricted to the aqueous medium. Surfactant molecules such as Aerosol OT (diisooctylsulfosuccinate):

$$CH_2CH_3$$
$$CH_3CH(CH_2)_4 OCOCH_2$$
$$CH_3CH(CH_2)_4 OCOCH—SO_3^- Na^+ \qquad \text{Aerosol OT}$$
$$CH_2CH_3$$

aggregate in nonpolar solvents such as octane in an inverted manner with polar head groups clustered inside and hydrophobic tails extending into the alkane solvent (cf. Fig. 1.2). These inverted micelles can solubilize large quantities of water in their inner core.[24] The inverted micelles with the inner water pool and normal micelles swollen with large quantities of long-chain alcohols solubilized in their hydrophobic core constitute the so-called water-in-oil (W/O) and oil-in-water (O/W) microemulsions. These are typical examples of three-component systems (surfactant, a long-chain alcohol, and water).[25-27] Chapter 5 describes recent photochemical studies in these systems.

In addition to the simple two-component systems, micelles and vesicles, there are also more sophisticated forms of two-component systems. Typical examples are monolayer and multilayer assemblies, black lipid membranes and liquid crystalline solvents.[28-37] From a practical point of view, these are more complex (delicate?) systems to handle and require utmost care and cleanliness. They are all based on the ability of lipids to form monolayer or multilayer arrays and/or membraneous films at the interfaces. An interesting feature is that these monolayers can be transferred to glass or some other support and manipulated. A brief discussion of photochemical studies in these systems can be found in Chapter 7.

With the enormous success obtained in the photochemical probing of static and dynamic properties of these organized molecular assemblies, attention is being focused on other organic or inorganic polymeric systems. Unlike the molecular aggregated systems, the host is composed of repeating units of inorganic or organic groups arranged in a linear (layer-type) or cyclic (cage) fashion. Due to their early industrial applications, photochemical studies of organic polymers such as polystyrene, polyvinylcarbazole, and polymethyl-

Polystyrene   Polyvinylcarbazole

acrylate has been active for more than a decade.[38-42] Studies of their ionic counterparts, polyelectrolytes[43-45] (e.g., polyvinylsulfate), however, has been

Polyvinylsulfate (PVS)

very recent. Aspects of molecular photophysics and photochemistry in these organic polymeric systems are reviewed in Chapter 8.

Inorganic polymeric systems provide a new class of microheterogeneous systems available for detailed investigations. Though the existence of linear, cyclic and colloidal polysilicates (for example zeolites,[46-48] clays,[49-50] and silica[51]) have been known for sometime, it is only recently that systematic studies have begun in these systems. Polysugars with carbohydrate repeating units (for example cyclodextrins, starch, and cellulose)[52,53] are another group of polymers of this type. Mainly due to the ingenuity of synthetic chemists, there is available now a third class of inorganic cyclic polymer units, composed of the ether linkage (—O—) as the repeating unit (e.g., polyethers such as crown ethers and cryptands[53-56]). Cyclic polymers such as zeolites, cyclodextrins, and crown ethers all have hydrophobic cavities that can "include" a variety of molecules, often with some shape and size selectivity. Chapter 9 is devoted to an outline of the type of photophysical and photochemical studies that are currently underway in these host–guest systems.

The last chapter is devoted to yet another form of microheterogeneous photochemistry that is gaining increasing interest. This concerns photo-chemistry of molecules in the *adsorbed state* (adsorbed on "inert" supports such as silica, alumina, or clays either in the dry state or in colloidal solutions of these oxides). Adsorption of molecules on surfaces leads to significant changes in the observed photophysics and in the reaction course of photoreactions.

1.2   Excited-State Processes and Reactions, an Overview

Let us consider various pathways available for electronically excited states of organic molecules. (The majority of probe molecules that we will be dealing with will be condensed aromatic hydrocarbons with or without polar substituents.) Following the initial act of light absorption, the molecule is raised to an electronically excited state in the singlet manifold. Schemes such as those in the Jablonski diagram summarize various unimolecular pathways available. (With inorganic ions and metal complexes, depending on the metal ions and its oxidation state, excited-state manifold can involve singlets, quartets, doublets, and triplet states). A very rapid relaxation ($\leqslant 10^{-12}$ sec) (internal conversion) leads to the arrival in the lowest singlet excited state S_1. The singlet excited state S_1 can either return to the ground state S_0 via nonradiative or radiative transitions (fluorescence) or intersystem crossing to a triplet state in the triplet manifold. Due to the spin restrictions, the mole-cules in the triplet state slowly return to the ground state S_0 again via a nonradiative or radiative transition (phosphorescence) (Fig. 1.5). In photo-physical studies we attempt to obtain quantitative information on all of these interrelated processes.

1.2.1   Medium Effects

In condensed media, the solvent molecules constituting the medium (in microheterogeneous systems the medium is referred to, more loosely, as "the environment") can intervene in some of the preceding excited-state processes. Usually solvents interact strongly with excited states that have higher dipole character compared to the ground state. The solute–solvent interactions can affect the relative populations at various vibrational levels of the excited state or the relative positions of the Franck–Condon envelope of the excited state with respect to the ground state. The former manifests itself as variations in the vibronic band intensites and the latter in large red or blue shifts in the emission maxima, often accompanied by changes in fluorescence yields. Molecules in

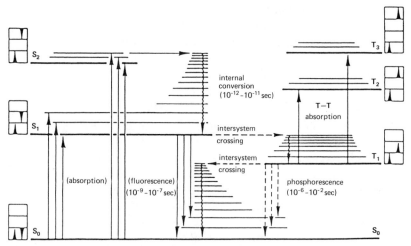

FIG. 1.5. Schematic representation of various excited electronic states of an organic molecule, their energy level relationships, and interconversion processes.

the excited state can rapidly protonate (or deprotonate) or even solvate to a relaxed state far from the initially formed Franck–Condon state. The emission from each of these "modified" states can often be different. The polarity and viscosity of the medium can influence the orientation and mobility of the probe molecule in the medium. This can lead to differences in the degree of depolarization of fluorescence. Thus, the solvent effects on emission occur in many ways, either as specific solute–solvent interactions or bulk effects as in solvent viscosity or polarity. Though the origin of these effects can be complex, the major attraction of luminescence probe analysis has been in the empirical utilization of these medium effects.

Arylaminonaphthalene sulfonates and substituted naphthylamines are a typical class of compounds that exhibit large solvent effects.[57-59] In this class of compounds, the fluorescence lifetime and quantum yields increase sharply with decreasing polarity of the medium. Detailed investigations of the excited-state photophysics have led to a rather complex picture of solvent relaxation in the excited state. According to these studies such probes as anilinonaphthalenesulfonate (ANS), on excitation to the singlet excited state S_1, undergo intramolecular charge transfer to form a charge-transfer singlet state S_{1CT}. Emission can occur from either of these two states. Solvents of high polarity promote the charge-transfer transition $S_1 \rightarrow S_{1CT}$ and rate of deactivation of S_{1CT} and decrease the frequency of the $S_{1CT} \rightarrow S_0$ fluorescence. However, the rates of the two processes $(S_1 \rightarrow S_{1CT})$ and $(S_{1CT} \rightarrow S_0)$ are controlled not only by the polarity of the medium but also by the viscosity of the medium.

In microheterogeneous systems, changes in the nature of the environment experienced by the probe on transfer from the aqueous medium to the host (or migration from one region to another within the host) are readily detected in the emission properties of the fluorescence probes. We shall discuss several applications of this effect in the study of microheterogeneous systems: onset of aggregation, phase transitions, binding constants, polarity of interfaces, gegenion binding, vesicle fusion, etc.

1.2.2 Acid–Base Equilibria in the Excited State

Most of the molecules in the excited state (singlet, triplet, or other spin states) are chemically new species with very different properties in several areas such as redox potentials, acid–base strength, and dipole moment.[60-62] Organic molecules with ionizable groups such as hydroxyl (—OH as in naphthols), carboxyl (—COOH as in naphthoic acid) or amines (—N=NH as in azobenzenes) often show large differences in the pK_a values between the ground and excited states. Excited molecules can undergo reactions without electronic deactivation. The pK_a values of the excited states can be evaluated from spectral measurements (absorption and luminescence) of the concentration of each form of the excited state species (protonated R—OH* and deprotonated R—O$^-$*):

$$
\begin{array}{ccc}
\text{R—OH*} & \xrightleftharpoons{\ pK_a^*\ } & \text{R—O}^-\text{*} + \text{H}^+ \\[2pt]
hv \Big\Updownarrow \text{fluor.} & & hv \Big\Updownarrow \text{fluor.} \\[2pt]
\text{R—OH} & \xrightleftharpoons{\ pK_a\ } & \text{R—O}^- + \text{H}^+
\end{array}
\tag{1.1}
$$

Commonly, the singlet excited states have substantially different pK_a values (differences often up to 8 pH units) than the ground state S_0 or the lowest triplet state T_1; pH indicators in the ground and excited states find interesting applications in the study of surface properties, pH gradient in the host interiors, and proton transport in biomembranes.

1.2.3 Depolarization of Fluorescence

Depolarization of fluorescence is an important property widely used in the study of macromolecules and microheterogeneous systems.[2,63,64] When a fluorescent molecule is excited by polarized light, its emission will be maximally polarized if, during its excited state lifetime, the probe does not change its position or orientation as in a viscous medium. However, if the molecule is not rigidly held, brownian motions will tend to remove the orientation imposed by the polarized excitation.

If the probe is excited with a plane-polarized light, then the polarization of fluorescence p and the fluorescence anisotropy r are given by

$$p = [I_v - I_h]/[I_v + I_h] \tag{1.2}$$

$$r = [I_v - I_h]/[I_v + 2I_h] \tag{1.3}$$

where I_v and I_h are the fluorescence intensities observed through a polarizer oriented vertically or horizontally to the plane of the polarization of the exciting beam. For randomly oriented molecules, p takes the characteristic value p_0:

$$p_0 = [3 \cos^2 \theta - 1]/[\cos^2 \theta + 3] \tag{1.4}$$

where θ is the angle between the emission and absorption oscillators. Commonly $\theta = 0$ and $p_0 = 0.5$.

In the absence of any motion, some fluorescence anisotropy exists which is called the fundamental anisotropy, r_0:

$$r_0 = [3 \cos^2 \theta - 1]/5 \tag{1.5}$$

If during the excited state lifetime, the probe undergoes motion, then the fluorescence anisotropy will depend on the motion amplitude:

$$r(t) = [I_v(t) - I_h(t)]/[I_v(t) + 2I_h(t)] \tag{1.6}$$

$$= r_0 P_2(t) = r_0 \langle [(3 \cos^2 \theta(t) - 1)/2] \rangle \tag{1.7}$$

where $\theta(t)$ is the angle through which the emission transition moment rotates in time t, the $\langle \rangle$ brackets mean an average over all the fluorescent molecules, and $P_2(t)$ is the second-order Legendre polynomial and corresponds to the orientation autocorrelation function of motion undergone by the fluorescent molecules.

Under continuous illumination, measurements lead to the mean fluorescence anisotropy, $\bar{r}$:

$$\bar{r} = r_0(\tau)^{-1} \int_0^\infty P_2(t) \exp\left(\frac{-t}{\tau}\right) dt \tag{1.8}$$

where τ is the mean fluorescence lifetime. Continuous illumination measurements are the easiest to perform, but they provide only limited information on the molecular motion. Fluorescence anisotropy decay curves can be obtained by time-dependent measurements carried out with narrow light pulses ($\simeq 2$ nsec) and a single photon counting system. We now briefly consider a motional model which has been used to describe the orientational autocorrelation functions as it corresponds in homogeneous solvents.

Isotropic Motion

Isotropic motion describes the motion of a rigid sphere of volume V in a medium of viscosity η. For rigid spheres, Einstein has shown that $P_2(t)$ is a single exponential:

$$r(t) = r_0 \exp(-6Dt/\tau) \tag{1.9}$$

where D is the rotational diffusion coefficient. This relation also leads to

Perrin's relationship:

$$1/r = (1/r_0)[1 + (3\tau_{fl}/\rho)] \tag{1.10}$$

or expressed in terms of the (mean) degree of polarization $\bar{p}$:

$$(1/\bar{p}) - \tfrac{1}{3} = [(1/p_0) - \tfrac{1}{3}][1 + (3\tau_{fl}/\rho)] \tag{1.11}$$

where τ_{fl} is the fluorescence decay time of the probe. If τ_{fl} is known, then a measurement of the single rotational relaxation time ρ can be made if time-averaged degree of polarization is measured under continuous excitation. In order to determine p_0, use is made of the Stokes relation:

$$\rho = 3V\eta/kt \tag{1.12}$$

where η is the viscosity of the medium. Measurement of $\bar{p}$ in various solvents of different viscosity η leads to evaluation of V and p_0. A popular, empirical application has been in the determination of microviscosity of host interiors. More rigorous applications are in the study of dynamics of solute diffusion in microheterogeneous systems via time-resolved depolarization studies.

Phase shift methods involving differential polarised phase fluorimetry can also be used to determine the time-dependent anisotropies. Discussions on this point can be found in Chapter 6 (Section 6.6) and in references 2a, 2b, and 64.

1.2.4 Excited-State Quenching

In competition with the normal decay pathways described earlier, the excited states can also be rapidly deactivated (quenched) in reactions involving suitable "quencher" molecules:

$$S^* + Q \longrightarrow S + Q \tag{1.13}$$

In addition to the external reagents present in the solution, the probe molecules S themselves can act as quenchers in the ground, excited states, a process identified as "self-quenching." In homogeneous solutions, the quenching process occurs via diffusion of molecules relative to each other and resulting bimolecular collisions. Hence, studies of fluorescence quenching have been a testing ground for various theories of diffusion-controlled reactions. The quenching rate constant k_q for reaction (1.13) follows Stern–Volmer kinetics:

$$\Phi_0/\Phi = 1 + k_q\tau_0[Q] \tag{1.14}$$

where Φ_0 and Φ are the fluorescence quantum yields in the absence and in the presence of the quencher whose concentration is given by $[Q]$ and τ_0 is the lifetime in the absence of quenchers. Since $\Phi = \tau_0/\tau$, the Stern–Volmer equation can also be written as

$$\tau_0/\tau = 1 + k_q\tau_0[Q] \tag{1.15}$$

This description of the collisional quenching constitute the so-called dynamic quenching case. Additional quenching can occur with increasing solute concentrations if the added solute forms a nonfluorescent complex with the ground-state probe. In this case, the quenching process is identified as static quenching. In the latter process, no molecular collisions are involved and, hence, the process is viscosity independent. Also in this case, we observe a drop in the total fluorescence intensity without any change in lifetime.

The kinetic aspects of dynamic quenching are explained fairly well in terms of a simple picture of diffusion-controlled reactions developed by Smolouchowski, Perrin, and others. Various aspects of diffusion-controlled reactions have been reviewed by Noyes[65] and by Birks *et al.*[66] According to Smolouchowski's diffusion model, the reaction rate will be controlled solely by the rate at which the reactants diffuse together, in which case, the bimolecular rate constant k for collisions is given by

$$k = (8RT/3000\eta)(pa/b) \tag{1.16}$$

where η is the macroscopic viscosity of the solvent, a and b are the interaction radii, and p is the reaction probability per collision, usually assumed to be equal to unity. Thus, dynamic quenching approaching the diffusion limit is strongly dependent on the solvent viscosity and temperature.

Analysis of the observance of the Stern–Volmer relationship is widespread in studies of excited-state quenching. Positive deviations from the Stern–Volmer relation is often attributed to cases in which the quencher happens to be present in the immediate vicinity of the probe at the instant of excitation (i.e., within the reaction sphere). A transient-effect model, a sphere of action model, as well as a dark complex model have all been proposed as static quenching mechanisms. A time-dependent rate constant has been derived for the transient case:[67a,b]

$$k'_d = \underset{\substack{\text{steady-} \\ \text{state term}}}{4\pi RD} + \underset{\substack{\text{transient} \\ \text{term}}}{4R^2(\pi D)^2 t^{1/2}} \tag{1.17}$$

According to Perrin's sphere of action model, the quenching is purely static. The probability p that an excited probe will lie within the active sphere is given by

$$p = \exp(-nv/V) \tag{1.18}$$

where V is the volume of the solution, v the volume of each active sphere, and n the number of quencher molecules per cubic centimeter. Unlike the earlier static quenching case (k independent of $[Q]$), in this case, the rate decreases exponentially with $[Q]$:

$$\Phi_0/\Phi = \exp VN[Q] \tag{1.19}$$

Studies of the efficiency of the quenching of excited states of solubilized probes in various microheterogeneous systems provide a simple and quantitative means of evaluating various properties such as extent of accessibility of the probes to other molecules, their mobility, influence of additives in affecting the internal structure of the host aggregates, surface charges and their charge density, counterion binding, and their exchange.

1.2.5 Excimers and Exciplexes

In suitable cases of fluorescence quenching, the probe and the quenchers can form a transient luminescent complex, as shown in reactions (1.21) and (1.22):

$$S^* \longrightarrow S + h\nu \qquad \text{(monomer emission)} \qquad (1.20)$$

$$S^* + S \longrightarrow [S.S]^* \longrightarrow S + S + h\nu' \qquad \text{(excimer emission)} \qquad (1.21)$$

$$S^* + Q \longrightarrow [S.Q]^* \longrightarrow S + Q + h\nu'' \qquad \text{(exciplex emission)} \qquad (1.22)$$

Formation of excimers at high probe concentrations is a common behavior with condensed aromatic hydrocarbons such as pyrene. At high probe concentrations, in addition to the structured monomer fluorescence, we also observe a broad, structureless, red-shifted emission attributable to the excimers.[68−72] The kinetics and mechanisms involved in excimer formation in homogeneous solutions have been extensively investigated. Forster and Kasper, who first discovered the excimer formation process, have proposed the following scheme:

$$P \xrightarrow{(1)} P^* + P \overset{(2)}{\rightleftharpoons} [PP]^*$$

$$\downarrow{(3)} \qquad\qquad\qquad \downarrow{(4)}$$

$$P + h\nu \quad P \qquad\qquad PP + h\nu' \quad PP$$

In homogeneous solvents such as cyclohexane or methanol, step (2), the excimer formation step, is diffusion controlled. In order to account for the diffuse, broad, and structureless nature of the excimer emission, it has been proposed that the excimer ground state is unstable. Based on the crystal structure studies, a sandwich configuration for pyrene excimers has been proposed in which the excimer is formed by mutual approach of two pyrene molecules with their molecular planes parallel:

 Pyrene excimers

The interplanar distance is about 3.53 Å. The entropy and enthalpy of dissociation for excimers is 18.5 ± 2.0 cal/deg mol and 9.5 ± 1.0 kcal/mol, respectively. These numbers indicate a strong binding and a rigid configuration for the pyrene excimers.

Molecules such as pyrene also form mixed or heteroexcimers (also called as exciplexes with other solutes such as substituted amines.[68,69] The stability of exciplexes is often restricted to nonpolar solvents. In polar solvents, due to the stabilization energy available via solvation, the transient complex dissociates into radical ions:

$$
\begin{array}{ccccc}
S^* + Q & \longrightarrow & [S^- \cdot \cdot Q^+]^* & \xrightarrow[\text{solvent}]{\text{nonpolar}} & S + Q + hv' \\
\downarrow & & \downarrow \text{polar} & & \text{exciplex emission} \\
& & \text{solvent} & & \\
S + hv & & S^-_{\text{solv}} + Q^+_{\text{solv}} & &
\end{array}
$$

As with the excimers, in homogeneous solvents, exciplex formation is diffusion controlled.

The formation of an excimer or exciplex can be intermolecular as between two isolated (freely moving) probe molecules or intramolecular as with two chromophores covalently linked to each other via one or more methylene or ether linkages:

$$
\begin{array}{cc}
Py^* + Py & Py^*\text{---}(CH_2)_n\text{---}Py \\
\text{intermolecular} & \text{intramolecular}
\end{array}
$$

In the latter case, the covalent linkages restrict the relative motion of the chromophore units toward each other. With diarylalkanes, $Ar\text{---}(CH_2)_n\text{---}Ar$, the efficiency of excimer/exciplex formation has been found to depend on the connecting chain length n, with optimal results obtained for $n = 3$. The diffusion-controlled nature of the formation of excimers finds a wide variety of applications in the study of microheterogeneous systems: statistics of solute distribution in the host aggregates, phase transitions, lipid exchange, vesicle fusion, cyclization dynamics in macromolecular systems, etc.

1.2.6 Excitation Energy Transfer

Under certain circumstances, quenching of an electronically excited molecule by a quencher can result in the formation of the quencher excited state:[1,73–75c]

$$
S^* + Q \longrightarrow S + Q^* \tag{1.23}
$$

For example, with suitable acceptor molecules whose absorption spectra overlap with the fluorescence of the excited molecule, excitation energy can be transferred to the acceptor molecule radiatively or nonradiatively. The process of reabsorption of the emitted fluorescence of the donor by the acceptor is known as the "trivial" process. In addition to the trivial process, there is transfer of excitation energy by coulombic and/or electronic exchange interactions. The processes involving isoenergetic radiationless transitions are known as resonance transfer or inductive resonance:

$$^1D^* + {}^1A \longrightarrow {}^1D + {}^1A^* \tag{1.24}$$

The energy transfer is adiabatic, irreversible, and slow compared to vibrational relaxation. The difference in the electronic energy of the donor and acceptor states is distributed as vibrational energy between the two final states. The process requires overlap of the donor fluorescence spectrum with the absorption of the acceptor. The energy transfer probability depends on the extent of overlap.

The coulombic interaction is usually expressed as multipole–multipole expansion, the leading term being the dipole–dipole. The theory for dipole–dipole coulombic energy transfer has been developed by Förster, who derived the relation

$$k = \frac{9000(\ln_{10})\kappa^2\Phi_D}{128\pi^5 n^4 r^6 N\tau_D} \int_0^\infty \frac{F_D(v)\varepsilon_A(v)\,dv}{v^4} \tag{1.25}$$

where n is the refractive index of the medium at the wavelength of excitation, N Avagadro's number, ε_A the molar extinction coefficients of the acceptor, κ the orientation factor, Φ_D the quantum yield of the donor fluorescence, τ_D the mean lifetime of the donor excited state, $F_D(v)$ the spectral distribution of fluorescence of the donor (in quanta normalised to unity), and r the interchromophore distance. An equivalent form of this equation is

$$k = (1/\tau_D)(R_0/r)^6 \tag{1.26}$$

where R_0 is the critical distance at which the energy transfer and donor excited state deactivation by fluorescence or internal quenching are of equal probability. The dipole–quadrupole interaction is proportional to r^{-8} and higher-order terms decrease as a higher power of interchromophore distance r. The coulombic (dipole–dipole) energy transfer occurs over long distances (20–60 Å) while the electron exchange interactions are over shorter distances (6–15 Å). The excitation energy transfer by inductive resonance mechanisms are highly selective between the donor and acceptor chromophores, occurring over specific distances. Hence, they serve as a spectroscopic ruler to measure interchromophore distances in macromolecules and larger bioaggregates.

1.2.7 Photoionization and Photoredox Reactions

Molecular excited states, either singlets or triplets, can ionize to give radical ions or undergo electron transfer reactions with added solutes as indicated in reactions (1.28) and (1.29)

$$S^* \;\rightleftharpoons\; S^+ + e^-_{solv} \tag{1.27}$$

$$S^* + Q \;\rightleftharpoons\; S^+ + Q^- \tag{1.28}$$

$$S^* + Q \;\rightleftharpoons\; S^- + Q^+ \tag{1.29}$$

Focusing our attention on the molecule S that absorbs the light, reactions (1.28) and (1.29) are identified as oxidative and reductive quenching, respectively. The photoionization process [reaction (1.27)] is often observed in polar solvents due to the enthalpy gain available from the solvation of the photoproducts and can occur either via monophotonic or biphotonic pathways.[76-78] With the rapid reduction/oxidation of one of the products in reactions (1.28) and (1.29), photoredox reactions are readily adapted to photosensitized redox processes:[79-81]

$$D + A \;\underset{\text{sensitizer}\,S}{\overset{h\nu}{\rightleftharpoons}}\; D^+ + A^- \tag{1.30}$$

Due to their implications in a variety of photobiological processes (photosynthesis, vision, etc.) and potential applications in the field of photochemical conversion of solar energy, these processes are being examined widely in microheterogeneous systems.

Practical utility of reactions such as (1.28) and (1.29) in homogeneous systems is often limited due to the very rapid recombination of the redox products in a reverse reaction. The presence of distinct hydrophobic and hydrophilic domains and charged interfaces provide a potential means of controlling these processes. Spatial separation of reactants or products, for example, is possible by ejection of one of the products from the host aggregate or by preferential solubilization of one of the products into the host aggregate. Electrical potential fields present in microheterogeneous systems can aid in this process. Significant progress has been made in this area. Later on, on several occasions, we will indicate how this is achieved.

1.2.8 Organic Photoreactions

Among the many areas of photochemistry that have seen a tremendous upsurge of interest and activity, the photochemistry of organic molecules has largely matured. There are numerous organic photoreactions known today that are extremely medium sensitive and whose mechanistic, and stereochemical aspects have been elucidated to finer details. These are ideally suited for examination in microheterogeneous systems. Typical examples are photoreactions of ketones such as Norrish type I (photofragmentation), type II (γH abstraction), and photocycloadditions.[1]

Consider Norrish type I reactions of ketones (α cleavage), for example. Photolysis of asymmetrical ketones such as R_1—CO—R_2 leads to homolytic cleavage and formation of the radical pair $[R_1-CO\cdot\cdot R_2]$. In principle, the radical pair can recombine in the primary cage [reaction (1.31)] to give the starting materials or rearranged products $R_1-CO-R'_2$:

$$[R_1-CO\cdot\cdot R_2] \xrightarrow{\text{recombination}} R_1-CO-R_2 \text{ or } R_1-CO-R'_1 \qquad (1.31)$$

The radical pair also separate and decarbonylate to give new products as shown in reaction (1.32), a process that can involve a new radical pair:

$$[R_1-CO\cdot\cdot R_2] \xrightarrow{-CO} [R_1\cdot\cdot R_2] \longrightarrow R_1H + R_2H + R_1-R_2 \qquad (1.32)$$

In homogeneous solvents, in the absence of a rigid cage, the radical pair intermediates are short lived and recombination products are rarely observed. Generation of the critical radical pairs in the inner cavities of microheterogeneous systems can provide the required restrictions on mobility. This has been observed in several systems, and novel applications have been found.

Norrish type II processes (γ H abstraction) generally proceed via a critical six-membered cyclic intermediate:

1,4-biradical

If the nature of the host medium is such that it hinders the formation of the intermediate, then the efficiency of the photoreactions can be lowered. Thus, host restrictions on the guest molecule mobility can be deduced.

The structural details of the host aggregate can also impose restrictions on the number of solutes that can be solubilized and also their mode of solubilization. In such cases, the efficiency and stereochemical course of reactions can be affected. Photocycloaddition reactions provide typical example of these effects which we shall discuss in various chapters.

1.3 Photochemistry in Microheterogeneous Systems

The main attractive feature of the photochemical methods is the range of time scales that we can probe. This extends from a few picoseconds to several seconds, an impressive range of 10^{12} in time. For dynamic systems such as micelles or processes such as solubilization of molecules in the cavity of a host aggregate, several events can occur over different time scales. Relaxation processes in the singlet and triplet excited states occur in very different time domains (the former in $\leqslant 10^{-9}$ sec and the latter lasting even several hundred milliseconds). Hence, fluorescence techniques are restricted to the study of fast kinetic processes, and triplet, or phosphorescence methods probe much slower processes.

A major discovery in photochemical studies of various microhetero-geneous systems has been that, irrespective of the finer details of the host architecture, the observed photophysics and photochemical behavior are unique and quite different from those observed in homogeneous solvent media. Throughout this monograph, we will illustrate numerous specific examples to support this contention. Reaction kinetics as developed for homogeneous systems need extensive revision. Supplementation by other physical methods, development of new theoretical models and kinetic schemes in the heterogeneous case are required for several reasons. Some of these include:

(1) *Probe Distribution.* Any quantitative interpretation of photo-induced processes (efficiency and kinetics) requires knowledge on the local concentration of the reactants in the reaction zone. While the concentration term is well defined in homogeneous solvents, it is less easily defined in heterogeneous systems. Due to the microheterogeneity of the system, the probe distribution often is not uniform. Multiple-site distribution would result clearly in kinetics and effects far different from those in simple solutions. In systems with a finite number of hosts (molecules or aggregates), description of the solute distribution using statistical models is a useful approached in this context.

(2) *Probe Movement.* Unlike in simple homogeneous systems, the probe molecules (or reactants) can find themselves in pockets or cages where their movement is restricted to a limited volume or even to restricted dimensions. Most of the phenomenalogical theories of reaction kinetics and diffusion in homogeneous media are based on uniform distribution of solutes and isotropic diffusion (translational and rotational) over an infinite volume. Anisotropic (hindered) rotational diffusion in a cone or a cylinder may be the only type of movement the probe can execute. Neutral molecules located in the lipid bilayers or ionic reactants adsorbed onto the surfaces of ionic micelles may be forced to diffuse in restricted dimensions. Lateral diffusion in the former case is a unique situation to encounter.

(3) *Conformational Restraints.* In environments where there are re-strictions on the probe mobility and diffusion, the molecules may be forced to assume a different orientation/conformation atypical of homogeneous media, and this can influence strongly the efficiency of those photoprocesses that require reactants to assume a preferred geometric orientation or con-formation, e.g., excimer formation and photocycloadditions. The local structures and the mode of solubilization can promote or inhibit these processes.

Reaction Kinetics in Microheterogeneous Systems[82-113]

An outstanding feature of the microheterogeneous systems is that, by careful design of the host–guest system and making use of the mode of distribution, location of solutes in the host aggregates, it is possible to organize reactants at the microscopic (molecular) level. This allows chemical reactions between reactants to occur under situations that are unique to these systems— diffusion and dimensionality so different to necessitate alternate kinetic treatments. For example, it is possible to pick conditions in which we can readily distinguish intraaggregate processes (reactions occurring between pairs of reactants in/on the host aggregate) from interaggregate ones:

$$A + B \longrightarrow \text{products} \tag{1.35}$$

The former follows a pseudo-first-order rate law (resembling a reaction between two bound labels in a macromolecule) while the interaggregate reactions follow normal second-order (bimolecular) kinetics as in homogeneous solvents.

Intraaggregate electron or energy transfer reactions in micelles, vesicles, microemulsions, or polyelectrolytes, for example, can occur in one of the three ways: (a) hydrophobic donors solubilized in the interior reacting with acceptors that are located at the surface, (b) both donors and acceptors (neutral, hydrophobic) cosolubilized in the interior, or (c) both (ionic species) confined to diffuse on the host ionic surface. Theoretical analysis of these situations have been made for each, using continuum theory of reaction diffusion kinetics [in cases (a) and (b) diffusion of the donor towards the acceptor can be described as a random walk process] and interpretations provided for the observed rate laws. A more general stochastic master equation approach has also been described recently that incorporates both physical diffusion and chemical reactions. Lattice statistical theories have also been utilized.

As mentioned earlier, the solubilization/inclusion of solute molecules in the interiors/cavities of the host aggregates can significantly restrict their mobility in each of the three directions. There is a growing appreciation for the role of dimensionality and spatial extent in influencing reactions in nonhomogeneous media.[82,84] It has been proposed, for example, that reduction in dimensionality in fact favors efficiency of chemical and biological processes.[82] The effect of confinement of reactants to a limited volume or restricted dimensions is being examined theoretically by several authors. Gösele *et al.*, for example, have made computer simulations and derived approximate solutions to the diffusion problem for two spherical molecules confined to a spherical isotropic region.[88] The results have been presented as quenching curves. One of the molecules was assumed to be excited and the other (quencher) quenched the luminescence on the first encounter. The logarithm of the remaining fraction of excited species is presented as a function of time for three cases: the excited molecule fixed at (a) the center of the sphere, (b) at the surface, or (c) diffusing freely within the sphere. The quencher is diffusing freely in all cases. Results are given for two values of the pertinent size parameter $R/r = 4$ and 10. Here R/r is the ratio of the aggregate and encounter radii. Of particular interest is the fact that the computer simulations show surprisingly good agreement with the equation for the encounter rate constant k in homogeneous solutions at conditions where a steady-state diffusion gradient has time and space to evolve:

$$k = 4\pi(D_A + D_B)R_{AB} \qquad (1.36)$$

The deviation in case (c) is only about 20% but substantially greater in case (a) or (b). On decreasing the size of confinement, an increase in the second-order rate constant would be expected both because of the transient effect, which also appears in homogeneous solutions at short times, and because of the spatial requirements of steady-state diffusion gradient. Astumian and Schelly[89] have also recently examined geometric effects of reduction of dimensionality in interfacial reactions by using collision, transition-state, and diffusion-control theories.

Berg and von Hippel have provided an illuminating overview of the kinetic aspects of diffusion controlled processes in macromolecular systems.[84] Diffusion under various situations (spherical case with and without molecular interactions between reactants, rod-like molecules, flat surfaces, and other geometries) as well as segmental mobility and reduction in dimensionality are considered. We conclude discussions of this chapter with a general overview of the structural, dynamical aspects of micellar aggregates, a model system we deal with at length in the next three chapters.

1.4 Structural and Dynamical Aspects of Micellar Aggregates[11–16,114–126]

Above a certain concentration range, the surfactant molecules aggregate in aqueous solution to form particles of colloidal dimensions, called micelles. The micelle formation takes place over a narrow range of surfactant concentration, around the critical micelle concentration (CMC) and is accompanied by distinct changes in various physical properties: light scattering, viscosity, electrical conductivity, surface tension, osmotic pressure, and solubilization capacity for a wide variety of solutes. Critical micelle concentration is one of the most thoroughly studied properties in the micellization process. Figure 1.6 presents the simplest and the most popular model of an ionic micelle, first proposed by Hartley in 1936. Below and up to the CMC, Hartley considers the monomeric surfactant as a strong electrolyte,

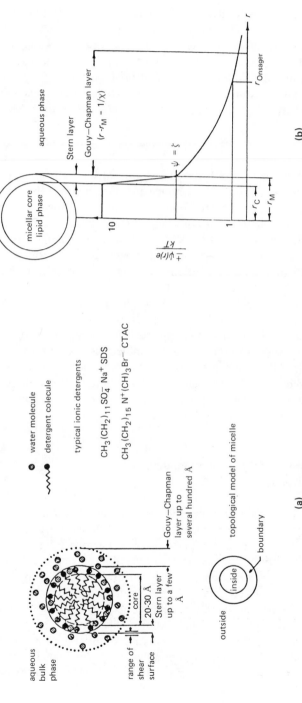

FIG. 1.6. (a) Schematic representation of a spherical ionic micelle and (b) the potential–distance function in the vicinity of a spherical micelle.

fully dissociated and unaggregated. At the critical concentration, aggregation begins, at first with the formation of relatively small micelles which grow rapidly over a very limited concentration range to a size which for a given surfactant remains approximately constant with further increase in surfactant concentration. Micelles are pictured as liquid-like, spherical in shape (globular?), their interior approximating that of a random distribution of liquid hydrocarbons but with the hydrophilic end of the chains constrained to the micellar surface. Above the CMC, increases in the surfactant concentration lead to an increase in the number of micelles, with little if any, increase in the number of monomeric surfactants.

Over the years, most of the major features of Hartley's model have been verified, though recently there have been attempts to provide alternate pictures to explain the observed properties (cf. discussions on this latter subject). Low-angle light, X-ray, and neutron scattering studies all indicate a roughly globular shape for ionic and nonionic micelles. For some recent structural studies consult references 127–133 (small-angle neutron scattering), references 134–139 (light scattering), and references cited therein. Ionic micelles typically have radii in the range of 15–30 Å and contain about 40–100 monomers. Depending on the head-group structure, ionic micelles have been found to be either (a) spherical at all concentrations, (b) rod shaped at all concentrations, or (c) spherical at low concentrations and rod shaped at high concentrations. Type (c) appears to be the most general behavior for a wide variety of ionic surfactants. Higher-order structures are also formed even at low surfactant concentrations on addition of electrolytes. The magnitude of the repulsive forces between the amphiphilic groups profoundly influences the micellar size and the conditions required for phase transitions to larger micelles. While micellar size of ionic surfactants is relatively insensitive to temperature, molecular weights of nonionic micelles increase rapidly with increasing temperature. Table 1.1 provides a collection of data on the CMC and aggregation number for a few common surfactants in aqueous solution at room temperature.

With the hydrocarbon chains forming the inner hydrophobic core of the spherical micelles, the polar head groups of ionic micelles are aligned together in a compact manner at the micellar surface. Significant advances have been made in the last decade in our understanding of the nature of the micellar core and the micellar surface. The ionic head groups of the surfactants and a portion of the counterions (gegenions) form a compact "Stern" layer at the micellar surface, in which about 60–75% of the micellar charge is believed to be neutralized. The remaining gegenions form a diffuse Gouy–Chapman layer. (cf. Fig. 1.6). Various estimates of the mean surface potential (zeta potential) for the compact region of the micellar double layer have been made and are mostly based on electrophoretic mobility measurements. Typical

TABLE 1.1. Micellar Parameters for a Few Common Surfactants in Aqueous Solution at
            Room Temperature

Surfactant	CMC	Aggregation number	Name
Anionic			
SD	2.4×10^{-2}	56	Sodium dodecanoate (laurate)
SDeS	3.3×10^{-2}	50	Sodium decylsulfate
SDS	8.0×10^{-3}	62	Sodium dodecyl(lauryl)sulfate (SLS or NaLS)
Cationic			
DAC	1.5×10^{-2}	52	Dodecylammonium chloride
DTAC	2.1×10^{-2}	56	Dodecyltrimethylammonium chloride
CTAC	1.4×10^{-3}	105	Cetyl(hexadecyl)trimethylammonium chloride
DeTAB	6.5×10^{-2}	48	Decyltrimethylammonium bromide
CTAB	9.2×10^{-4}	60	Cetyl(hexadecyl)trimethylammonium bromide
TTAB	3.6×10^{-3}		Tetradecyltrimethylammonium bromide
CTA-OH	1.5×10^{-3}	46	Cetyltrimethylammonium hydroxide
TTA-OH	7.0×10^{-3}	42	Tetradecyltrimethylammonium hydroxide
DTA-OH	3.1×10^{-2}	20	Dodecyltrimethylammonium hydroxide
CPyC	8.0×10^{-4}		Cetylpyridinium chloride
DPyB	1.1×10^{-2}	86	Dodecylpyridinium bromide
TPyB	2.9×10^{-3}		Tetradecylpyridinium bromide
Neutral			
Triton X-100	2.6×10^{-4}	143	Polyoxyethylene(E9-10)t-octylphenol
Igepal CO-630	4.6×10^{-5}	150	Polyoxyethylene(E9-10)nonylphenol
Igepal CO-730	2.0×10^{-4}		Polyoxyethylene(E12)nonylphenol
Brij 35	6.0×10^{-5}	40	Polyoxyethylene(E23)laurylether
Zwitterionic			
DB	2.0×10^{-3}	40	N-dodecyl-N,N'-dimethylbetaine
DDAPS	1.2×10^{-3}		Dodecyldimelhylammonium propylsulfonate

estimates are in the range of 50–100 mV. For a micellar double layer thickness of 3–5 Å, this would correspond to a surface dielectric constant of approximately 40. The drop in the micellar charge (zeta potential) on moving farther away from the Stern layer has been described in terms of solutions to the reduced Poisson–Boltmann equation.

Closely related to the studies of the nature of the micellar double layer are the studies on the nature and extent of counterion binding and their mobility at the micellar–water interfaces. These studies include EMF, conductance, ESR, and NMR relaxation techniques. The binding of counterions to the micellar surface has been pictured in chemical terms as a mass-action type of ion pairing and, in physical terms, as a distribution of counterions in the Gouy–Chapman electrical double layer. Counterions possessing the lowest energy of hydration and largest amount of coordinated water are associated most strongly with the micelle. Thus, the relative binding efficiency of various ions to the micellar surface follows a typical lyotropic series. Since the relative amounts of coordinated water molecules for cations are not as large as for the anions, the nature of the counterion is not as critical for anionic surfactants as for the cationic surfactants. For cationic surfactants, a Langmuir-type adsorption isotherm for counterion adsorption has been demonstrated. For the majority of counterions, the affinity of the micelle for the counterions is entropy controlled. NMR relaxation studies indicate binding of counterions as a type of ion pairing.

In recent years, considerable attention has been devoted to studies on the nature of the micellar inner core as well. Hartley visualizes micelles as liquid-like. The hydrocarbon chains in micelles are generally regarded as disordered so that the hydrophobic core is, in effect, a small drop of liquid hydrocarbon. The earliest evidence for the liquid-like nature of the micellar core comes from the ability of micelles to dissolve (solubilize) a wide variety of hydrophobic solutes and from their ability to form mixed micelles. Recently, more detailed information on the micellar core has been obtained from spectroscopic relaxation methods, which measure the freedom of motion of the hydrocarbon chains in the micelle. Thermodynamic data for transfer of small hydrocarbons from water to micelles are strikingly similar to the corresponding data for transfer from water to liquid hydrocarbons. Although all results indicate that the micellar core is, in general, fluid, there are several indications that it is less fluid (higher viscosities) than a comparable hydrocarbon solvent of the same chain length as the surfactant. NMR relaxation studies, in fact, indicate a gradient in the segmental mobility of the hydrocarbon chains. Those near the head group are comparatively rigid, with increasing freedom of motion as we move farther away into the interiors. Raman scattering studies are also in accord with such descriptions.

The currently accepted model of an ionic micelle consists of a core of radius about equal to the length of the fully extended alkyl chain of the detergents, surrounded by the Stern layer containing water, the head groups and more than half the counterions and this in turn is surrounded by the Gouy–Chapman layer containing water and remaining counterions. It is also generally agreed that micelles are heavily hydrated species. Hydrodynamic data estimates indicate that as many as 10–12 water molecules are bound per surfactant ion, corresponding to about 40% of the micellar volume. These water molecules are bound firmly enough to move as a kinetic part of the micelle during a diffusion or sedimentation process. However, there is no agreement as to the exact location of these water molecules in the micelle. Some water molecules may be entrapped by the micelle and under certain circumstances, part of the hydrocarbon chains may extend into the aqueous phase. Ample evidence exists however for extensive hydration of the micellar surface. The amount of water in the micellar interior varies from surfactant to surfactant, but at present, water is considered to penetrate the micellar surface up to distances of approximately six carbon atoms. In contrast to ionic micelles, the location and extent of water penetration in nonionic micelles appear to be fairly clear. There are evidences that suggest that the inner hydrophobic core is free of water and that the outer mantile of ethylene oxide units is heavily hydrated. On the average, there are about four water molecules per ethylene oxide unit and the water concentration increases on moving farther toward the outer surface of the micelle.

Micelles are dynamical structures in equilibrium with monomers and other forms of aggregates in solution. Depending on the experimental condition (surfactant concentration, temperature, presence of additives, etc.), the micellar size distribution for one form or the other may dominate. The number average distributions, however, are not unduly broad. During phase changes, the molecular weights level off on both sides indicating that small spherical and large rod-shaped micelles may be discrete populations and that micelles of intermediate size have relatively low stability.

Micellar association equilibria have been subject to detailed investigations by stopped-flow, temperature-jump, pressure-jump, and ultrasonic relaxation techniques. These studies show in general, the presence of two relaxation processes, characterized by two relaxation times τ_1 and τ_2 with a ratio of τ_1/τ_2 greater than 100. Both relaxation times depend on the concentration of the surfactant and on the ionic strength of the medium. The fast process is of the order of a few microseconds and the slow process extends over several milliseconds. The fast process has been assigned to the exchange of a monomeric surfactant ion between the micellar phase and the bulk solvent:

$$\text{(M)} + \xi \rightleftharpoons \text{(M')} + \xi \qquad (1.37)$$

The slow process has been assigned to the micellization–dissolution equilibrium:

$$\text{(M)} \rightleftharpoons m\ \xi \qquad (1.38)$$

The classical model of ionic micelles, originally proposed by Hartley, has been subject to much debate in recent years, especially with regard to the question of water penetration in the micelle. Hartley's model cannot be reconciled with any appreciable water-to-hydrocarbon contact. Experimental studies do indicate that depending on the micelle type, water appears to penetrate the micelle to various depths. In recent years various models have been proposed that allow different degrees of water penetration into the micelle. In Menger's model,[121] the micelles are pictured as porous clusters of surfactants which provide considerable water penetration. Fromherz[124] has rationalized micellar structures in terms of a surfactant block model that allows wetting of entire hydrocarbon chains on a time average. Dill and Flory[125] have discussed molecular organization of surfactants in terms of a statistical theory by using lattice models. This model provides a substantial probability for methylene groups, even in the middle of the chain, to be in the outer layer of the surface of the micelle.

Distribution of Solutes Among the Micelles[97–106]

Given the discrete number of micelles present at a given total surfactant concentration, introduction of solutes (probes and quenchers) leads to their solubilization in the available number of micelles. Knowledge of this distribution of solutes among the micelles is essential for the interpretation of various excited-state bimolecular processes such as fluorescence quenching, excimer formation, energy, and electron transfer.

Given the average number of solutes per micelle $\bar{n}$ [$\bar{n} = N/[M]$, where N is the total number of solutes introduced and $[M]$ the concentration of micelles] the determination of the probability P_i of finding i solutes in a given micelle can be an interesting exercise in statistical mechanics. Considering the solutes as simple balls N in number to be thrown into a given number M of micelles (boxes), there are several distribution laws that we can derive under various conditions: random or geometric, binomial, and poissonian. For indistinguishable balls, the number of ways of arranging them among the micelle is given by (Bose statistics)

$$W = M! \bigg/ \prod_{i=0} M_i! \qquad (1.39)$$

where M_i represents micelles with i solutes. Maximizing W under the conditions $\sum M_i = M$ and $\sum i M_i = N$, we obtain the geometric (or random) distribution:

$$P_i = \bar{n}^i/(1 + \bar{n})^{i+1} \qquad (1.40)$$

where P_i is the probability of finding a micelle that contains i solutes. For a poisson distribution, we have

$$W = (N!M!) \bigg/ \prod_{i=0} [M_i!(i!)^{M_i}] \qquad (1.41)$$

and

$$P_i = (\bar{n}^i e^{-\bar{n}})/i! \qquad (1.42)$$

Binomial distribution is characterized by:

$$P_i = \binom{i}{m}\left(\tfrac{\bar{n}}{m}\right)^i [1 - (\bar{n}/m)]^{m-i} \qquad (1.43)$$

where m is the maximum number of solutes that can be solubilized in a given micelle ($i = 0, 1, 2, \ldots, m$). When $m \to \infty$, the binomial distribution reduces to poissonian.

Poissonian distribution is the most widely used model for the distribution of solutes and micelles and other aggregate/molecular host systems. The poissonian distribution can be shown to be a normal consequence of the dynamic model of the micellar association and solubilization process. The dynamic equilibrium of the solute S with the micelle can be represented as:

$$S_w + \left(M_i \right) \underset{(1+i)k'}{\overset{n}{\rightleftharpoons}} \left(M_{i+1} \right) \tag{1.44}$$

where S_w represents a solute molecule in water and $M_{i'}$, a micelle containing i solute molecules. The rate n is a second-order rate constant which describes the rate of entry of solutes in the micelles. It is assumed to be independent of the number of solutes in the micelle. The k' is the exit rate of one solute and is assumed to be independent of the occupation number i. The rate at which the solutes leave micelles containing several of them is assumed to be linearly dependent on the number of solutes, i.e., rate $= (i + 1)k'$. There is no limit to the maximum number of solutes that may occupy a given micelle. Writing equilibrium expressions [Eq. (1.44)] for all values of i, it is easy to obtain the fraction of micelles that are occupied by i solute molecules as

$$[M_i]/[M] = (\bar{n}^i e^{-\bar{n}})/i! \tag{1.45}$$

where $\bar{n}$ is the average number of solutes per micelle and $[M]$, the concentration of micelles. This equation shows that the distribution of solubilizates in micelles to be governed by a poisson distribution. The same conclusion can also be reached by considering that each micelle is a small fraction v of a total micellar volume V offered to the solubilizates.

Figure 1.7 presents schematically the variation of P_i values for various mean occupancy number $\bar{n}$. It is worthwhile examining these P_i values to appreciate the influence of the poisson distribution. At a low value of $\bar{n} = 0.1$, most of the micelles are empty and only 10% of the micelles contain one or more solutes ($P_0 = 0.90$, $P_1 = 0.09$, and $P_n = 0$ for $n \geqslant 2$). When $\bar{n} = 1.0$, i.e., there is equal concentrations of solutes and micelles, 37% of the micelles are still empty, but about 26% of them contain two or more solutes ($P_0 = 0.37$, $P_1 = 0.37$, and $P_n = 0.26$ for $n \geqslant 2$). Thus, if we want to avoid bimolecular self-quenching of the excited state (reactions such as excimer formation, and triplet–triplet annihilation), then $\bar{n}$ should be much less than 0.5.

In the next two chapters dealing with singlet and triplet excited-state processes of molecules in micelles, we shall illustrate on several occasions the importance/necessity of considering a statistical distribution picture to interpret the experimental results. It will be shown that poisson distribution adequately explains the results in most cases. For excimers, Dorrance and Hunter have claimed that geometric distribution better describes the experimental results. When i is large, both poisson and geometric distributions predict roughly the same results. The geometric distribution, however, is an anamolous result, for it predicts that, at large values of mean occupancy

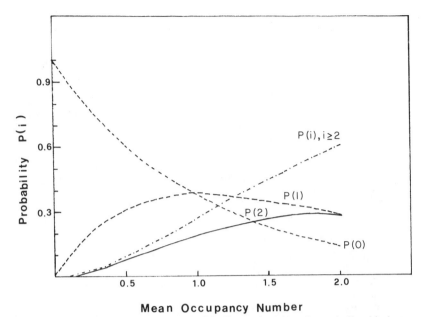

Mean Occupancy Number

Fig. 1.7. Variation in the probability P_i of finding i solutes in a given micelle with the mean occupancy number N of solutes in a given micellar solution, calculated according to poisson statistics.

number $\bar{n}$, there will be more unoccupied micelles than those containing $\bar{n}$ solubilized molecules. Miller, Hauser, and Klein have used binomial distribution in their analysis of excimer kinetics. Since the number of solutes that can be accommodated in a given micelle cannot be infinite, binomial distribution is also a good representation. Its deviation from a poisson distribution is significant only when the maximum limit m is greater than the mean occupancy number $\bar{n}$. Moroi has commented that we can also use a gaussian distribution. However, the variable of a gaussian distribution is continuous whereas the number of reactants in a given micelle is an integer!

It should be pointed out that in assessing the potential micellar effects on photoprocesses, consideration of the solubilization sites and conformational restraints on the probe movement are just as important as their statistical distribution.

References

1. N. J. Turro, "Modern Molecular Photochemistry." Benjamin, New York, 1978.
2. (a) J. R. Lakowicz, "Principles of Fluorescence Spectroscopy." Plenum, New York, 1984. (b) D. V. O'Connor and D. Phillips, "Time-Correlated Single Photon Counting." Academic Press, Orlando, Florida, 1984. (c) J. N. Demas, "Excited State Lifetime Measurements." Academic Press, New York, 1983.

3. J. B. Birks, "Photophysics of Organic Molecules." Wiley, New York, 1970.

4. J. B. Birks, ed., "Organic Molecular Photophysics," Vols. 1 and 2. Wiley, New York, 1975.

5. C. A. Parker, "Photoluminescence of Solutions." Elsevier, Amsterdam, 1968.

6. D. O. Cowan and R. L. Drisko, "Elements of Organic Photochemistry." Plenum, New York, 1976.

7. E. L. Wehry, ed., "Modern Fluorescence Spectroscopy," Vols. 1–4. Plenum, New York, 1976.

8. V. Balzani and V. Carasitti, "Photochemistry of Coordination Compounds." Academic Press, New York, 1970.

9. A. W. Adamson and P. D. Fleischauer, "Concepts of Inorganic Photochemistry." Wiley (Interscience), New York, 1975.

10. J. H. Fendler, "Membrane Mimetic Chemistry." Wiley, New York, 1982.

11. C. Tanford, "The Hydrophobic Effect: Formation of Micelles and Biological Membranes," 2nd Ed. Wiley, New York, 1980.

12. P. H. Elworthy, A. T. Florence, and C. B. McFarlane, "Solubilisation by Surface Active Agents." Chapman & Hall, London, 1968.

13. K. L. Mittal, ed., "Micellisation, Solubilisation and Microemulsions," Vols. 1 and 2. Plenum, New York, 1977.

14. K. L. Mittal, ed., "Solution Chemistry of Surfactants," Vols. 1 and 2. Plenum, New York, 1979.

15. E. J. Fendler and K. L. Mittal, eds., "Solution Behaviour of Surfactants," Vols. 3 and 4. Plenum, New York, 1982.

16. J. H. Fendler and E. J. Fendler, "Catalysis in Micellar and Macromolecular Systems." Academic Press, New York, 1975.

17. E. E. Bittar, ed., "Membrane Structure and Function." Wiley, New York, 1980.

18. D. Papahadjopoulos, ed., "Liposomes and Their Uses in Biology and Medicine," Ann. N. Y. Acad. Sci., Vol. 308. New York, 1978.

19. G. Gregoriadis and A. C. Allison, eds., "Liposomes in Biological Systems." Wiley, New York, 1980.

20. M. K. Jain and R. C. Wagner, "Introduction to Biological Membranes." Wiley (Interscience), New York, 1980.

21. F. Szoka and D. Papahadjopoulos, $Annu.$ $Rev.$ $Biophys.$ $Bioeng.$ $\mathbf{9}$, 467 (1980).

22. E. Grell, ed., "Membrane Spectroscopy," Mol. Biol. Biochem. Biophys. Ser., Vol. 31. Springer–Verlag, Berlin and New York, 1983.

23. D. Chapman, ed., "Biomembrane Structure and Function." Macmillan, New York, 1983.

24. (a) P. Luisi, ed., "Reversed Micelles: Biological and Technological Relevance of Amphiphilic Structures in Apolar Media." Plenum, New York, 1984. (b) H. F. Eicke, $Top.$ $Curr.$ $Chem.$ $\mathbf{87}$, 85 (1980).

25. L. M. Prince, ed., "Microemulsions." Academic Press, New York, 1977.

26. I. D. Robb, ed., "Microemulsions." Plenum, New York, 1982.

27. D. O. Shah, ed., "Macro and Microemulsions." Plenum, New York, 1983.

28. G. L. Gaines, Jr., "Insoluble Monolayers at Liquid–Gas Interfaces." Wiley, New York, 1966.

29. G. L. Gaines, Jr., in "Surface Chemistry and Colloids" (M. Kerker, ed.), MTP Int. Rev. Sci., Vol. 7, pp. 1–24. Butterworth, London, 1972.

30. H. Kuhn, D. Mobius, and H. Bücher, in "Physical Methods of Chemistry" (A. Weissberger and B. W. Rossiter, eds.), Vol. I, Part IIIB, pp. 577–701. Wiley (Interscience), New York.

31. H. Ti Tien, "Bilayer Lipid Membranes (BLM): Theory and Practice." Dekker, New York, 1974.

32. H. Ti Tien, in "Photosynthesis in Relation to Model Systems" (J. Barber, ed.). pp. 115–140. Elsevier, Amsterdam, 1979.

33. M. K. Jain, "The Bimolecular Lipid Membrane: a System." Van Nostrand–Reinhold, New York, 1972.

34. F. D. Saeva, ed., "Liquid Crystals: The Fourth State of Matter." Dekker, New York, 1983.

35. S. Chandrasekar, "Liquid Crystals." Cambridge Univ. Press, London and New York, 1977.

36. P. G. de Gennes, "The Physics of Liquid Crystals." Oxford Univ. Press, London and New York, 1974.

37. J. J. Wolken and G. H. Brown, "Liquid Crystals and Biological Systems." Academic Press, New York, 1980.

38. P. C. Hiemenz, "Polymer Chemistry: Basic Concepts." Dekker, New York, 1984.

39. R. B. Seymour, "Polymer Chemistry: An Introduction." Dekker, New York, 1981.

40. F. W. Billmeyer, "Textbook of Polymer Science," 3rd Ed., Wiley, New York, 1984.

41. H. Morawetz, "Macromolecules in Solution." 2nd Ed., Wiley, New York, 1975.

42. C. Tanford, "Physical Chemistry of Macromolecules." Wiley, New York, 1975.

43. S. A. Rice and M. Nagasawa, "Polyelectrolyte Solutions." Academic Press, New York, 1961.

44. A. Rembaum and E. Selegny, eds., "Polyelectrolytes and Their Applications." Reidel, Boston, Massachusetts, 1975.

45. F. Oosawa, "Polyelectrolytes." Dekker, New York, 1971.

46. D. W. Breck, "Zeolite Molecular Sieves," 2nd Ed. Wiley, New York, 1984.

47. J. A. Rabo, ed., "Zeolite Chemistry and Catalysis," ACS Monogr. No. 171. Am. Chem. Soc., Washington, D.C., 1976.

48. G. D. Stucky and F. G. Dwyer, eds., "Intrazeolite Chemistry," ACS Symp. Ser. No. 218. Am. Chem. Soc., Washington, D.C., 1983.

49. H. Van Olphen, "An Introduction to Clay Colloid Chemistry," 2nd Ed. Wiley, New York, 1977.

50. B. K. G. Theng, "The Chemistry of Clay Organic Reactions." Wiley, New York, 1974.

51. R. K. Iler, "The Chemistry of Silica," 2nd Ed. Wiley, New York, 1979.

52. M. L. Bender and M. Komiyama, "Cyclodextrin Chemistry," React. Struct. Concepts Org. Chem. Ser. Vol. 6. Springer-Verlag, Berlin and New York, 1978.

53. J. Szejtli, "Cyclodextrins and Their Inclusion Compounds." Akadémiai Kiadó, Budapest, 1982.

54. L. J. Mathies and C. W. Carrehar, eds., "Crown Ether and Phase Transfer Catalysis in Polymer Science." Plenum, New York, 1984.

55. M. Hiraoka, "Crown Compounds: Their Characteristics and Applications." Elsevier, New York, 1982.

56. J. L. Atwood, J. E. D. Davis, and D. D. McNicol, eds., "Inclusion Compounds," Vols. 1–3. Academic Press, New York, 1984.

57. S. R. Meech, D. V. O'Connor, D. Phillips, and A. G. Lee, *J.C.S. Faraday II* **79**, 1563 (1983), and references cited therein.

58. E. M. Kosower, H. Kanety, H. Dodiuk, G. Striker, T. Jovin, H. Boni, and D. Huppert, *J. Phys. Chem.* **87**, 2479 (1983), and references cited therein.

59. J. Slavik, *Biochim. Biophys. Acta* **694**, 1 (1982).

60. J. F. Ireland and P. A. H. Wyatt, *Adv. Phys. Org. Chem.* **12**, 131 (1976).

61. W. Klöpfer, *Adv. Photochem.* **10**, 311 (1977).

62. E. Vander Donckt, *Prog. React. Kinet.* **5**, 273 (1970).

63. G. Weber, *Annu. Rev. Biophys. Bioeng.* **1**, 553 (1972).

64. R. B. Cundall and R. E. Dale, eds., "Time Resolved Fluorescence Spectroscopy in Chemistry and Biology," NATO ASI, Vol. A69. Plenum, New York, 1980.

65. R. M. Noyes, *Prog. React. Kinet.* **1**, 129 (1961).

66. A. H. Alwatter, M. D. Lumb, and J. B. Birks, *in* "Organic Molecular Photophysics" (J. B. Birks, ed.), Vol. 1, p. 403. Wiley, New York, 1973.

67a. T. L. Nemzek and W. R. Ware, *J. Chem. Phys.* **62,** 477 (1975).
67b. L. Monchick, *J. Chem. Phys.* **62,** 1907 (1975).
68.   J. B. Birks, *Prog. React. Kinet.* **5,** 181 (1970).
69.   H. Beens and A. Weller, *in* "Organic Molecular Photophysics" (J. B. Birks, ed.), Vol. 2, p. 159. Wiley, New York, 1975.
70.   F. C. DeSchryver, N. Boens, and J. Put, *Adv. Photochem.* **10,** 359 (1977).
71.   M. Gordon and W. R. Ware, eds., "The Exciplex." Academic Press, New York, 1975.
72.   W. Klopfer, *in* "Organic Molecular Photophysics" (J. B. Birks, ed.), Vol. 1, p. 357. Wiley, New York, 1973.
73.   F. Wilkinson, *Adv. Photochem.* **3,** 241 (1964).
74.   A. A. Lamola and N. J. Turro, eds., "Energy Transfer and Organic Photochemistry." (Interscience), New York, 1969.
75a. T. Förster, *in* "Energetics and Mechanisms in Radiation Biology" (G. O. Phillips, ed.), p. 183. Academic Press, New York, 1968.
75b. T. Förster, *Discuss Faraday Soc.* **27,** 1 (1959).
75c. T. Förster, *in* "Comparative Effects of Radiation." (M. Burton, J. S. Kirby-Smith, and J. L. Magee, eds.), p. 300. Wiley, New York, 1960.
76.   R. C. Jarnagin, *Acc. Chem. Res.* **4,** 420 (1971).
77.   P. Lesclaux and J. Jousset–Dubien, *in* "Organic Molecular Photophysics" (J. B. Birks, ed.), Vol. 1, p. 457. Wiley, New York, 1973.
78.   P. L. Piciulo and J. K. Thomas, *J. Chem. Phys.* **68,** 3260 (1978).
79.   V. Balzani, F. Boletta, M. T. Gandolfi, and M. Maestri, *Top. Curr. Chem.* **75,** 1 (1978).
80.   N. Sutin, *Prog. Inorg. Chem.* **30,** 441 (1983).
81.   M. Grätzel, *in* "Modern Aspects of Electrochemistry" (R. E. White and J. O. M. Bockris, eds.), Vol. 15, p. 83. Plenum, New York, 1983.
82.   G. Adam and M. Delbruck, *in* "Structural Chemistry and Molecular Biology" (A. Rich and N. Davidson, eds.), p. 198. Freeman, San Francisco, California, 1968.
83.   D. F. Calef and J. M. Deutsch, *Annu. Rev. Phys. Chem.* **34,** 493 (1983).
84.   O. G. Berg and S. P. von Hippel, *Annu. Rev. Biophys. Bioeng.* **14,** 131 (1985).
85.   J. H. Fendler, *Annu. Rev. Phys. Chem.* **35,** 137 (1984).
86.   M. K. Musho and J. J. Kozak, *J. Chem. Phys.* **80,** 159 (1984), and references cited therein.
87.   P. Lee and J. J. Kozak, *J. Chem. Phys.* **80,** 705 (1984), and references cited therein.
88.   U. Gösele, V. K. A. Klein, and M. Houser, *Chem. Phys. Lett.* **68,** 291 (1979).
89.   R. D. Astumian and Z. A. Schelly, *J. Am. Chem. Soc.* **106,** 304 (1984).
90.   C. S. Owen, *J. Chem. Phys.* **62,** 3204 (1975).
91.   M. Tomkiewicz and G. A. Corker, *Chem. Phys. Lett.* **37,** 537 (1976).
92.   K. Razi Naqvi, J-P. Behr, and D. Chapman, *Chem. Phys. Lett.* **26,** 440 (1974).
93.   S. Kondo and Y. Takano, *Int. J. Chem. Kinet.* **8,** 481 (1976).
94.   P. H. Richter and M. Eigen, *Biophys. Chem.* **2,** 255 (1974).
95.   H. C. Berg, "Random Walks in Biology." Princeton Univ. Press, Princeton, New Jersey, 1984.
96.   A. J. Frank, M. Grätzel, and J. J. Kozak, *J. Am. Chem. Soc.* **98,** 3317 (1976).
97.   P. P. Infelta, M. Grätzel, and J. K. Thomas, *J. Phys. Chem.* **78,** 190 (1974).
98.   P. P. Infelta and M. Grätzel, *J. Chem. Phys.* **70,** 179 (1979); **78,** 5280 (1983).
99.   M. Tachiya, *Chem. Phys. Lett.* **33,** 289 (1975); *J. Chem. Phys.* **78,** 5282 (1983).
100. M. Tachiya, *J. Chem. Phys.* **76,** 340 (1982); *Radiat. Phys. Chem.* **21,** 167 (1983).
101. H. Sano and M. Tachiya, *J. Chem. Phys.* **75,** 2870 (1981).
102. H. Sano, *J. Chem. Phys.* **74,** 1394 (1980).
103. R. C. Dorrance and T. F. Hunter, *J.C.S. Faraday I* **70,** 1572 (1974).
104. T. F. Hunter, *Chem. Phys. Lett.* **75,** 152 (1980).
105. D. J. Miller, *J. Chem. Educ.* **55,** 776 (1978).

106. Y. Moroi, *J. Phys. Chem.* **84**, 2186 (1980).
107. J. J. Keizer, *J. Am. Chem. Soc.* **105**, 1494 (1983).
108. J. E. McCarthy and J. J. Kozak, *J. Chem. Phys.* **77**, 2214 (1982).
109. K. Razi Naqvi, S. Waldenstrom, and K. J. Mork, *J. Phys. Chem.* **86**, 4750 (1982).
110. M. D. Hatlee, J. J. Kozak, G. Rothenberger, P. P. Infelta, and M. Grätzel, *J. Phys. Chem.* **84**, 1508 (1980).
111. M. D. Hatlee, J. J. Kozak, and J. J. Kozak, *Ber. Bunsenges. Phys. Chem.* **86**, 157 (1981).
112. M. D. Hatlee and J. J. Kozak, *J. Chem. Phys.* **72**, 4358 (1980); **74**, 1098 (1981).
113. M. D. Hatlee and J. J. Kozak, *J. Chem. Phys.* **74**, 5627 (1981).
114. B. Lindman and H. Wennerstrom, *Top. Curr. Chem.* **87**, 1 (1980), and references cites therein.
115. P. Mukerjee, *Adv. Colloid Interface Sci.* **1**, 241 (1967).
116. T. Nakagawa, *Colloid Polym. Sci.* **252**, 56 (1974).
117. E. A. G. Aniansson, S. N. Wall, M. Almgren, H. Hoffman, I. Kielmann, W. Ulbricht, R. Zana, J. Lang, and C. Tondre, *J. Phys. Chem.* **80**, 905 (1976), and references cited therein.
118. E. Cordes, ed., "Reaction Kinetics in Micelles." Plenum, New York, 1973.
119. I. V. Berezin, K. Martinek, and A. K. Yatsimirskii, *Russ. Chem. Rev.* (*Engl. Transl.*) **42**, 787 (1973).
120. K. L. Mittal and B. Lindman, eds., "Surfactants in Solution." Plenum, New York, 1984.
121. F. M. Menger, *Acc. Chem. Res.* **12**, 111 (1979).
122. F. M. Menger and B. J. Boyer, *J. Am. Chem. Soc.* **102**, 5936 (1980).
123. F. M. Menger, J. M. Jerkunica and J. C. Johnston, *J. Am. Chem. Soc.* **100**, 4676 (1978).
124. P. Fromherz, *Chem. Phys. Lett.* **77**, 460 (1980).
125. K. A. Dill and P. J. Flory, *Proc. Natl. Acad. Sci. USA.* **77**, 3115 (1980); **78**, 676 (1981).
126. K. A. Dill, *J. Phys. Chem.* **86**, 1498 (1982).
127. R. Triolo, L. J. Magid, J. S. Johnson, and H. R. Child, *J. Phys. Chem.* **86**, 3689 (1982).
128. R. Triolo, J. B. Hayter, L. J. Magid, and J. S. Johnson, *J. Chem. Phys.* **79**, 1977 (1983).
129. L. J. Magid, R. Triolo, and J. S. Johnson, *J. Phys. Chem.* **88**, 5730 (1984).
130. M. Zulauf and J. B. Hayter, *Coll. Polym. Sci.* **260**, 1023 (1982).
131. M. Zulauf and J. P. Rosenbusch, *J. Phys. Chem.* **87**, 856 (1983).
132. J. C. Ravey, *J. Coll. Int. Sci.* **94**, 289 (1983).
133. D. J. Cebula and R. H. Ottewill, *Coll. Polym. Sci.* **260**, 1118 (1982).
134. M. Corti and V. Deghiorghio *in* "Light Scattering in Liquids and Macromolecule Solutions" (V. Deghiorghio, M. Corti, and M. Giglio, eds.), p. 111, Plenum, New York, 1980.
135. N. A. Mazer, G. B. Banedek, and M. C. Carey, *J. Phys. Chem.* **80**, 1075 (1976).
136. M. Corti and V. Deghiorghio, *J. Phys. Chem.* **85**, 711, 1442 (1981).
137. M. Corti, C. Minero, and V. Deghiorghio, *J. Phys. Chem.* **88**, 309 (1984).
138. W. Brown, R. Johnson, P. Stilbs, and B. Lindman, *J. Phys. Chem.* **87**, 4548 (1983).
139. V. Deghiorghio and M. Corti in reference 120, p. 441.

Chapter 2

Micellar Photophysics—Singlet-State Reactions

2.1 Introduction

To start our discussions on the micellar photophysics and photochemistry a few general remarks on micellar effects on the distribution of the reactant molecules are in order. Earlier brief discussions on the structural aspects of micellar aggregates (Chapter 1) pointed out three features of these systems, all of which have important consequences in influencing the photophysics and chemistry of solubilized molecules. These are their heterogeneous character with hydrophobic and hydrophilic domains separated by a charged interface, the discrete number of the micellar aggregates, and the dynamic nature of their association equilibria.

Due to the microheterogeneous character of the micellar aggregates, the distribution of reactant partners (probe and the quencher molecules) is not uniform throughout the entire volume of the aqueous solution wherein the micelles are present. Neutral solutes can be associated solely with the micellar phase, partitioned with aqueous phase, or even dispersed solely in the aqueous phase, depending on the nature of their hydrophobic/hydrophilic character. Even when they are asssociated solely with the aggregates, a probe can reside solely in the inner hydrocarbon core, be sandwiched near the surface between the polar headgroups, or just adsorbed at the surface by electrostatic interactions. The mode of solubilization has pronounced effects on the observed photophysical and chemical properties. For reactions involving ionic reactants and products, the presence of the charged interface in ionic micelles leads to strong electrostatic effects. Depending on the sign of the charges on the micelle and the quencher ions, the latter may all be collected at the micellar surface (e.g., metal cations in anionic micelles) or kept away from the solubilized probe molecules. In redox reactions where charged products are formed from neutral species, the micellar interface can bring about enhanced charge separation by stabilizing one of the redox products and repelling the other.

The discrete number of micellar aggregates also influences the distribution among the micelles. Depending on the relative number of solutes and the micelles to solubilize them, the solute molecules can either all crowd together in the micelles leading to a locally high concentration or be well separated from each other. Thus, by varying the micelle concentration we can promote or totally inhibit bimolecular reactions of the excited probe. Various possible solute distribution modes were outlined in Chapter 1 (Section 1.4). As we shall see later, the distribution of solutes among the micelles occurs statistically and a poisson distribution is often found to be adequate to describe it.

The dynamic nature of the micellar association has its effects in either bringing together reactant partners in a transient manner or keeping them apart. For very hydrophobic (water-insoluble) solutes, the residence time in the micelle can approach the lifetime of the micelle, while for those which partition significantly between the micellar and aqueous phase the solute exit rates can be much higher. The implications are that for short-lived singlet excited state reactions of hydrophobic solutes ($< 1 \; \mu sec$) the picture is a "static" one. The micelles are intact species, and no exit of the probe occurs during the quenching process. For the longer-lived triplet states, the probe during its lifetime can exit a micelle and be quenched in the aqueous phase. All these points are important ones to bear in mind throughout this and the following chapters.

2.2   Luminescence Probe Analysis with Solvent-Sensitive Probes

The simplest approach is to introduce a luminescence probe into the micellar aggregate and from the observed luminescence changes and the known behavior of the probe in various homogeneous environments, attempt to map out the static picture of the aggregate.[1-4] Applications of fluorescence probes in micellar systems include:

(1) *Critical micelle concentration (CMC) determination.* The solubilization process of the transfer of the probe from water to the micellar interior/surface is often accompanied by large changes in the probe luminescence properties. Hence, monitoring of the probe luminescence as a function of the surfactant concentration provides a simple method for the determination of the critical micelle concentration.

(2) *Influence of additives.* In a similar manner if the micellar structure undergoes significant changes in the presence of additives such as electrolytes and alcohols, these can be followed readily.

(3) *Interface properties and polarity.* From the observed luminescence features of surface-adsorbed probes we can attempt to estimate the polarity (effective dielectric constant) or surface electrical potential.

Table 2.1 provides a summary of various fluorescence probes that have been used in the study of micellar systems.[5-50] In the table we have indicated the possible origin of the spectral changes observed for various probes, and these are to be taken as tentative assignments. Application of fluorescence probe techniques require utmost care, and we would like to draw attention to our earlier cautionery remarks (Section 1.3). The need for supplementary knowledge on the probe photophysics in homogeneous systems, its location, and possible perturbations of the aggregated system are also pertinent and

TABLE 2.1. Fluorescence Probes in the Study of Micellar Assemblies

Medium-dependent property	Origin of solvent effects	Probe examples	Reference
Intensities of vibronic bands in absorption and fluorescence	Ham effect (vibronic coupling)	Pyrene and other aromatic hydrocarbons	5–13 14,15
Fluorescence maximum (λ_{max}) and quantum yields (ϕ_{Fl})	$(n, \pi^*)-(\pi - \pi^*)$ state shift and polarity of (π, π^*) state	Pyrenecarboxaldehyde, 7-alkoxy coumarin, alkylaminobenzylidine malonitrile	16–20 21 22
Fluorescence maximum (λ_{max}) and quantum yields (ϕ_{Fl})	$^1S^*$ high dipole moment (polar)	Cyanoalkoxybiphenyl	23
ϕ_{Fl} and τ_{Fl}	H-bonding interactions in the excited state	Xanthene dyes (Rose bengal, acridine, acridine orange)	24–28
λ_{max} and ϕ_{Fl}	Solvent interaction in the CT excited state	ANS, TNS	29–31
τ_{Fl} and λ_{max}	Solvent relaxation in the CT excited state	ML_3^{2+}, M = Ru, Os L = Substituted bpy, phen	32–40
Rate constant for cis–trans isomerization	Polar transition state	Azobenzenes	41
ϕ_{Fl} and τ_{Fl}	?	Indoles, Dansylglycine, perylene	42–45 46 30,47
λ_{max}, ϕ_{Fl}, and τ_{Fl}	(Changes in rates radiationless process)	Cyanines	49
λ_{max}, and ϕ_{Fl}	Solvent interactions in the excited state?	Anthroyl, fatty acids	50
Relative intensity of normal and anamolous fluorescence	Two conformationally distinct singlets (planar and twisted CT states?)	Dimethylamino benzonitrile (DABN)	48

necessary. For the interpretation of the photophysical and chemical properties, it is often desirable to obtain as much information on the probe environment in the aggregate from the unquenched luminescence itself.

An interesting case of "medium effects" where the solvent interactions perturb mainly the relative intensities of the vibrational fine structures of the fluorescence spectra can be found in the Ham effect.[5-15] In aromatic molecules such as benzene or pyrene with a minimum D_{2h} symmetry, the absorption and fluorescence spectra show mixed polarization owing to the vibronic coupling between the first (S_1) and second (S_2) singlet excited states. The first singlet absorption $(S_0 \rightarrow S_1)$ is symmetry forbidden and is weak. In the Ham effect, the forbidden vibronic bands in weak electronic transitions show marked intensity enhancements under the influence of solvent polarity. Though the number of aromatic hydrocarbons that exhibit this effect is

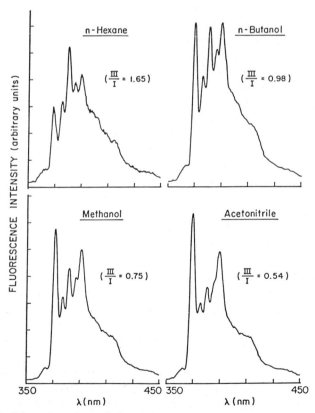

FIG. 2.1. Solvent dependence of vibronic band intensities in pyrene monomer fluorescence at room temperature. The concentration of pyrene is $2 \times 10^{-6}\ M$ and $\lambda_{\text{excit}} = 310$ nm. (From Kalyanasundaram and Thomas.[5] Copyright 1977 American Chemical Society.)

numerous, the most popular probe has been pyrene. Figure 2.1 illustrates typical changes in the fluorescence spectrum of pyrene in homogeneous solvents. If we number I–V the principal vibronic bands observed in the room temperature fluorescence in solution, then band III $(0-737\ \mathrm{cm}^{-1})$ at 328.9 nm is strong (allowed) and shows minimal variation in intensity. Band I $(0-0)$ located at 372.4 nm shows significant intensity enhancement in polar solvents. Thus, the peak intensity I/III in the normal fluorescence spectra can serve as a measure of the polarity of the environment. Table 2.2 provides some quantitative data on the I/III ratio in various neat solvents. Various quantitative correlations of the intensity variation with single or multiple solvent polarity parameters have been attempted with limited success.[30] For recent studies of the Ham effect phenomenon in homogeneous solvents see reference 5,7–14a,17a,b,18.

In early studies, the pyrene I/III ratio has been used as a probe to determine the critical micelle concentration and the extent of water penetration in various micelles.[5] Pyrene fluorescence undergoes significant changes both in I/III ratio and the lifetime so that we can readily detect the CMC by monitoring either of these as a function of the surfactant concentration. The measured CMC values are in very good agreement with those determined by other physical methods. Interestingly the measured I/III ratios in various micellar solutions are quite different, depending more on the nature of the head group and somewhat less on the length of the hydrocarbon chain and the nature of the counterions. Presumably pyrene is solubilized in slightly different regions of the micelle and I/III ratios merely reflect the nature of the immediate environment around the probe.

Some interesting correlations have been observed recently[12] between the pyrene I/III ratio, fluorescence lifetime, solubilization time, and the efficiency of excimer formation. The I/III ratio is now used widely as a structural probe in many applications. High I/III peak ratios observed for surfactants with quarternary ammonium head groups as compared to sulfate head groups (~ 1.40 for C_n—$\overset{+}{N}(Me)_3$ versus ~ 1.20 for C_n—OSO_3^-) have been interpreted as due to the affinity of arenes to the former. Offen and Turley[13] have used it to monitor changes that occur when micellar solutions are subjected to high pressures. The intensity ratio is nearly unaltered in a compressed micellar solution but shows a discontinuous change at the pressure-induced solubility limit where the micelle is transformed to a hydrated solid. From comparison of the peak ratio with those in neat solvents it has been inferred that some water is present in the solvation shell of pyrene, and this could possibly be achieved by locating pyrene near the micelle–water interface.

More familiar examples of fluorescence probes involve solvent-induced spectral shifts in the fluorescence maxima of molecules of the arylaminonaphthalenes type [e.g., N-phenylnaphthylamine (NPN), 1-anilinonaphthalene-8-

TABLE 2.2. Variation in the Relative Band Intensities (I/III) in Pyrene Monomer Fluorescence in Homogeneous Solvents and in Aqueous Micellar Solutions

Solvent	Band ratios (I/III)		Solvent	Band ratios (I/III)	
	Kalyanasundaram and Thomas[a]	Dong and Winnik[b]		Kalyanasundaram and Thomas[a]	Dong and Winnik[b]
Neat Solvents			n-Propanol	1.09	1.09
Perfluorodecalin		0.52	Ethanol	1.10	1.18
Perfluoromethylcyclohexane	0.50	0.52	Benzyl alcohol	1.22	1.24
Hexane	0.61	0.58	Chloroform	1.22	1.24
Cyclohexane	0.57	0.58	Tetrahydrofuran	1.33	1.35
Methylcyclohexane	0.55	0.58	Methanol	1.30	1.35
Dodecane	0.60	0.59	Acetic acid		1.37
Isooctane	0.60		Ethyl acetate	1.45	1.37
n-Butylether		0.84	Pyridine		1.42
Isopropylether		0.93	Dimethoxyethane		1.48
1,2,4-Trichlorobenzene		0.97	Dioxane		1.50
o-Dichlorobenzene	0.92	1.02	Formamide		1.57
n-Pentanol	0.93	1.02	Diglyme		1.60
i-Butanol	0.98	1.02	Ethelene glycol	1.64	1.64
Ethyl ether			Formic acid	1.64	1.69
Toluene	1.04	1.11	Acetonitrile	1.75	1.75
Benzene	1.14	1.05	N,N-dimethylformamide	1.82	1.81
n-Butanol	1.02	1.06	Water	(1.84)	1.87
Chlorobenzene	1.09	1.08	Dimethyl sulfoxide	1.88	1.95
Micellar Solutions			Igepal Co-630	1.30	
SD	1.04		DeTAB	1.28	
DAC	1.05		CTAB	1.30	
SDS	1.14		CTAC	1.35	
Brij 35	1.18		DTAC	1.37	
Triton X-100	1.32				

[a] K. Kalyanasundaram and J. K. Thomas, J. Am. Chem. Soc. 99, 2039 (1977).
[b] D. Dong and M. A. Winnik, Photochem. Photobiol. 35, 17 (1982).

sulfonate (ANS)], indole, and coumarin:

NPN   ANS   Indole

These compounds show large spectral shifts with increasing polarity (generally red shifts) accompanied by significant changes in the fluorescence lifetimes and quantum yields. Pyrene-3-carboxaldehyde,[16–20] p-N,N-dialkylamino-benzylidene malononitrile (DABN)[22] and the 7-alkoxycoumarins[21] belong to a class of neutral probe molecules whose emission properties change due to a shift in the nature of the emitting state from (n, π^*) to (π, π^*) upon increasing solvent polarity:

Py–CHO   DABN   7-Alkoxycoumarin

The resulting (π, π^*) state being very polar shows large red shifts. For recent discussions of solvent effects on the photophysical properties of these probes in homogeneous solvents, the reader should consult the following sources: pyrenecarboxaldehyde,[16–18] ANS.[51–53]

Barring a few exceptions, over a large solvent dielectric constant range (all protic solvents; $\varepsilon = 10$–80), pyrenecarboxaldehyde fluorescence shows a linearity between the fluorescence maximum λ_{max} and ε as shown in Fig. 2.2. The data can be fitted with a linear relationship of the form

$$\lambda_{max}(nm) = 0.52\varepsilon + 431.5 \tag{2.1}$$

Due to the presence of polar carbonyl group, pyrenecarboxaldehyde is solubilized near the micelle–water interface, and, hence, the fluorescence spectral shifts have been interpreted in terms of polarity at the micelle–water interfaces.[16,19] Table 2.3 contains a collection of ε estimates determined using this and several other probe molecules. The polarity estimates, in general, are in good agreement with similar estimates from electrophoretic measurements.

Traditionally the polarity (or effective dielectric constant ε) of probe binding sites is determined using the spectral shifts in the ground-state charge transfer absorption bands of various dye molecules.[54–59] In alkylpyridinium halides the absorption maxima (Kosower's Z value, for example) correspond to transitions from a predominantly dipolar ground state to an excited state of considerably reduced polarity. With increasing polarity of the medium the

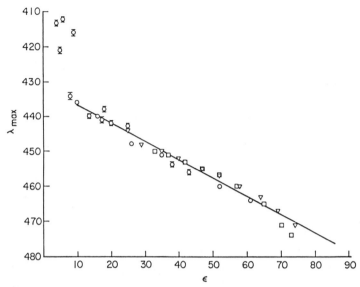

FIG. 2.2. Variation in the fluorescence maximum of pyrenecarboxaldehyde with the solvent dielectric constant: (○), methanol–water mixtures; (□), ethanol–water mixtures; (▽), dioxane–water mixtures, and (◌̄), neat solvents. (From Kalyanasundaram and Thomas.[16] Copyright 1977 American Chemical Society.)

dipolar ground state is more stabilized, and this causes a red shift in the absorption maximum, i.e., Z values increase with increasing polarity. We also use dyes which show large spectral shifts due to changes in the ground-state acid–base equilibria, as in phenols and coumarins. There have been numerous reports on ε estimates of binding sites using ground-state absorption changes in various micelles including alkylpyridinium iodide,[59] bromophenol blue,[58] alkylcoumarins,[56] pyridinium N-phenolbetaine,[55] aromatic ketones,[54] and simple arenes such as benzene and naphthalene.[57] In the majority of cases the ε estimates are in the range of 20–50, neither that of bulk water nor that of a medium exclusively made up of alkanes. The implications are that a good majority of aromatic solutes are solubilized at the micelle–water interface.

Otruba and Whitten[59a] have explored the solvent sensitivity of the absorption maximum of the charge transfer complex formed between the hydrocarbon t-stilbene and methyl viologen, as an indicator of the "solubilization" in micelles

$$HC \cdot MV^{2+} \xrightarrow[hv_{CT}]{} (HC^{+} \cdot MV^{+})^{*}$$

The charge transfer absorption blue shifts with increasing polarity (e.g., the maximum located at 492 nm in EtOH versus 445 nm in acetonitrile for t-4-hydroxystilbene). Efforts to correlate the hv_{CT} with the common single parameters of solvent polarity (e.g., Kosower's Z or Reichardt–Dimroth E_{T}

TABLE 2.3. Estimates on the Polarity at the Micelle–Water Interface using Different Medium-Sensitive Probes [a]

| | Fluorescence | | | | | Absorption | | | | |
| Micellar system | PyCHO | | DABN | ANS | | PhCOPh | | NPB | | Coumarin |
	ε	Z	ε	ε	Z	$E_T(30)$	Z	$E_T(30)$	ε	ε
SDS	45	88	40	25(?)	82(?)	58	88	57.5	51	32
SD	39		(40)							
DAC	26	83								
CTAB	16	80	36	36	85.5	51.5	79	53.4	31	32
Triton X-100	15	66	28	35	85.5	42.5	66	53.0	30	32
Igepal Co-630	12	66		35	85.0					
Brij 35								52.8	28	
CPyC	24	82								
CTAC	24									
References	15–19		22	29–31		54		55		56

[a] PyCHO, pyrene-1-carboxaldehyde; DABN, p-N,N-dialkylaminobenzylidinemalonitrile; ANS, 1-anilino-8-naphthalenesulfonate; NPB, N-phenylbetaine; and coumarin, 7-hydroxyalkylcouramin.

values) had only limited success, due to the sensitivity of these parameters to H-bonding effects. Interestingly good linear relationship of hv_{CT} with (I_1/I_3) values of pyrene monomer fluorescence was found. For several substituted stilbenes in SDS (C_{12}) and SDeS (C_{10}) micelles, a wide range of apparent polarity values Py_{eff} was obtained.

Demas and co-workers have examined the excited-state lifetime of a series of polypyridyl complexes of Ru(II) and Os(II) in various micellar solutions.[32–37] The measured lifetimes in H_2O and D_2O changed substantially on introduction of surfactants. Depending on the nature of the metal complex, the increases in lifetimes were either monotonic with increasing concentration of the surfactant (no clear CMC indication) or a sudden increase was observed at concentrations at or near the known CMC. To understand the observed effects the data have been analyzed in terms of a kinetic scheme involving association of the metal complex (D) with the surfactant monomers (S) and micellar assemblies (M):

$$D \xrightarrow{hv} D^* \xrightarrow{\tau_w} D + hv' \qquad \text{(water)} \qquad (2.2)$$

$$D + S \rightleftharpoons DS \xrightarrow{hv} DS^* \xrightarrow{\tau_s} DS + hv'' \qquad \text{(monomer)} \qquad (2.3)$$

$$D + M \rightleftharpoons DM \xrightarrow{hv} DM^* \xrightarrow{\tau_m} DM + hv'' \qquad \text{(micelle)} \qquad (2.4)$$

τ_w, τ_s, and τ_m denote the lifetime of the metal complex in water, in the associated complex with the surfactant monomer and micelle, respectively. The observed lifetime is given as the weighted sum

$$\tau_{obs} = \sum_i f_i \tau_i \qquad (2.5)$$

where the f_i represent the fraction of species in each category.

Using the lifetime data, a solvent exposure factor F (fraction of the surface of the sensitizer that is exposed to the aqueous medium when it is solubilized in the micelle) has been deduced:

$$F = \left(\frac{\tau[H(m)]^{-1} - \tau[D(m)]^{-1}}{\tau[H(s)]^{-1} - \tau[D(s)]^{-1}} \right) \qquad (2.6)$$

Here $\tau[H(s)]$ and $\tau[D(s)]$ refer to the lifetime of the complex in surfactant–free H_2O and D_2O, respectively, and $\tau[H(m)]$ and $\tau[D(m)]$ are the lifetimes of the micelle-bound complex in H_2O and D_2O, respectively. Based on the magnitude of F and the spectral shifts in the emission maxima, the metal complexes have been classified into four classes:

(1) large solvent exposures ($F > 0.5$) and negligible spectral shifts ($\Delta\lambda < 1$ nm)

(2) large solvent exposures ($F > 0.5$) and appreciable spectral shifts ($\Delta\lambda < 1$ nm)

(3)   low solvent exposures ($F < 0.5$) and appreciable spectral shifts ($\Delta\lambda >$ 1 nm)

(4)   small solvent exposures ($F < 0.5$) and negligible spectral shifts ($\Delta\lambda <$ 1 nm)

Anionic surfactants such as sodium dodecyl sulfate (SDS) are expected to interact substantially with cationic probes such as $Ru(bpy)_3^{2+}$ both in the micellar and monomeric forms. Examination of emission decays in SDS solutions at concentrations below the CMC do show a fast component preceeding the normal exponential decay.[38–40] Earlier the fast decay had been attributed to triplet–triplet annihilation reactions at high local concentration of the probes.[38] It is now believed that below the CMC, $Ru(bpy)_3^{2+}$ forms clusters with SDS anions which incorporate several emitter molecules (pre-micellar aggregates).[39,40] At SDS concentrations below the CMC, these $Ru(bpy)_3^{2+}$–SDS clusters behave as micelles in that they incorporate aromatic molecules which act as quenchers of $Ru(bpy)_3^{2+*}$. The concentration of the clusters have been found to be proportional to the probe concentration and decrease as SDS concentration increases. Similar pre-micellar interactions between monomeric surfactants and dye molecules have also been observed in acridine orange–SDS, and 1-anilinonaphthalene-8-sulfonate–SDS systems. The ANS however shows normal behaviour in nonionic micellar solutions.

2.3   Fluorescence Depolarization Studies

An early application of the fluorescence polarization measurements in the steady state has been in the determination of microviscosity of the micellar interior. The term *microviscosity* is used to distinguish the viscosity of the probe environment in the interior of the aggregate from that of the bulk solvent medium in which the aggregates are present. Some prefer the term *microfluidity*. The classical model of micelles pictures the interior of the micelles to be liquid-like, and, hence, it is of interest to measure directly the microviscosity η in different micellar systems.[60–67] Weber *et al.*[60] have shown that η can be derived from the measurements of mean degree of polarization ($\bar{p}$) in the steady state using Perrin's relation (cf. Section 1.2.3):

$$r_0/r = [(1/\bar{p}) - \tfrac{1}{3}]/[(1/p_0) - \tfrac{1}{3}] = 1 + (kT\tau/\eta V_0) \qquad (2.7)$$

Molecules such as 2-methylanthracene, perylene, and diphenylhexatriene (DPH) have been used as probes. All these probes are very hydrophobic and, hence, are expected to be solubilized in the inner hydrophobic regions. Typical estimates of η values for the micellar interior determined using various probes are (η values in centipoise are shown in parentheses): 2-methylanthracene:[60] DTAB (26), TTAB (32), CTAB (30); perylene:[60,66] DTAB (17), TTAB (21),

CTAB (19), SDS (10), CTAC (18); dansylamine:[61] CTAB (17), TTAB (16), cetylbetaine (31); and 6-indole-11:[66] CTAB (20) (cf. Table 1.1 for the abbreviation of various surfactants).

The measured microviscosities, as can be seen, are in the order of 15–30 cP for the normal micelles while the values are much higher ($\geqslant 100$ cP) for bile acid micelles such as sodium taurocholate (STC). Although these values are high compared to 1–2 cP observed for pure hydrocarbon liquids, they nevertheless stress the fluid nature of the micellar core. Steady-state measurements of fluorescence polarization can also be used to monitor structural changes that occur on addition of various molecules. Microviscosity changes have been recorded during the addition of alcohols, cholesterol, NaCl, and others to CTAB[60,61] and STC[63] micelles. Addition of electrolytes to micellar solutions of dodecylammonium chloride (DAC) causes phase transitions of normal spherical micelles to larger rod-shaped aggregates.[64] In the latter aggregates, hydrocarbon chains are expected to assume a bilayer configuration. The resulting increases in microviscosity are readily detected by inner probes such as 2-methylanthracene but not by surface probes such as ANS.

A few time-resolved fluorescence depolarization studies of rotational diffusion of probes in micelles have been made.[68–71] Picosecond transient absorption,[69] fluorescence,[71] and phase fluorometric techniques[70] have been used. In micellar media the measured rotational relaxation times τ_{rot} can originate from one or more of the following situations:

(1) unsolubilized probes in water,

(2) dye molecules are attached to the micelle (in the interior or at the surface) with some degree of freedom for reorientation of the dye while the micelles themselves undergo slow isotropic rotational diffusion,

(3) the dye molecules are rigidly bound or immobilized within, leading to rotation of the micelles themselves.

Based on the relative size/volume of the rotating dye molecules and of micelles in low-viscosity solvents (water) we would anticipate τ_{rot} to be in the range of few hundred picoseconds for the former and about 8–10 nanoseconds for the latter. Values in between these ranges would represent rotational diffusion of dye in/on the micellar aggregate.

When the dyes are not associated with the micelles for electrostatic reasons (e.g., rhodamine 6G–cetyldimethylammonium chloride and erythrosin B–SDS systems), the τ_{rot} values measured in micellar solutions are very close to those observed in water. Table 2.4 contains data on τ_{rot} values determined in micellar media. In the majority of cases we observe τ_{rot} in the range of 2–5 nsec in micellar media. Rose bengal in CTAB micelles showed a minor long-lived anisotropic decay component ($\tau_{rot} = 10.6 \pm 2.0$ nsec) which would correspond to case (3) mentioned above.[71] The intermediate τ_{rot} values have been

TABLE 2.4. Rotational Relaxation Times for Dye Molecules in Micellar Media

Dye	Medium	Temperature	$\tau_{rot}{}^a$	Reference
Rhodamine 6G	SDS	20°C	0.543, 2.17	68
Rose bengal	CTAB	25°C	2.40, 10.6	71
Rose bengal	SDS	25°C	1.27	71
Acridine orange	SDS	25°C	1.4	69
Acridine orange	Triton X-100	25°C	2.7	69
Crystal violet	SDS	25°C	1.75	69
Crystal violet	Triton X-100	25°C	4.1	69
Rhodamine 6G	SDS	25°C	2.2	69
Rhodamine 6G	Triton X-100	25°C	6.7	69
Rose bengal	Triton X-100	25°C	8.7	69
Cyanosine	Triton X-100	25°C	7.5	69

^{a} Values given in nanoseconds.

interpreted as anisotropic rotational diffusion with a 40° angle between the rotation axis and the transition dipole or isotropic diffusion under stick boundary conditions. Discussions on some of the inherent problems in measuring and interpreting the fluorescence polarizations can be found in later chapters (e.g., Sections 6.6 and 8.1.3).

2.4   Micellar Effects on Acid–Base Equilibria

Solubilization or adsorption of dye molecules in aqueous micellar solutions influences significantly the acid–base properties of the dye.[72-82] Table 2.5 presents a collection of data on the micellar effects on the ground state pK_a of various dye molecules.[56,72-78] The pK_a of acid–base reactions in micellar media is influenced first by the effective dielectric constant at the site of solubilization and second by the effect of charge distribution of the counterions in the double layer. The dielectric effect can be separated by using a nonionic surfactant such as Triton X-100 and noting the shift in the pK_a with respect to water. The lower dielectric constant of the micelle–water interface as compared to water favors the uncharged species (e.g., acid form of phenols and basic form of amines). Hence the pK_a measured in Triton X-100 is higher than that measured in water for, say, 2-naphthol but lower for acridine. The change in pK_a can be correlated with homogeneous solvents (taking care to include the species H_3O^+ in this case) to give a value for the apparent dielectric constant for the probe environment.

The charge effect in ionic micelles depends on the distribution of counterions: anionic micelles concentrate protons (and other positive ions) in the double layer and suppress the ionization of a proton from the probe, while with cationic micelles the reverse situation is true. Hence, anionics will raise

TABLE 2.5. Micellar Effects on the Ground-State pK_a of Dye Molecules

Dye	Micelle	Water	Micelle	ΔpK_a	Reference
7-Hydroxycoumarin (umbelliferone)	DTAB	7.75	6.85	−0.90	72
7-Hydroxycoumarin	Triton X-100	7.75	7.80	+0.05	72
7-Hydroxycoumarin	SDS	7.75	8.25	+0.50	72
Bromothymol blue	DTAB	7.18	6.45	−0.73	72
Bromothymol blue	SDS	7.18	8.35	+1.17	72
Methyl red	DTAB	4.95	3.68	−1.27	72
Methyl red	Triton X-100	4.95	5.18	+0.23	72
Methyl red	SDS	4.95	6.63	+1.68	72
Methylcobalamin	CTAB	2.63	2.65	+0.02	73
Methylcobalamin	SDS	2.63	5.73	+3.10	73
Methylcobalamin	Triton X-100	2.63	2.65	+0.02	73
Cyanocobalamin	CTAB	0	0	0	73
Cyanocobalamin	Triton X-100	0	0	0	73
Cyanocobalamin	SDS	0	3.2	+3.2	73
C_{11}-(7-Hydroxy)coumarin (alkylumbelliferone)	CTAB	7.75	6.35	−1.4	56
C_{11}-(7-Hydroxy)coumarin	Triton X-100	7.75	8.85	+1.1	56
C_{11}-(7-Hydroxy)coumarin	SDS	7.75	11.15	+3.4	56
C_{17}-(7-Amino)coumarin	Triton X-100	2.35	1.25	−1.1	56
C_{17}-(7-Amino)coumarin	SDS	2.35	3.55	+1.2	56
Dimethyl yellow	SDS	3.35	4.85	+1.5	74
Pyridine-2-azo-p-dimethylaniline	SDS	4.5	6.0	+1.5	74
2-Naphthol	SDS	9.4	10.7	+1.3	75
Proflavine	SDS	9.5	12.5	+3.0	76
Bromocresol green	CTAB	4.8	3.1	−1.7	77
Bromocresol green	Brij 58	4.8	6.45	+1.65	77
Bromocresol green	Triton X-100	4.8	6.2	+1.40	77
Bromocresol green	Tween 80	4.8	6.0	+1.20	77
Bromocresol green	CTAB	4.60	3.35	−1.25	78
Bromophenol red	CTAB	5.95	4.90	−1.05	78
Bromothymol blue	CTAB	7.10	6.20	−0.90	78
Thymol blue	CTAB	8.60	8.40	−0.2	78

the pK_a of the dyes (with respect to Triton X-100), and cationic ones will lower it. The charge effect is much larger than the dielectric effect. The shift in pK_a (ΔpK_a) upon solubilization/adsorption to anionic micelles can be as much as 3 pH units.

The large shifts in the ground state pK_a of dye molecules upon solubilization in micelles leads to novel effects of deprotonation/protonation of the dye under certain pH ranges while no such reactions occur in aqueous solutions. At a given solution/bulk pH, the fluorescence of a dye can be

from the acid form (ROH or RNH_3^+) when it is in the micelles even if the
fluorescence in aqueous solution is from the basic form (RO^- or RNH_2):

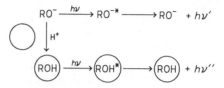

In amines RNH_2 is a better quencher than RNH_3^+. With these above the
protonations can lead to inhibitive effects on the fluorescence quenching
efficiency of amines. This has been shown to be the case for linked anthracene–
dimethylaniline (Anth–$(CH_2)_n$–DMA) in SDS micelles.[81]

Solubilization of molecules in micellar media can also affect the excited-
state acid–base reactions as has been observed in dyes such as naphthols.
2-naphthol has a pK_a of 9.4 in water. In the first excited singlet state it is a
very strong acid (pK_a^* drops to 2.8). In aqueous solutions at $pK_a^* < pH < pK_a$,
the proton equilibrium, however, is not set up completely during the lifetime
of the excited state, so that fluorescence is observed from both NpOH* and
NpO^-*:

$$NpOH^* \underset{pK_a}{\rightleftharpoons} NpO^-* + H^+ \tag{2.8}$$

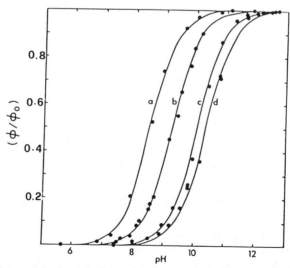

FIG. 2.3. Fluorescence titration of 1-naphthol (9.8×10^{-6} M) against pH in aqueous and
various micellar solutions: (a) CTAB (cationic) (5.5×10^{-3} M), $pK_s = 8.53 \pm 0.10$; (b) water,
$pK = 9.23 \pm 0.09$, (c) SDS (anionic) (4.9×10^{-2} M), $pK_s = 10.04 \pm 0.11$; and (d) 16A16
(nonionic) (4.3×10^{-3} M), $pK_i = 10.30 \pm 0.14$. (From Selinger and Harris.[82a])

In aqueous micellar SDS solution 2-naphthol distributes between water and the micelles. While the part of 2-naphthol in water shows the pH-dependent fluorescence change, the part dissolved in the micellar phase does not show the fluorescence of the naphtholate anion even when the surrounding pH is basic. Increasing the micelle concentration increases the fluorescence of the undissociated (NOH*) forms.[79–81]

Molecules such as 1-naphthol and pyrene-1-amine in aqueous solution show complete equilibration for excited-state reactions, but in the presence of SDS micelles only partial equilibration occurs.[81,82] Since the protonation/deprotonation reactions of the excited state are diffusion controlled (viscosity dependent), these changes have all been interpreted as due to the micellar viscosity effect. Figure 2.3 presents fluorescence titration curves for 1-naphthol in various micellar media. Pileni and Grätzel have made an interesting study of micellar effects (SDS) on the pK_a of the ground, singlet, and triplet states of proflavine.[76] The scheme below outlines the shifts in pK_a for ground and excited states on micellar solubilization:

H_2N—[structure]—N—[structure]—NH_2 $\xrightarrow{H^+}$ H_2N—[structure]—$\overset{+}{N}\!\!-\!\!H$—[structure]—NH_2 $\xrightarrow{H^+}$

(PF)   (PFH$^+$)

H_2N—[structure]—$\overset{+}{N}\!\!-\!\!H$—[structure]—$\overset{+}{N}H_3$

(PFH$_2{}^{2+}$)

			pK (H$_2$O)	pK (SDS)
PF $\underset{\longleftarrow}{\xrightarrow{H^+}}$ PFH$^+$ $\underset{\longleftarrow}{\xrightarrow{H^+}}$ PFH$_2{}^{2+}$			0.2, 9.5	?, 12.5
$\downarrow h\nu$ $\downarrow h\nu$ $\downarrow h\nu$				
^{1}PF* $\underset{\longleftarrow}{\xrightarrow{H^+}}$ ^{1}PFH^{+*} $\underset{\longleftarrow}{\xrightarrow{H^+}}$ ^{1}PFH$_2{}^{2+*}$			1.5, 12.5	3.5, 11.5
$\downarrow$				
^{3}PF* $\underset{\longleftarrow}{\xrightarrow{H^+}}$ ^{3}PFH^{+*} $\underset{\longleftarrow}{\xrightarrow{H^+}}$ ^{3}PFH$_2{}^{2+*}$			4.0	3.5

Closely related to these are the recent studies of laser pH jump initiated proton transfer on charged micellar surfaces. The technique utilizes a proton donor which is a very much stronger acid in its singlet excited state than in the ground state (i.e., $pK^* \ll pK$). Excitation by a strong but short laser pulse (1–10 mJ, 2–15 nsec) at a suitable pH ($pK^* < pH < pK$) results in proton ejection. The ejected proton can, in turn, react with appropriate indicator dyes.

TABLE 2.8. Studies of Fluorescence Quenching of Solubilized Probes by Cationic Quenchers in Anionic Micelles (SDS)

Probe	Quencher	Reference
Pyrene	Cu^{2+}	129
Pyrene	Cu^{2+}, Tl^+	130
RuL_3^{2+}, L=phen	C_n-V^{2+} (viologen)	131
9-Methylanthracene	Tl^+	132
Perylene	Aryldiazonium salts	133
Pyrene	Cu^{2+}, Hg^{2+}, Eu^{3+}	134
Anthracenes	Aryldiazonium salts	135
$Ru(bpy)_3^{2+}$	Cu^{2+}, Cr^{3+}, Fe^{3+}	136
Pyrene	Ag^+, Tl^+, Ni^+, Sm^{3+}	137
$Ru(bpy)_2(CN)_2$, RuL_3^{2+}	Cu^{2+}, MV^{2+}	138
C_n-Stilbene, DPB	MV^{2+}	139
Methylpyrene, pyrene	Tl^+, Cu^{2+}, Eu^{3+}	140
C_n-Stilbene	MV^{2+}	141
Pyrene	MV^{2+}	142
Pyrene	Cs^+, Cu^{2+}, Tl^+	143
$Ru(bpy)_2L^{2+}$, L = C_n-bpy	MV^{2+}	144
RuL_3^{4-}, L = bpy—$(COO^-)_2$	Cu^{2+}	145
Pyrene	Tl^+, Ag^+, Cu^{2+}, Eu^{3+}	146
Naphthalene	Ni^{2+}	147
$Ru(bpy)_3^{2+}$	MV^{2+}	148
ω-(α-naphthyl)dodecanoic acid	Cu^{2+}, Eu^{2+}	149
Pyrene	Cu^{2+}	128
Pyrene	Cu^{2+}, MV^{2+}, CPyC	150
Anthracenes	Cu^{2+}	96
Indole derivative (6-In-11)	Co^{2+}	117
Naphthalene	Cu^{2+}	120
$Ru(bpy)_2L^{2+}$	Cu^{2+}, MV^{2+}	100

(mean occupancy number per micelle ≪ 1) exhibit single exponential fluorescence decay and lifetime much longer than that observed in water, thus suggesting their total association with the micellar aggregates.

In some cases though there can be a significant partitioning of the probe between the micellar pseudo-phase and the aqueous solution as has been observed with naphthalene in SDS micelles.[127] The fluorescence decay can then be deconvoluted into two distinct decays τ_1 and τ_2 (with τ_2 close to the value observed in water)

The relative fraction of the two components to the total fluorescence reflect the partitioning of the probe between the two phases. Quantitative analysis in this case can lead to evaluation of the partition coefficient for the probe in the two-phase system.

For the rest of the discussion we will consider *exclusively* the cases in which the probe is quantitatively associated with the micellar aggregates at very low loading levels and the intensity of the excitation source is kept quite low. Under such conditions, we have mostly empty micelles and monomer surfactants with some of the micelles containing at most one fluorescent probe. [Multiple occupancy of probes such as pyrene can lead to excimer formation. This can be used to obtain the mode of distribution (statistics) of probes among the micelles, as will be shown later.] Since micellar aggregation is a dynamic process these assumptions imply that the decay of the excited probe is much faster than the average residence time in the micelle. Phosphorescence studies of solute entry and exit rates (discussed in Chapter 3) indicate that for fluorescence this is indeed true.

With the introduction of the quenchers we can have several situations:

(1) like the probe molecules, the quenchers associate solely with the micelles,

(2) the quencher partitions significantly between the micellar and aqueous phases, and

(3) the quencher molecules exchange between the micelles in addition to partitioning with the aqueous phase.

In each of these cases the quenching interactions of the excited state with the quenchers (cohabitants of the micelle) can be static or dynamic. (In the static case only those fraction of P_m^*-containing micelles that are free of quenchers fluoresce. In the other, the lifetime of the probe P^* in a given micelle is dependent on the number of cohabitant quencher molecules. We will elaborate more on the distinction later.) Thus, we can identify at least five limiting situations which we will discuss in some detail. In all cases the distribution of the probes and quencher molecules among the micelles is assumed to be poissonian (cf. Section 1.4). It is also assumed that the rate of quenching in a micelle containing n quencher molecules is n times that in a micelle with just one quencher. That is, the quenching ability of a Q molecule is independent of the presence of other Q molecules in the same micelle.

For quencher molecules which partition significantly between the micellar and aqueous phases, we can define an equilibrium constant K in the following manner:

$$K = [Q_m][Q_w][M] = (k_+/k_-) \qquad (2.11)$$

where $[Q_m]$ and $[Q_w]$ refer to the concentration of the quencher in the micellar and aqueous phases, respectively and [M] is the total micelle concentration. This definition of K coupled with the following definitions of the mean occupancy number $\bar{n}$ and total quencher concentration $[Q_t]$,

$$\bar{n} = [Q_m]/[M] \qquad (2.12)$$

$$[Q_t] = [Q_m] + [Q_w] \qquad (2.13)$$

allow the existence of the following interrelationships which are useful in simplifying the kinetic expressions derived for the various quenching scenarios:

$$[Q_w] = [Q_m]/K[M] = \bar{n}/K \qquad (2.14)$$

$$[Q_w] = [Q_t] - [Q_m] = [Q_t]/(1 + K[M]) \qquad (2.15)$$

$$[Q_t] = [Q_m] + [Q_w] = \bar{n}([M] + 1/K) \qquad (2.16)$$

$$\bar{n} = K[Q_t]/(1 + K[M]) \qquad (2.17)$$

To describe the decay of an excited state probe in a micelle containing m quenchers and at most one probe molecule the following rate constants are introduced:

(1) $k_0 = (1/\tau_0)$, the reciprocal lifetime of the excited probe fluorescence (unquenched) (in $\sec^{-1}$),

(2) k_+, the entry of a quencher molecule into a micelle (in $M^{-1}\ \sec^{-1}$),

(3) k_-, the exit of a quencher molecule from a micelle (in $\sec^{-1}$), and

(4) k_q is the quenching rate of an excited probe by one quencher molecule in the micelle (in $\sec^{-1}$).

2.5.1 Static Quenching of Excited States

Case A: Static Quenching Where the Quencher is Totally Micellized

Here we consider a case in which both the probe and the quencher are solely associated with the micellar phase ($[Q_t] = [Q_w] + [Q_m] \simeq [Q_m]$). The quenching in or on a micelle containing an excited probe and a quencher is much faster than the fluorescent decay (static quenching) so that, in effect, fluorescence is observed only from micelles devoid of quenchers. Although the overall luminescence intensity is thus reduced, the measured lifetime τ remains independent of the added quencher concentration. Figure 2.4 presents schematically the shape of the fluorescence decay curves with and without the quenchers for this case:

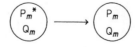

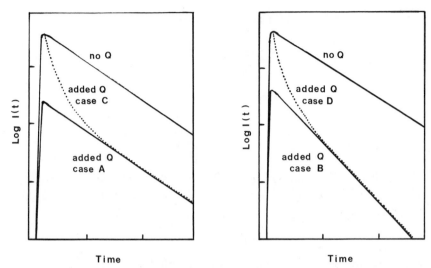

FIG. 2.4. Schematic representation of fluorescence decay curves of a probe under different quenching scenarios. Case A: static quenching where the quencher is totally micellized. Case B: static quenching where the quencher is partially micellized. Case C: dynamic quenching with totally micellized quencher. Case D: dynamic quenching where the quencher is partially micellized. (From Yekta et al.[153])

The relation between the fluorescence intensity I and the total quencher concentration $[Q_t]$ is given by

$$\ln(I/I_0) = [Q_t]/[M] = \bar{n}, \quad (\because [Q_t] \simeq [Q_m]) \qquad (2.18)$$

Recall that the micellar aggregation number N, the concentration of micelles $[M]$ and of the surfactant $[Surf]$, are related by

$$[M] = ([Surf] - [CMC])/N \qquad (2.19)$$

Equations (2.18) and (2.19) provide a simple steady-state method for the determination of the micellar aggregation number N. Quenching of $Ru(bpy)_3^{2+*}$ in SDS micelles by 2-methylanthracene and methylphenothiazine have been analyzed in this manner. We elaborate more on this point in the next section.

Case B: Static Quenching Where the Quencher is Partially Micellized

Here the case considered is similar to case A but with the relaxation of the requirement that the quencher be also fully associated with the micelle. The quenchers have some hydrophilic character so as to partition significantly between the micellar and the aqueous phases:

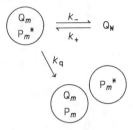

The lifetime of the emitting state P* in this case is reduced by the dynamic diffusional quenching of P* by water solubilized quencher molecules $[Q_w]$. This is expressed as

$$1/\tau = (1/\tau_0) + k_+[Q_w] \tag{2.20}$$

where the rate constant of quenching equals the forward rate constant k_+ since every association results in complete quenching. Figure 2.4 presents schematically the expected luminescence decay curves corresponding to case B. Substituting Eq. (2.15) for the term $[Q_w]$ in Eq. (2.20) and rearranging, we have

$$(1/\tau) - (1/\tau_0) = k_+[Q_t]/(1 + K[M]) \tag{2.21}$$

For steady-state irradiation the relative fluorescence intensities are given by

$$(I_0/I) = (1 + k_+\tau_0[Q_w]) \exp\{K[Q_t]/(1 + K[M])\} \tag{2.22}$$

According to Eq. (2.21), the observed luminescence lifetime will increase with increasing surfactant concentration (hence, of the micelles). This is simply the result of the fact that increasing the micelle concentration decreases the concentration of active quenchers that are in water.

2.5.2 Dynamic Quenching of Excited States

The most interesting cases however are those that involve dynamic quenching of the excited state. (The quenching process is identified as dynamic whenever there is a relative motion of probes and quenchers toward each other, and this process is affected by medium viscosity, temperature, etc.) We will consider three limiting cases here.

Case C: dynamic quenching with totally micellized, immobile quenchers,

Case D: dynamic quenching with partially micellized quenchers (quenchers which partition between the micellar and aqueous phases), and

Case E: dynamic quenching with ionic quenchers which partition between the micellar and aqueous phases and also exchange between micelles.

Case C: Dynamic Quenching with Totally Micellized,
Immobile Quenchers

We consider here as in case A micellar systems in which the quencher and the probe molecules are associated almost exclusively with the micellar phase, i.e., $[Q_m] \gg [Q_w]$ so that the total/bulk concentration of the quencher is given by $[Q_t] = [Q_m] + [Q_w] \simeq [Q_m]$. It is assumed that for a given micelle the quenching probability is proportional to the number of cohabiting quencher molecules. The decay rate of a fluorescent probe in a micelle with i quenchers is given by

$$k_0 = (k_0 + ik_{qm}) \qquad (2.23)$$

where k_0 is the rate constant for the unquenched probe decay and k_{qm} is the first-order rate constant for the quenching by one quencher molecule. The quenching process described is intramicellar and by analogy to intramolecular processes is assumed to be a first-order process (units are sec^{-1}). The observed fluorescence signal is given by the sum of contributions from micelles with different number of quenchers:

$$F(t) = \sum_{i=0} P_i F(0) \exp(-k_i t) \qquad (2.24)$$

where P_i is the probability of finding a probe molecule in a micelle with i quenchers. Assuming poissonian distribution for P_i prior to and during the quenching process,

$$P_i = (\bar{n}^i e^{-\bar{n}})/i! \qquad (2.25)$$

where $\bar{n}$ is the mean occupancy number of quenchers in a micelle, we have

$$F(t) = \sum_{i=0} (\bar{n}^i e^{-\bar{n}}/i!) F(0)[\exp - (k_0 + ik_q)t]$$

$$= F(0) \exp(-k_0 t) \sum_{i=0} (\bar{n} e^{-k_q t})^i / i!$$

$$= F(0) \exp[-k_0 t] + \bar{n}[\exp(-k_q t) - 1]$$

or

$$\ln[F(t)/F(0)] = -k_0 t + [\bar{n}(e^{-k_q t} - 1)] \qquad (2.26)$$

Equation (2.26) can be written in a more general form so that we can identify its functional form with those of other cases:

$$\ln[F(t)/F(0)] = -At + B(e^{-Ct} - 1) \qquad (2.27)$$

where $A = k_0$, $B = \bar{n}$, and $C = k_q$. Equation (2.27) describes the decay of fluorescence after a short pulse excitation and suggests that the fluorescence decay is nonexponential with a fast initial decay arising from the rapid quenching of P* in micelles which contain the quenchers. If $t \to \infty$ then

Eq. (2.27) reduces to

$$F(t) = F(0) \exp(-\bar{n} - k_0 t) \tag{2.28}$$

If $t \to 0$, then $\exp(-k_q t) = (1 - k_q t)$ and Eq. (2.27) reduces to

$$F(t) = F(0) \exp[-(k_0 + \bar{n} k_q)t] \tag{2.29}$$

Thus, in this model the logarithmic fluorescence decay curves for different values of $\bar{n}$ (i.e., different [Q]) should be a family of curves with equal slopes at long times. At short times, however, the slopes depend on [Q]. Figure 2.4 also presents schematically the decay curves for case C. For steady-state irradiation the relative fluorescence intensities are given by

$$(I/I_0) = e^{-\bar{n}} \sum_{i=0} \frac{\bar{n}^i}{i![1 + i(k_q \tau_0)]} \tag{2.30}$$

Case D: Dynamic Quenching with Mobile (Neutral) Quenchers
Which Partition between the Micellar and Aqueous Phases

We now consider a more general case in which the quencher molecules partition between the aqueous and micellar pseudo-phases with an associated equilibrium constant K. The scheme below outlines various processes that define the fluorescence decay of the probe molecules solubilized in the micelle:

Here S_m and S_{m-1} represent micelles each with one probe and with m or $(m-1)$ number of quencher molecules at the time of δ-pulse excitation. The rates of the exit, quenching processes are assumed to depend on the number m of quencher molecules in the micelle and are given by $(mk_-[S_m^*])$ and $(mk_q[S_m^*])$, respectively, where $[S_m^*]$ is the concentration of micelles with m quenchers and one excited probe.

The rate equation for this situation is given by

$$-\frac{d[S_m^*]}{dt} = (k_0 + k_+[Q_w] + mk_- + mk_q)[S_m^*] - (m+1)k_-[S_{m+1}^*]$$

$$+ k_+[Q_w][S_{m-1}^*], \qquad m = 0, 1, 2, \dots \tag{2.31}$$

If the concentration of the fluorescent molecules in the micelles and the intensity of exciting light are kept low, for a poisson distribution of quenchers, the following decay law for the fluorescence signal can be derived:

$$F(t) = F(0) \exp[-At + B(e^{-Ct} - 1)]$$

or

$$\ln[F(t)/F(0)] = [-At + B(e^{-Ct} - 1)] \tag{2.32}$$

where

$$A = k_0 + \alpha[Q_t] \qquad \text{with} \qquad \alpha = \frac{k_q k_+}{(k_q + k_-)(1 + K[M])} \tag{2.33}$$

$$B = \beta[Q_t] \qquad \text{with} \qquad \beta = \frac{(k_q^2 k_+)}{k_-(k_q + k_-)^2(1 + K[M])} \tag{2.34}$$

and

$$C = (k_q + k_-) \tag{2.35}$$

Using the relationship $\bar{n} = K[Q_t]/(1 + K[M])$ [Eq. (2.17)] the above expressions can also be written as

$$A = k_0 + \left[\frac{k_- k_q n}{(k_q + k_-)} \right] \tag{2.36}$$

$$B = n[k_q^2/(k_q + k_-)^2]$$

Note that the functional form of Eq. (2.32) is similar to that derived for case C [Eq. (2.27)]. From a detailed kinetic analysis of the time-resolved fluorescence decay curves under various conditions, it is possible to evaluate the rate constants of interest (k_q, k_+, and k_-) using these equations. At long times ($t \to \infty$) Eq. (2.32) reduces to

$$F(t) = F(0) \exp(-At - B) \tag{2.37}$$

Thus a plot of log $F(t)$ versus t will give a slope of A and intercept B from the emission curves at longer times. At short times ($t \to 0$) we have

$$F(t) = F(0) \exp -At \tag{2.38}$$

By integrating Eq. (2.31) over all times it is possible to derive the following expression for steady-state experiments:

$$\frac{I}{I_0} = \frac{\tau}{\tau_0} \exp(-n\alpha^2) \sum_{i=0} \frac{(n\alpha^2)^i}{i![1 + (k_q + k_-)]} \tag{2.39}$$

where α is given by

$$\alpha = [k_-/(k_- + k_q)] \tag{2.40}$$

Case E: Dynamic Quenching with Ionic Quenchers that Exchange between the Micellar and Aqueous Phases and also between Micelles

In recent studies of fast reactions in micellar solutions involving ions of charge opposite to that of the micelles, a dependence of the exit rates on micelle concentration has been observed. This has made it necessary to extend the previously formulated kinetic scheme (case D) to include a "hopping" process in which transfer of quencher occurs upon micellar collisions or close approach in addition to the normal migration in the aqueous phase:

$$S_m \quad \underset{(2)\ k_0 + mk_q}{\overset{(1)\ h\nu}{\rightleftarrows}} \quad S_m^* \quad \underset{(4)\ k_+\ [Q_w]}{\overset{(3)\ mk_-}{\rightleftarrows}} \quad S_{m-1}^* + Q_w$$

$$(6)\ k_{ex} \Big\updownarrow k_{ex}\ (5) \qquad Q_j$$

$$S_{m+1}^* \quad + \quad Q_{j-1}$$

As in case D, S_m, S_{m-1}, and S_{m+1} represent micelles each with one probe and with m, $(m-1)$ and $(m+1)$ quencher molecules, respectively; Q_w represents free quencher ions in the aqueous phase and $\left(Q_j \right)$ a micelle with j associated quencher ions. It is tacitly assumed that the entry and exit rate constants k_+ and k_- (and hence their ratio $K = k_+/k_-$) are independent of the micelle concentration. The concentration effects mentioned earlier are prescribed by processes (5) and (6). These hopping processes do not necessarily imply that real micellar collisions occur. A substantial increase in transfer rate could result from close approaches of two micelles. Analysis leads to essentially the same form of the rate law:

$$F(t) = F(0) \exp[-At + B(e^{-Ct} - 1)] \tag{2.41}$$

where

$$A = k_0 + \alpha[Q_t] \quad \text{with} \quad \alpha = \left\{ \frac{k_q(k_+ + k_{ex}K[M])}{(1 + K[M])(k_q + k_{ex}[M] + k_-)} \right\} \tag{2.42}$$

$$B = \beta[Q_t] \quad \text{with} \quad \beta = \left\{ \frac{k_q^2}{(k_q + k_{ex}[M] + k_-)^2(1 + K[M])(k_- + k_{ex}[M])} \right\} \tag{2.43}$$

$$C = (k_q + k_{ex}[M] + k_-) \tag{2.44}$$

At long times ($t \to \infty$) Eq. (2.41) reduces to

$$F(t) = F(0) \exp(-At - B) \tag{2.45}$$

Thus, a plot of log $F(t)$ versus t will give a slope of A and intercept B from the emission curves at longer times. At short time ($t \to 0$) we can use the approximation $\exp(-C) \simeq (1 - C)$ and Eq. (2.41) can be reduced to

$$F(t) = F(0) \exp - (A + BC)t \tag{2.46}$$

Tables 2.9 and 2.10 summarize the rate constant data obtained for the quenching process (k_q), solute entry (k_+), exit (k_-), and exchange (k_{ex}) for

TABLE 2.9. Rate Constant Data on the Micellar Entry (k_+), Exit (k_-), and Quenching (k_q) for Neutral Quenchers (Case D)

Quencher[a]	Probe/micelle	$k_+ (M^{-1} \text{sec}^{-1})$ $(\times 10^9)$	$k_- (\text{sec}^{-1})$ $(\times 10^5)$	$K (M^{-1})$ $(\times 10^4)$	$k_q (\text{s}^{-1})$ $(\times 10^7)$	References
RI, R = ethyl	Pyrene/SDS	9.7	83	0.12		90
RI, R = butyl	Pyrene/SDS	8.8	14	0.63		90
RI, R = hexyl	Pyrene/SDS	5.4	7.5	1.04		90
RI, R = octyl	Pyrene/SDS	5.9	4.0	1.65		90
m-DCB	1-Methylpyrene/CTAC		56	0.1865	3.1	83
m-DCB	1-Methylpyrene/SDS	10	50	0.19	3.5	88
m-DCB	1-Methylpyrene/SDS + 0–1 M NaCl	7.4	46	0.168	2.9	88
m-DCB	1-Methylpyrene/SDS + 0.3 M NaCl	8.6	60	0.14	2.6	88
m-DCB	Pyrene/SDS	11.5	76	0.15	2.7	93
p-CNT	DEII		61	0.19	7.5	93
Benzene	Pyrene			0.10		93
Toluene	Pyrene			0.53		93
CH$_2$I$_2$	Pyrene	25	95	0.26	7.5	102,152
O$_2$	Dimethylnaphthalene/SDS	14	530	0.026		104

[a] m-DCB = m-dicyanobenzene; DEII = dimethylindoloindole; p-CNT = p-cyanotoluene.

TABLE 2.10. Micellar Entry (k_+), Exit (k_-), Exchange (k_{ex}), and Quenching (k_q) Rate Constants for Ionic Quenchers (Case E)

Quencher	Probe/micelle	$k_+ (M^{-1}\ \text{sec}^{-1})$ ($\times 10^9$)	$k_- (\text{sec}^{-1})$ ($\times 10^5$)	$K (M^{-1})$ ($\times 10^4$)	$k_{ex} (M^{-1}\ \text{sec}^{-1})$ ($\times 10^8$)	$k_q (\text{sec}^{-1})$ ($\times 10^7$)	References
Cu^{2+}	1-Methylpyrene/SDS	1.2	1.2	1.0	6	2.7	128,140
Eu^{3+}	1-Methylpyrene/SDS	1	1	0.55	1	1.6	128,140
Cr^{3+}	1-Methylpyrene/SDS	1	1	0.59	1	0.98	128,140
Ni^{2+}	1-Methylpyrene/SDS	1	1	1.09	5	0.89	128,140
Co^{2+}	1-Methylpyrene/SDS	1	1	0.87	7	0.54	128,140
Pb^{2+}	1-Methylpyrene/SDS	7	3	2.77	3	0.91	128,140
Tl^+	1-Methylpyrene/SDS	29	76	0.38	55	1.9	128,140
Ag^+	1-Methylpyrene/SDS	16	80	0.2	80	1.6	128,140
Cs^+	1-Methylpyrene/SDS	20	20	1.0	(60)	0.027	128,140
I^-	Pyrene/DDTAC	50	24	2.0	6–10		113
Cu^{2+}	Pyrene/SDS	29	48	6.0	30	1.2	140
Ag^+	Pyrene/SDS	59	45	0.13		4.1	146
Tl^+	Pyrene/SDS			0.11			146
Eu^{3+}	Pyrene/SDS					0.9	146
Cu^{2+}	Pyrene/SDS					1.2	146
I^-	1-Methylpyrene/CTAC		36	0.7275	30	5.8	83
Cu^{2+}	Pyrene/SDS	1.2	1.2	1.0	4.3	2.5	130

various ionic and neutral quencher molecules in micellar media (cases D and E). It can be readily verified that case C is the limiting condition of cases D and E when $k_- \ll k_q$ and $(k_- + k_{ex}[M]) \ll k_q$, respectively. The entry rate constants k_+ are in the range of diffusion-controlled rates that we can calculate using the Debye–Smolouchowski equation. In all of the cases considered (A–E) the kinetic rate expressions given were all derived with the following principal assumptions: (a) the poisson distribution describes the statistics of distribution of solutes among the micelles; (b) there is no limit on the number of solutes that can occupy a given micelle; (c) the quenching rate constant in a micelle with m quenchers is (mk_q) where the first-order rate constant for quenching by one quencher molecule; and (d) the exit and entry rates are given by (mk_-) and $(k_+[Q_w])$, where k_- is the first-order exit rate of a single solute, k_+ the second-order rate constant for solute entry. Analysis in terms of distributions other than poissonian should also be considered[158] (cf. Section 1.4 for a discussion on other distribution laws). The following distribution law, for example,

$$P_i = [\bar{n}^i/(\bar{n} + 1)^{1+i}] \tag{2.47}$$

appears to fit better the fluorescence of pyrene in C_nTAB $(n = 10\text{--}16)$ micelles. Similarly the number of solutes that can be solubilized can be limited and a modified rate constant k'_+ considered:

$$k'_+ = k_+[1 - (n/m)], \qquad n = 0, 1, 2, \ldots, m \tag{2.48}$$

where m is the limit to the number of solubilized molecules. Tachiya has derived explicit kinetic expressions for several of the cases considered earlier.[155]

2.6 Fluorescence Quenching II: Applications

Fluorescence quenching studies carried out under a wide variety of conditions have provided detailed and quantitative information on the dynamic nature of the micellization process (for example, aggregation number, nature and extent of counterion binding and exchange, and influence of additives) and the solubilization process (entry and exit rates of solutes and partition coefficients). We consider some of these applications in this section. Depending on the nature of the solubilization of the probe and quenchers, the quenching process can occur mainly in the aqueous phase or on the micellar pseudo-phase either in the interior or at the surface. Kinetic treatments for the five limiting cases considered in the previous section provide a means of obtaining information of interest such as k_-, k_+, k_q, k_{ex}, and N. One such application involve determination of the micellar aggregation number using probes which are associated exclusively with the micellar phase.

2.6.1 Micellar Aggregation Number Determination

Method I

One popular method is based on the intramolecular quenching of an excited-state molecule (case C) by another ground-state molecule (self-quenching) as in the case of pyrene:

$$^1Py^* + Py \rightleftharpoons \{Py\cdot\cdot Py]^* \qquad 2\,Py \qquad (2.49)$$

At very low probe concentration (mean occupancy number $\bar{n} \ll 1$), we observe essentially the monomer fluorescence. With increasing $\bar{n}$, the concentration of multiply occupied micelles increases. While the pyrene in singly occupied micelles will exhibit normal fluorescence, those with multiply occupied probes will undergo dynamic quenching as represented by Eq. (2.49). Recall that under such conditions the observed monomer fluorescence is given by

$$\ln[F(t)/F(0)] = -k_0 t + \bar{n}[\exp(-k_q t) - 1] \qquad (2.31)$$

where $\bar{n}$ is the mean occupancy number of probes in the micelle and k_0 the rate constant for unquenched monomer fluorescence decay. The fluorescence decay thus exhibits a biphasic character with a initial fast component as illustrated in Fig. 2.5. At long times ($t \to \infty$) Eq. (2.31) reduces to the linear form

$$\ln[F(t)/F(0)] = -(\bar{n} + k_0 t) \qquad (2.33)$$

According to Eq. (2.33), the limiting slope of the plots of ln $F(t)/F(0)$ versus t is $-k_0$ and the extrapolation of this linear region to $t = 0$ will give $\bar{n}$. Since $\bar{n}$ is related to the micellar aggregation number N by the relation

$$\bar{n} = [Q_t]/[M] = [Q_t]N/([Surf] - [CMC]) \qquad (2.19)$$

knowing all the concentration terms, this method can be used to determine N.

It should be pointed out that this method based on excimer formation assumes (in addition to those cited under case C) that the product excimer Py_2^* dissociation is a process much less probable than the excimer decay and the excimer formation is intramicellar in nature. The method was originally proposed in 1979.[154,163] It is now used widely to determine micellar aggregation number N for new surfactant systems[164-167] and also to study the effect of additives (alcohols, electrolytes, etc)[168-172] on N. The method is very general and is not restricted to excimer-forming probes. Any intramicellar quenching conforming to case C can be analysed in this manner to derive N.

Method II:

In 1978 Turro and Yekta[98] proposed a fluorescence quenching method based on case A discussed earlier. Recall that in case A the fluorescence quenching in on the micelle is much faster than the fluorescence decay so that

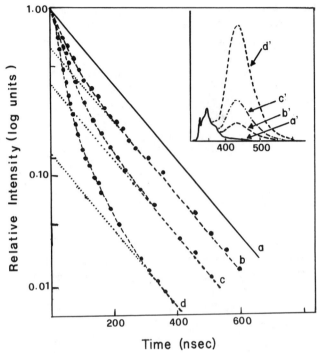

FIG. 2.5. Fluorescence decay curves for pyrene monomers in cetyltrimethylammonium chloride (CTAC) micellar solutions (10^{-2} M) at various pyrene concentrations: (a) 7.5×10^{-6} M, (b) 5.2×10^{-5} M, (c) 1.04×10^{-4} M and (d) 2.08×10^{-4} M. ($\bigcirc$), experimental points; (- - -), fitted curves according to Eq. (2.73). See text for details. Insert: Steady-state emission spectra of corresponding solutions, normalized to the monomer emission of solution (a) (7.5×10^{-6} M). (From Atik et al.[163])

fluorescence in effect is observed only from micelles devoid of quenchers. With the distribution of probes, quenchers taken as poissonian, the fluorescence intensity ratio with and without the quencher is simply related to $\bar{n}$, the mean occupancy number. The micellar aggregation number N is given by

$$\ln(I/I_0) = -[Q_t]/[M] = \bar{n} \qquad (2.18)$$

The method was originally applied to the quenching of $Ru(bpy)_3^{2+}$* by 2-methylanthracene in SDS micelles. The attraction of this method is that it is a simple steady-state method involving measurement of relative intensities of the probe luminescence with and without quenchers. Time-resolved studies by Almgren and Lofroth[94] and by Rodgers and Baxendale[95] have shown that for this probe–quencher pair the quenching process is best described by case C rather than by case A. It can be pointed out that the difference between cases A and C lies in the relative magnitudes of the fast decay component with respect to the total emission. In both cases, the long time decay is characterized by unquenched probe luminescence $(k = k_0)$. If the quenching process is extremely rapid as in static quenching, then the contribution from quenched probe molecule emission will be negligible and one observes in essence a single exponential decay (case A). The $Ru(bpy)_3^{2+}$ –methylanthracene system in SDS can be treated under case C along lines similar to that outlined for pyrene excimers. For cases which conform strictly to case A, the applicability of the method is quite valid. Time-resolved fluorescence measurements in the presence of quenchers will indicate the applicability of case A. In the absence of this technique, the steady-state method needs to applied with extreme caution.

2.6.2 Determination of Partition Coefficients (Binding Constants)

A major application of the fluorescence quenching studies has been in the determination of partition coefficients of various solutes in aqueous micellar media. In addition to providing quantitative data on the dynamic aspects of the micellar solubilization process, these studies allow a direct comparison of micelles with various polar and nonpolar solvents as a medium to dissolve or extract organic solutes. For neutral and ionic quenchers detailed kinetic analysis along the lines indicated earlier (cases D and E) provide rate constants data for K, k_+, and k_- (cf. Tables 2.9 and 2.10). Determination of the exit and entry rate constants requires time-resolved measurements on the probe fluorescence decay either by single-photon-counting techniques or by laser flash photolysis. There are, however, several simpler steady-state methods discussed in the literature which under suitable conditions can provide partition coefficient data for quenchers and probe molecules.

Binding Constants of Quencher Solutes

Encinas and Lissi[92] have proposed a steady-state method for the determination of K for the cases where the relative fluorescence intensity I_0/I with and without quenchers is simply related to the average concentration of the quencher in the micellar pseudophase as measured by the mean occupancy number $\bar{n}$. It may be recalled that the binding constant or partition coefficient K is characterized by equilibrium relations shown in Eqs. (2.11)–(2.19).

$$K = \bar{n}/[Q_w] = [Q_m]/[M][Q_w] \tag{2.14}$$

where $[Q_m]$ and $[Q_w]$ represent the micellized and aqueous quenchers and $[M]$ is the concentration of the micelles.

When the fluorescence quenching is solely controlled by $\bar{n}$, then the value of K can be obtained from a plot of total quencher concentration $[Q_t]$ necessary to lead to a fixed value of I_0/I against the micelle concentration $[M]$:

$$[Q_t] = [Q_m] + [Q_w] = \bar{n}[M] + (\bar{n}/K) \tag{2.50}$$

In practice, from the conventional Stern–Volmer plots of I_0/I versus $[Q_t]$ at several micellar concentrations $[M]$, the values of $[Q_t]$ mentioned earlier are obtained by cutting "horizontally" and subsequently replotting these $[Q_t]$ values against the micelle concentration M to conform to Eq. (2.50). Figure 2.6 illustrates the application of such a methodology. Table 2.11 presents a collection of partition coefficient K values of quenchers that have been determined in this manner.

Ziemiecki and Cherry[96] have proposed a fluorescence quenching method that uses an approximation of an expression originally derived by Atik and Singer for the determination of K of metal ions. The following expression has been derived from the description of the nonlinear Stern–Volmer plots of pyrene monomer fluorescence:

$$I_0/I = (1 + \beta k_q K \tau_0 [Q_t])/(1 + \alpha \beta k_q K \tau_0 [Q_t]) \tag{2.51}$$

Here I_0 and I are the fluorescence intensities in the absence and in the presence, respectively, of quencher Q at a total concentration $[Q_t]$, α is the fraction of micelles without a quencher and $\beta = (1 - K[M])^{-1}$. All other parameters have the usual meaning as described earlier.

As the total concentration of the quencher increases, the value of α approaches zero. Hence at high values of $[Q_t]$ Eq. (2.51) reduces to the usual form of the Stern–Volmer equation but with a different slope:

$$I_0/I = 1 + k_q K \tau_0 [Q_t] \tag{2.52}$$

The linear region of the quenching plot can then be used to obtain a value of K by simply evaluating the slope in the linear region as a function of the micelle concentration:

$$(1/\text{slope}) = (k_q k \tau_0)^{-1} + [M](k_q \tau_0) \tag{2.53}$$

The region of the Stern–Volmer plot where the condition $\alpha = 0$ is fulfilled can be readily recognized because the curve is linear and leads to an extrapolated intercept of 1.0. Cherry et al.[96,143] have used the method to determine binding constants of several monovalent and divalent cations onto SDS micelles. Opinion has been expressed[175] that the approximations used by Cherry are not valid under the conditions of his experiment.

Binding Constants of Probe Molecules

Method I. For probes that partition significantly between the micellar and the aqueous phases under certain conditions, the lifetime of fluorescence in water can be quite different from that in a micelle. Consider the case of naphthalene whose fluorescence can be quenched by Br^- ions. When this

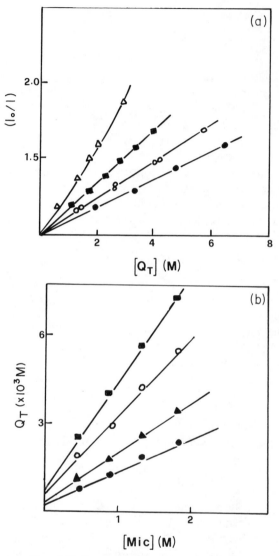

FIG. 2.6. Determination of binding constants from the fluorescence quenching of benzo-(a)pyrene by CCl_4 in CTAC micellar solutions. Intensity ratio I_0/I plotted against the total quencher concentration: ($\triangle$), 0.04 M; ($\blacksquare$), 0.075 M; ($\bigcirc$), 0.11 M; and ($\bullet$), 0.15 M. (b) Variation in the $[Q_T]$ values corresponding to a fixed I_0/I value [determined from plots shown in (a)] with micelle concentration. I_0/I values: ($\blacksquare$), 1.7; ($\bigcirc$), 1.5; ($\blacktriangle$), 1.3; and ($\bullet$), 1.2. (From Encinas and Lissi.[92])

TABLE 2.11. Partition Coefficient (Binding Constant) Data for Solutes in Aqueous Micellar Solutions using the Method of Encinas and Lissi

Solute/micelle	Probe used[a]	Mean occupancy number	$K(M^{-1})$ $(\times 10^3)$	Reference
2,5-Dimethyl-2,4-hexadiene/SDS	DMN	0.2–0.9	2.5	92
2,5-Dimethyl-2,4-hexadiene/SDS	Fluorene	1.0–8.9	2.5	92
2,5-Dimethyl-2,4-hexadiene/SDS	Pyrene	0.9–2.2	2.8	92
Di-t-butyl nitroxide/SDS	Benzopyrene	0.8	1.5	92
Di-benzoylperoxide/SDS	DMN	0.5	11.0	92
Methyl methacrylate/SDS	Pyrene	1.5	1.1	92
CHCl₃/SDS	DMN	8	0.57	92
3-Hydroxy-3-methyl-2-butanone/SDS	Biphenyl	2.6	0.29	92
Di-t-butyl peroxide/SDS	DMN	1–5	1.8	86
Di-t-butyl peroxide/CTAC	DMN	1–6	2.5	86
t-Butyl hydroperoxide/CTAC	DMN	1–5	0.56	86
t-Butyl hydroperoxide/SDS	DMN	0.5–4	0.44	86
Dibenzoyl peroxide/CTAC	DMN	0.5–0.6	12.0	86
H₂O₂/CTAC	DMN	1–15	0.15	86
H₂O₂/SDS	DMN	1–9	0.08	86
CCl₄/SDS	DMN	0.1–0.6	2.5	85
CCl₄/SDS	Benzopyrene	0.5–2.5	2.7	85
CCl₄/CTAC	Benzopyrene	0.5–3.5	5.2	85
CHCl₃/CTAC	DMN	3–10	2.35	85
CHCl₃/SDS	DMN	1.2–8.6	0.57	85
Benzene diazonium salt/SDS	Perylene	1.4–2.7	5.41	64
o-Toluene diazonium salt/SDS	Perylene	1.8–2.9	7.67	64
p-Methoxy benzene diazonium salt/SDS	Perylene	0.8–2.2	7.98	64
Acetone/SDS	Biphenyl		0.052	175
1,2-Dicyanoethylene/SDS	Pyrene		0.05	83
Acrylonitrile/SDS	Pyrene		0.37	87
Methyl methacrylate/SDS	Pyrene		1.0	87
Methyl methacrylate/CTAC	Pyrene		0.89	87
Tetrachloroethylene/SDS, CTAC	Pyrene		20.0	87
1,3-Octadiene/SDS, CTAC	Pyrene		20.0	87

[a] DMN = 2,3-Dimethylnaphthalene.

probe is incorporated in a CTAB micelle we observe a decrease in the total emission yield compared to that in water.[127] By changing the proportion of the "dissolved" to "free" solute, the partition coefficient of the probe can be obtained from the relationship

$$(I_\infty - I_0)/(I_t - I_0) = (1 + K[M])^{-1} \tag{2.54}$$

where I_∞, I_0, and I_t are the relative intensity of fluorescence at 100% solubilization, without surfactant present, and at intermediate amounts of surfactant concentrations, respectively. Almgren et al.,[173] have used this method to measure K values of naphthalene as follows: 3×10^4, 7×10^4 and $(2-2.8) \times 10^4 \ M^{-1}$ in SDS, CTAB, and $Zn(DS)_2$, respectively. The method requires that the fluorescence quantum yield inside the micelle and in the aqueous phase be independent of the surfactant concentration. Hence, the method cannot be strictly applicable to the naphthalene–CTAB system for it has been reported[127] that the naphthalene lifetime both in water and in CTAB decreases with increasing CTAB concentration.

Method II. An analogous fluorescence quenching method has been described based on the following scheme:[174]

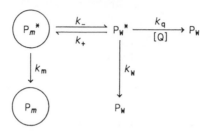

In this case the quencher and the micellar system have the same sign of charges so that the micellar surface repels the quenchers. The quenching effects can then be attributed solely to the probes in the free solution on a time scale where the escape routes do not interfere with the simple free-solution kinetics. With increasing quencher concentration we can selectively scavenge all of the excited-state probes that are in water leaving those that are micelle-bound. Using an appropriate kinetic analysis it is possible to obtain the fraction of the probes that are in micellar and aqueous phases. Quina and Toscano[174] first applied this method to the quenching of pyrenebutyrate (PBA) fluorescence by I^- ions in the presence of anionic SDS micelles and obtained $K(PBA) = 2.4 \pm 0.15 \times 10^4 \ M^{-1}$. Almgren *et al.*[173] later applied it to naphthalene* quenched by Br^- in the presence of SDS micelles and obtained $K(Naph) = 4.0 \times 10^4 \ M^{-1}$. Abuin and Lissi[147] have proposed a modification of this method. For the selective quenching of Naph* that are in the aqueous phase by Ni^{2+} in the presence of CTAB micelles, the quenching is described by:

$$(I_0 - I)/I = k_q \tau \alpha [Q]/[1 + [Q](k_q \tau)] \tag{2.55}$$

where α is the fraction of the probe emission from water at zero Ni^{2+} concentration.

Method III. A complementary version of Method II has also been studied using probes that partition between the aqueous and micellar phase (quarternary ammonium salts of pyrene or anthracene) and functional micelles. The latter were composed of surfactants having quenchers as the head group or as counterions (e.g., alkylpyridinium halides or surfactant

$$P = Ar-(CH_2)_n-NMe_3$$
$$Q = CH_3-(CH_2)_n-R$$

nitroxyls.[20,21,28,108] Due to the high local concentration of the quenchers in the functional micelles, micelle-solubilized probes are very efficiently quenched by static or pseudostatic quenching mechanisms. Those that are in water undergo inefficient dynamic quenching by the monomeric surfactant quencher molecules or by free counterions (e.g., Br^- in cetylpyridinium bromide). The evaluation of the concentration of the aqueous quenchers is not straightforward. For the surfactant quencher monomers, it can be taken as equal to the critical micelle concentration and for the counterions it is the fraction of the surfactant concentration not bound to the micelle. Quantitative analysis in terms of this model has shown that K for the amphiphilic probe molecules is very sensitive to the concentration of detergent and added salts present.

2.6.3 Studies of Counterion Binding and Exchange

Yet another application of fluorescence quenching is in the determination of the nature and extent of counterion binding and on the exchange of bound ions with other ions present in the bulk aqueous phase. Quenching of solubilized probe molecules by halide ions (Br^-, I^-, etc) in cationic alkyl trimethylammonium (CTA) micelles and by metal ions (Cu^{2+}, Eu^{2+}, etc) in anionic sodium dodecylsulfate (SDS) have been the principal systems examined in several laboratories.

For a quantitative analysis, the binding of a quencher cation Q of charge Z^+ to an anionic micelle (or added anion to a cationic micelle) has been described either as an adsorption or an an exchange with another ion of the aqueous bulk. In the first case the process is represented as

$$Q_W + \boxed{M} \underset{k_-}{\overset{k_+}{\rightleftharpoons}} \boxed{Q_m} \qquad (2.56)$$

The binding constant K is given by

$$K = [Q_m]/[Q_w][M] \qquad (2.57)$$

where $[Q_m]$ and $[Q_w]$ denote the concentration of bound and free-ion concentration, respectively, and $[M]$ is the concentration of micelles. On the other hand if the process is considered as an exchange between the added ion and surfactant anion A, the process is represented as

$$(Z_Q/Z_A)A_m + Q_w \rightleftharpoons (Z_Q/Z_A)A_w + Q_m \qquad (2.58)$$

The exchange constant K_{ex} is then given by

$$K_{ex} = [Q_m][A_w]^{(Z_Q/Z_A)}/[Q_w][A_m]^{(Z_Q/Z_A)} \qquad (2.59)$$

where $[Q_m]$ and $[A_m]$ are measured in terms of their local concentrations (for example, in terms of number ions per micelle or per micellized surfactant). This exchange constant K_{ex} is different from the exchange rate constant k_{ex} referred to earlier (under case E in Section 2.4) in the discussion of various quenching scenarios. The latter refers to the exchange between the bound ions of different micelles.

Comparison of Eqs. (2.57) and (2.59) shows that both equations can be employed equivalently for a given surfactant concentration when $[Q_m] > [A_m]$ and hence (A_w/A_m) is nearly a constant. By monitoring the solubilized probe fluorescence lifetime when the nonquencher counterions are replaced by quencher counterions, it is possible to estimate the exchange constants K_{ex}. Abuin, Lissi, Quina, and co-workers have developed extensive kinetic analysis schemes to evaluate such K_{ex} values from the fluorescence quenching data. Table 2.12 provides a summary of evaluated counterion exchange coefficients $K_{(X/Y)}$ in various aqueous, ionic micellar solutions:

$$\boxed{Y_m} + X_w \rightleftharpoons \boxed{X_m} + Y_w \qquad K_{(X/X)} \qquad (2.60)$$

Generally the counterion distribution or exchange in ionic micelles appears to be a very rapid process. The basic experimental criterion for fast counterion redistribution is single exponential decay of the probe emission in the presence of quencher species, e.g, pyrene fluorescence decay in CTAB or Cu(DS)$_2$ micelles. Abuin et al.[107] have shown that in the limit of fast counterion redistribution, the quenching of fluorescence of micelle-bound species is a direct function of the average fractional coverage of the micellar surface by the quencher counterions and, hence, of the local quencher concentration in the vicinity of the micellar surface.

Foreman et al.[145] have used an extra micellar probe RuL_3^{2+} (L = 4,4'-dicarboxybipyridine) to monitor the binding capacity of quencher cations

TABLE 2.12. Counterion Exchange Coefficients $K_{X/Y}$ for Various Quencher Ions in Aqueous Ionic Micellar Solutions (ions/micelle)

$$Y_m + X_w \rightleftharpoons X_m + Y_w, \qquad (K_{X/Y})$$

Replacing ion/Ion replaced (X/Y)	Micelle	Probe	$K_{X/Y}$	Reference
Br^-/Cl^-	CTAY	Biphenyl,	4.2	107
		naphthalene	5.0	
I^-/Br^-	CTAY	Biphenyl, naphthalene	13.0	107
$S_2O_3^-/SO_4^-$	CTAY	Biphenyl, naphthalene	2.5	107
IAc^-/Be^-	CTAY	Biphenyl, naphthalene	1.0	107
Br^-/NO_3^-	CTAY	Biphenyl, naphthalene	0.90	107
Tl^+/Na^+	SDS	9-Methyl anthracene	3.8	132
$S_2O_3^-/Br^-$	CTAY	Pyrene, perylene	5.0	105
$S_2O_3^-/Cl^-$	CTAY	Pyrene, perylene	1.3	105
$S_2O_3^-/NO_3^-$	CTAY	Pyrene, perylene	17.0	105
SO_4^-/Br^-	CTAY	Pyrene, perylene	74–250	105
Br^-/Cl^-	CTAY	Fluoranthene	11.0	115,116
I^-/Cl^-	CTAY	Fluoranthene	25	115,116
Br^-/F^-	CTAY	Fluoranthene	70	115,116

$(Cu^{2+}, MV^{2+}, BV^{2+},$ and $CBP^+)$ onto SDS micelles:

In aqueous solution devoid of surfactants the luminescence of $RuL_3^{4-}*$ is quenched by Cu^{2+} by normal Stern–Volmer kinetics. Introduction of anionic micelles removes Cu^{2+} from the solution by binding them to the micellar surface and attenuates its quenching activity. Detailed analysis allows determination of the binding capacity of these cations in terms of cations per micelle. The estimated values in SDS micelles are 6.0, 3.2, 3.0, and 2.0 for Cu^{2+}, MV^{2+}, benzylviologen (BV^{2+}), and CBP^+, respectively. The higher binding capacity coupled with the existence of a limited capacity suggest that hydrophobic cations such as viologens bind at different sites as compared to metal ions or neutral organic molecules. Miola et al.[108] have used a similar procedure to monitor the incorporation of amphiphilic anthracene [Anthr—$(CH_2)_n$—NMe_3^+] into cetylpyridinium halide micelles using $Ru(bpy)_3^{2+}$ as the extramicellar probe.

The differences due to the presence of quencher and nonquencher species as counterions in ionic micelles on the photophysical properties of solubilized probe molecules have been the subject of numerous investigations (cf. Tables 2.7 and 2.8 for specific references). The reduced fluorescence lifetimes of probes in cetyltrimethylammonium bromide (CTAB) as compared to that in cetyltrimethylammonium chloride (CTAC) micelles was noted as early as 1973 and was attributed to the quenching of excited states by the locally high concentration of bound bromide ions.[47,62,127] Using the fluorescence lifetimes of a surface-bound probe pyrene-3-sulfonic acid (PSA) in DTAC and CTAB micelles as a guide, the surface concentration of Br$^-$ in CTAB micelles was estimated to be near 0.52 M.[125] Laser photolysis studies of these solutions indicated that in CTAB micelles the fluorescence quenching is accompanied by the formation of triplets in rather high yields as compared to that in DTAC micelles. Similar results have been obtained by Wolff[109,110] who measured the quantum yields of fluorescence and triplet formation of a series of aromatic hydrocarbons in ethanol, CTAC, and CTAB micelles. For several arenes there is a significant decrease in Φ_{fl} with a corresponding increase in Φ_T on going from CTAC to CTAB micelles. For example with anthracene, the Φ_{fl} and Φ_T values are 0.26 and 0.74 in CTAC and 0.08 and 1.00 in CTAB, respectively.

Kira and co-workers have similarly measured yields of pyrene triplet and cation radicals during the quenching of pyrene solubilized in SDS micelles by various metal ion quenchers.[129,134,137] For heavy atom metal ion quenchers there is a pronounced increase in triplet yields (mainly due to the external heavy atom effects) without any appreciable yield of pyrene cations. Pyrene triplet yields in SDS micelles in the presence ($\simeq 10^{-2}$ M) of Ag$^+$, Tl$^+$, Ni^{2+}, Co^{2+}, Tb^{3+}, Sm^{3+}, and Dy^{3+} and in the absence of these heavy atom salts are 0.66, 0.61, 0.55, 0.46, 0.35, 0.33, 0.30, and 0.23 (SDS), respectively.[137] Quenching by Cu^{2+} ions occurs via electron transfer involving both singlet and triplet excited states ($\Phi_{P+} \simeq 0.4$ and $\Phi_T \simeq 0.01$). Using a simple kinetic model the quenching rate constants have been correlated with the yield of transient species. Observation of room-temperature phosphorescence from aromatic hydrocarbons in SDS micelles in the presence of heavy atom salts such as Tl^{3+}, a topic we will elaborate in Chapter 3, is another manifestation of these novel counterion effects.

Fluorescence quenching studies using counterions as quenchers can also be used to verify the co-existence of two types of micellar structures in concentrated ($\geqslant 0.3$ M) aqueous solutions of mixtures of CTAB and CTAC.[119] It is known that at high concentrations CTAB exists as rod-shaped micelles while CTAC retains its spherical globular shape. The fluorescence decay of pyrene in these mixtures is a double exponential. The short-lived component is ascribed to pyrene solubilized in rodlike micelles with Br$^-$ as the principal counterion and the longer-lived component to pyrene in globular micelles with Cl$^-$ as the principal counterions. The co-existence is suggested to be a consequence of a higher preference for Br$^-$ as counterions in the case of rod-like structures. The electrolyte-induced phase transition from spherical to rod-shaped aggregates in aqueous micellar solutions of dodecylammonium chloride (DAC) has also been studied by similar fluorescence quenching methods.[124]

2.6.4 Studies on the Influence of Additives

The scope and limitations of fluorescence quenching methods in the determination of micellar aggregation number N are best illustrated with the studies of influence of additives (electrolytes, alcohols, etc.) to aqueous micellar solutions.[84,89,91,94,168-172,176] Consider, for example, the addition of NaCl to aqueous micellar solutions of SDS. Light scattering studies have established that introduction of electrolytes leads to relatively small increases in N initially (i.e., [NaCl] $\leqslant$ 0.4 M). At high concentrations ([NaCl] > 0.4 M) N increases sharply indicating formation of larger rod-shaped micelles. Aggregation numbers of 60,150, and 1000 have been estimated at 0, 0.4, and 0.6 M NaCl, respectively. Determination of N via fluorescence quenching methods I and II referred to earlier faithfully reproduce moderate increase in N at low NaCl concentration. At higher salt concentrations these methods estimate rather low increases in N (by a factor of two). Clearly the fluorescence methods are as reliable as any other method for the determination of N, such as light scattering, osmometry, and centrifugation, provided $N \leqslant 200$. (For methods based on static quenching, e.g, $Ru(bpy)_3^{2+}$ –methylanthracene, $N \leqslant 110$). The reasons for the deviation at large N are not clear. Almgren has suggested that the quasi-elastic light scattering methods yield a weight-averaged aggregation number N_w whereas fluorescence quenching methods yield the number-average value N_n. A ratio of $N_w/N_n = 2$ is expected for larger rod-shaped micelles. Also for larger aggregates the models based on poisson statistics may not be valid.

The use of aggregation number as a guide to detect structural changes in the aggregate has been reported in several studies of the influence of solubilization of alcohols, hydrocarbons, and multivalent and hydrophobic counterions to aqueous ionic micellar systems. For the addition of medium-chain-length alcohols (C_5-C_7) fluorescence studies have confirmed the occurrence of phase transitions from spherical to near-spherical mixed micelles (surfactant + alcohol) above a certain concentration ratio. Addition of hydrophobic compounds to SDS micelles results in a growth of the micelles with an increase in aggregation number N that is large enough to keep the charge density at the surface a constant.

2.7 Aspects of Excimer Formation

An interesting mode of deactivation of an excited-state probe is via quenching by other ground-state probe molecules themselves (self-quenching) In aromatic hydrocarbons such as pyrene this process is accompanied by the formation of luminescent excited state dimers called *excimers*:

$$^1Py^* + Py \underset{k_2}{\overset{k_1C}{\rightleftharpoons}} {}^1Py\cdot\cdot Py^*$$

$$hv \downarrow\downarrow \tau_{mon} \qquad\qquad \downarrow \tau_{ex}$$

$$Py, hv' \qquad\qquad 2Py, hv''$$

These are characterized by their structureless fluorescence emission occurring at wavelengths red-shifted with respect to that of the monomer. In micellar media excimer formation requires the presence of two probe molecules in a given micelle at the time of excitation. Hence, photochemical studies of excimer formation are of interest in the exploration of the influence of statistical distribution of solutes among the micelles.[3,67,80,154,158,163,171,177–192]

Pyrene was introduced as a probe for detergent micelles by Förster and Selinger in 1964.[177] From the dependence of the efficiency of excimer formation on the detergent concentration, these authors recognized the existence of a random distribution of pyrene molecules among the micelles. Pownall and Smith[179] tried to interpret the micellar effects on the relative intensities of monomer to excimer using a simple kinetic model from homogeneous solutions and derived values for the microviscosity prevailing in the micellar interior assuming normal diffusion-controlled kinetics for the excimer formation. Unfortunately, the statistical effects of solute distribution were not properly taken into account. Monomer fluorescence intensities were overestimated and abnormally high values for the microviscosity were obtained.

In micellar media the efficiency of excimer formation depends on the probability of an excited pyrene finding a ground-state pyrene as a cohabitant in the same micelle. This, in turn, depends on the relative number of micelles present to solubilize a given number of probe molecules. With the availability of a large number of micelles it is possible to obtain conditions where excimer formation is totally inhibited at relatively high concentration of pyrene (e.g., 5×10^{-4} M). Conversely excimer formation can be enhanced even at low pyrene concentrations (e.g., 10^{-5} M) by decreasing the number of host micelles and forcing the pyrene molecules to occupy the fewer micelles with multiple occupancies. From the mean occupancy number $\bar{n}$ (i.e., the ratio of number of probe molecules to the number of host micelles) we can use various statistical models to calculate the relative probability of finding singly and multiply occupied micelles (cf. Section 1.4 for a discussion on this point).[154,180,181,184] There have been numerous quantitative studies[180–192] both by steady-state and time-resolved studies of excimer formation in micellar media. The majority of these studies are in favor of treating the allocation of solutes among the micelles as conforming to a poisson distribution.

Kinetic analysis of fluorescence growth and decay curves of monomers and excimers are usually carried out using the scheme indicated above. For a δ-pulse excitation in homogeneous solution, analysis of the scheme shows that the fluorescence decay of monomer $I_M(t)$ to be a sum of two exponentials while that of the excimer $I_D(t)$ as a difference of the two exponential terms:

$$I_M(t) = A_1 \exp(-\lambda_1 t) + A_2 \exp(-\lambda_2 t) \qquad (2.61)$$

$$I_D(t) = A_3 \exp(-\lambda_1 t) - \exp(-\lambda_2 t) \qquad (2.62)$$

where A_1, A_2, and A_3 are constants. This gives rise to monomer fluorescence decay exhibiting a dual exponentiality and the excimer decay a rise followed by an exponential decay.

In micellar media excimer formation comes from micelles with discrete occupation numbers and hence depend additionally on the type of solubilizate distribution. For kinetic analysis each micelle can be considered as a subsystem to which Eq. (2.62) can be applied. The occupancy number i determines the effective intramicellar concentration C_i and, hence, a pair of time constants $\lambda_{1,i}$ and $\lambda_{2,i}$. If P_i is the probability of finding a micelle with i solutes, then the excimer fluorescence can be given as a weighted sum of biexponentials:

$$I'_D(t) = A'_3 \sum_{i=2} I_D(t) P_i \qquad (2.63)$$

Note that the summation starts at $i = 2$. There has been partial success in quantitative analysis of the excimer kinetics with P_i given either by poisson, binomial, or geometric distributions. The biexponential nature of the monomer fluorescence (the fast component arising from the self-quenching reaction) can be analyzed quantitatively using a kinetic scheme discussed earlier (case C of quenching scenarios, Section 2.4). Micellar aggregation numbers can be determined in this way, and we have already dealt with this aspect of the application in Section 2.5.

The foregoing discussions lead to the conclusion that in studies of microviscosity for the probe environment in the micellar interior it is advantageous to use intramolecular excimer-forming probes (e.g., dipyrenylalkanes) instead of intermolecular ones (e.g., isolated pyrene molecules). With intramolecular excimer-forming bichromophoric probes the excimer emission is exclusively unimolecular, and furthermore at low solute concentrations the statistical factors related to the probe distribution can be safely ignored. Studies of diarylalkanes (Ar—$(CH_2)_n$—Ar) have shown that geometric optimization (parallel sandwich structures for the arenes) is achieved best with $n = 3$. Hence various 1,3-diarylpropanes (aryl = pyrenyl, napthyl, phenyl, phenanthryl, etc.) are widely used as intramolecular excimer forming probes.[193-202]

Table 2.13 summarizes data on the microviscosity η for the probe environment (interior of the micelles?) estimated using several 1,3-diarylalkanes. Clearly there is a distinct dependence of η on the size of the probe. This could simply be a reflection of the non-isotropic nature of the micellar architecture with different sized probes sampling different regions. This situation is not unique in that we finds similar situations in studies of larger aggregates such as lipid vesicles and polymers. Estimates of η via fluorescence depolarization methods using probes such as 2-methylanthracene are in the same range as those obtained with dinaphthylpropanes.

Turro and co-workers have used the intramolecular excimer formation in dinaphthylpropanes to study microviscosity changes when ionic micellar solutions are subjected to high pressure and in the presence of additives.[196] In CTAB micelles η increases from ~ 27 to 107 cP on going from atmospheric pressure to 2610 bar. Addition of NaCl, Na_2SO_4, or ethanol to SDS was found to generally decrease the η values. There is extensive discussion in the literature on the photophysics of excimer formation of diarylalkanes even in neat solvents. In covalently linked bichromophoric systems there are *conformational restraints* that influence efficiency of excimer formation (cf. discussions on excimers in polymers, Section 8.1.4). In studies of microviscosity using diarylalkanes caution needs to be exercised in the choice of solvent system that one uses to determine the calibration curves and in the utilization of calibration curves determined elsewhere.[199–202]

Micelles composed of surfactants containing a phenyl group at the end or middle of the alkyl chain show, in addition to the normal phenyl chromophore emission (max. at 280 nm), an excimerlike emission from adjacent phenyl groups. This is an unusual feature arising from the high local concentration of

TABLE 2.13. Microviscosity Data for the Micellar Interior Estimated using Intramolecular Excimer-Forming Probes [1,3-Diarylpropanes in Aqueous Micellar Solutions at Room Temperature (20°C)]

Probe	Temperature (°C)	Microviscosity η (cP)			Reference
		SDS	CTAC	CTAB	
1,3-Diphenylpropane	20	4			193
1,3-Di(1-naphthyl)propane	20	5	10		194
1,3-Di(1-naphthyl)propane	20	9	31	39	195
1,3-Di(1-naphthyl)propane	25	11.5	27	47	196
1,3-Di(pyrenyl)propane	20	19			193
1,3-Di(4-biphenyl)propane	20	10	21	42	194
Pyrene[a]	22	33			190
Pyrene[a]	22	193		151	179

[a] Intermolecular excimers.

phenyl chromophores in the micelle.[67,178,187] Such excimerlike emission (max. at 340 nm) has also been observed in concentrated micellar solutions of non-ionic surfactants such as Triton X-100.[185] Since the absorption spectrum is not independent of the surfactant concentration it appears that some form of dimeric species are responsible for these new features.

2.8 Dynamics of Excited-State Charge Transfer Complexes (Exciplexes)

Formation of excited-state charge transfer complexes (exciplexes) is often observed during the quenching of aromatic hydrocarbons by donor molecules such as amines and cyanocompounds. In nonpolar solvents the exciplexes are stable and luminesce but in polar solvents they rapidly dissociate into molecular ions:

$$A + D \xrightarrow{h\nu} A^* + D \rightleftharpoons [A \cdots D]^* \begin{cases} [A^- \cdots D^+]^* - A + D + h\nu' \\ \text{exciplex} \\ \\ A^-_{solv} + D^+_{solv} \end{cases}$$

Studies in micellar media with one or both of the donor/acceptor pair solubilized in the micelle provide an elegant method to probe heterogeneities in the micellar architecture and the effect of micelle on the stability of the complex. Both intra- and intermolecular quenching have been investigated in various ionic and non-ionic micellar systems.[203-210] Let us consider first the intermolecular case.

The singlet excited state of pyrene is efficiently quenched by amines in a charge transfer process to form the ionic exciplex $(Py^- \cdots A^+)^*$ whose fate depend critically on the nature of the micelle. In cationic micelles, following the quenching, the donor cation D^+ is expelled from the micellar surface while the pyrene anion (Py^-) is retained (stabilized) leading to fairly long lifetimes for the pyrene anions. In anionic micelles the micellar surface traps D^+ and enhances geminate-iron combination. Due to the trapping of some of D^+ at the surface of the micelle, some of the pyrene anions escape recombination and decay with lifetimes much longer than those observed in homogeneous solvents. Addition of inert electrolytes have little effect on the ion recombination process:

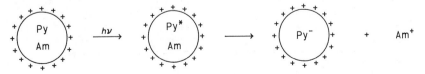

The efficiency of the charge separation process has been examined using several pyrene derivatives [pyrene butyric acid (PBA), pyrene sulfonic acid (PSA), pyrene carboxaldehyde (PyCHO), and pyrene dodecanoic acid (PDA)].[204] With surface probes such as PyCHO, PSA, or pyrenetetrasulfonate (PTSA) in cationic micelles the pyrene anion lifetimes approach a millisecond. Table 2.14 presents typical half-life data $(t_{1/2})$ for the pyrene anion radicals dissociated from the pyrene–dimethylaniline exciplex system. In all cases no emission attributable to an exciplex has been observed. Matsuo et al.[208] have demonstrated similar beneficial effects of micelles on charge separation in the quenching of surfactant derivatives of $Ru(bpy)_3^{2+}$ by dimethylaniline.

Analysis of the quenching showed that the steady-state data cannot be adequately explained by normal Stern–Volmer kinetics. Plots of I/I_0 versus $[Q]$ show large curvatures. Monitoring of the pyrene fluorescence decay in the presence of quenchers showed (instead of a single exponential as is observed during the amine quenching in homogeneous solvents) a fast component whose contribution increases with increasing quencher concentration. The quenching kinetics are readily understood in terms of case D of the quenching scenarios discussed in Section 2.4. Amines such as dimethylaniline (DMA) do have some solubility in water. The binding constant K for DMA in CTAB micelles has been deduced to be $4.8 \times 10^3 \ M^{-1}$. Figure 2.7 presents spectra

TABLE 2.14. Effect of Micellar Phase and Additives on the Decay (First Half-Life $t_{1/2}$) of Pyrene Anion Dissociated from a Pyrene and Its Derivatives–Dimethylaniline Exciplex System[a]

Medium	Acceptor	$t_{1/2}(\mu sec)$
CH$_3$OH	Pyrene	6.0
CTAB (0.02 M)	Pyrene	500
CTAB (0.20 M) + 0.2 mM NaOH	Pyrene	71.4
CTAB (0.02 M) + 0.1 M NaCl	Pyrene	230
CTAB (0.02 M) + 0.02 M Na$_2$SO$_4$	Pyrene	190
NaLS (0.02 M)	Pyrene	66.1
Igepal (0.01 M)	Pyrene	13.1
CTAB (0.02 M)	PBA	466
CTAB (0.02 M)	PDA	22
CTAB (0.02 M)	PSA	960
CTAB (0.02 M)	P-CHO	1030
CTAB (0.02 M)	PTSA	906
CTBAB (0.02 M)	PSA	935
DTAC (0.02 M)	PSA	400
CTAC (0.02 M)	PSA	392
CAC (0.02 M)	PSA	0.2

[a] Data from B. Katusin-Razem et al.[204]

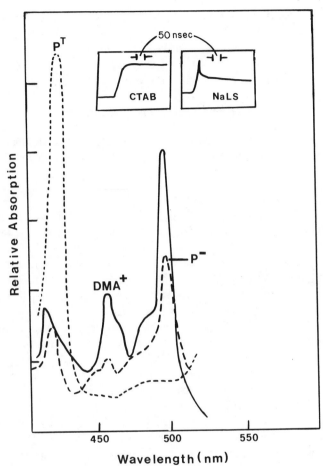

FIG. 2.7. Spectra of short-lived transients formed during the quenching of pyrene singlet excited state by dimethylaniline in cyclohexane (· · ·), methanol (– – –), and CTAB micelles (——). Insets: Oscilloscope time traces of pyrene anion (P⁻) in CTAB and NaLS micelles. (From J. K. Thomas, *ACS Adv. Chem. Ser.* **184,** 97 (1979). Copyright 1979 American Chemical Society.)

of short-lived intermediates that are formed during the DMA quenching of pyrene fluorescence as well as transient absorption decay curves for pyrene anion Py⁻ demonstrating the charge separation process. Amines such as dibutylaniline bind quantitatively with the micellar phase, and hence its quenching process is analyzed according to case C.

In covalently linked systems such as pyrene—$(CH_2)_3$—DMA, emission from the intramolecular exciplex and the formation of pyrene anion to varying degree in different micelles have been reported.[204] The exciplex yield and lifetimes decrease as we go through the series hexane, Igepal CO-630 (neutral), methanol, CTAB (cationic), and SDS (anionic). Mataga et al.[203] in their studies with a series of linked compounds py—$(CH_2)_n$—DMA, however, have failed to observe any exciplex emission in these systems though they do observe formation of ion products by laser flash photolysis.

2.9 Studies of Excitation Energy Transfer Processes

Under certain conditions an excited molecule D* can transfer its excitation energy to a suitable acceptor A over distances much longer than the collisional diameters (e.g., 50–100 Å):

$$^1D^* + A \rightarrow D + {}^1A^* \qquad (2.64)$$

Such long-range (Förster-type) energy transfer is of much interest in studies of macromolecules and microheterogeneous systems as a spectroscopic ruler in the determination of interchromophore distances. Studies in micellar systems are attractive for several reasons. Due to their rather small dimensions we can construct a model system in which the donor–acceptor pairs are in well-defined geometry and proximity. By varying the solute-to-micelle ratio it is possible to have fairly high local concentrations of the solutes inside the micelle and study self-quenching and bimolecular energy transfer phenomena. Table 2.15 provides a listing of various energy transfer processes that have been explored in micellar systems.[211–227]

The rate constant for energy transfer by resonance mechanisms is strongly dependent on the distance of separation between the donor–acceptor pair R ($k_{DA} \propto (1/R^6)$). Hence, in homogeneous solvents, concentrations of $10^{-3} M$ are required for resonance transfer processes to be of importance. In micellar media using spatial proximity and locally high concentrations it is possible to observe novel energy transfer processes. Enhanced efficiencies in energy transfer have indeed been observed for several donor–acceptor pairs cosolubilized in micelles. These include thionine–methylene blue,[211] and rhodamine 6G–coumarin I.[213] Efficient transfer of excitation energy from the built-in phenyl group in micelles composed of phenyl undecanoate and Igepal CO-630 to solubilized aromatic molecules (e.g., naphthalene and pyrene) has been advanced as evidence for the solubilization of these aromatics inside the micelle closer to the phenyl group.[185,212]

An intriguing feature of the green-plant photosynthesis has been the efficient transfer of excitation energy among the antenna chlorophyll in chloroplasts where the in-vivo concentration levels are estimated to be ~0.1 M. There have been several attempts in model systems to achieve this efficiency without encountering intense self-quenching. There have been studies of energy transfer by chlorophyll a in non-ionic micelles composed of Triton X-100. Fluorescence polarization studies do indicate that local

TABLE 2.15. Excitation Energy Transfer Studies in Micellar Meida

Donor	Acceptor	Micelle	Reference
Thionine	Methylene blue	SDS	211
"Phenyl" (micellar)	Naphthalene	Phenyl undecanoate	212
Coumarin-I	Rhodamine 6G	SDS, CTAB, and Triton X-100	213
Chlorophyll a	Chlorophyll a	Triton X-100	214
"Phenyl" (micellar)	Pyrene	Igepal Co-630	185
Naphthalene	Tb^{3+}	SDS	215
N-Methylphenothiazine	t-Stilbene and naphthalene	CTAB	216
Naphthalane, bromonaphthalene, biphenyl, and phenanthrene	Tb^{3+}, Eu^{3+}	SDS	217
Rhodamine 6G	3,3'-Diethylthiacarbocyanine iodide	SDS	218
Acridine orange	Methylene blue	SDS	219
Chrysene	t-Stilbene	CTAB	220
N,N-Dimethyldihydroindolo-carbazole(DI)	Cu^{2+}	$Cu(DS)_2$	220
1,3-Dioctylalloxazine	Proflavine and acriflavine	SDS	221
Coumarin-I	Rhodamine 6G	Triton X-100	222
Diphenyl acetylene	Pyrene	SDS	223
$Ru(bpy)_3^{2+}$	Cresyl violet	SDS	224
Acetophenone, propiophenone, and isobutyrophenone	r-Methylvalerophenone	SDS	225
N-alkylcarbazole	ZnTPP	SDS	226
Rhodamine 6G	Pinacyanol	SDS	227

concentration of chlorophyll a in the micelles to be higher by 1–3 orders of magnitude as compared to bulk concentration.[214] Self energy transfer between built-in naphthyl chromophores has also been studied in micelles formed by cetylaminonaphthalene sulfonates.[212]

While there is no energy transfer observed between aromatic donors (e.g., naphthalene) and heavy metal salts such as Eu^{3+} or Tb^{3+} in homogeneous solvents, very efficient energy transfer has been demonstrated in aqueous anionic micelles.[215] Here again the role of micelles is to allow compartmentalization of no more than one donor per micelle and at the same time concentrate a large number of acceptors at the surface. In homogeneous solutions triplet–triplet annihilation reactions rapidly deactivate the donor triplets before they can transfer their excitation energy. Kinetic aspects of these specialized energy transfers using molecular organization have been studied.[217] Formal second-order rate constants for energy transfer from bromonaphthalene and biphenyl to Tb^{3+} have been evaluated to be

$2 \times 10^4 \, M^{-1} \sec^{-1}$ and $5 \times 10^4 \, M^{-1} \sec^{-1}$, respectively. These strikingly low energy transfer rate constants while explaining the absence of any energy transfer in homogeneous solvents also illustrate the remarkable ability of micelles to organize reactants so as to achieve even such inefficient energy transfer processes. This work also provides a semiquantitative approach for the conversion of the intramicellar pseudo-first-order rate constants to formal second-order rate constants that are appropriate for homogeneous solvent systems.

Under suitable conditions energy transfer processes can also be used to determine micellar aggregation numbers. The overall efficiency of energy transfer depends on two factors: (1) the occupation statistics, i.e., the probability that n acceptor molecules share a micelle with an arbitrary donor and (2) intramicellar energy transfer efficiency. The latter depends on the size of the micelle with respect to the Förster radius. If the micelle diameter is less than the Förster radius the energy transfer will be very efficient. The overall efficiency is then governed by the occupation statistics alone and problems concerned with the solubilization site, the orientation factor, or diffusion do not arise. Photostationery measurements then can enable determination of the fraction of acceptor-occupied micelles and, hence, the aggregation number.

When the number of acceptor molecules per micelle is small so that we need to consider only micelles with zero or one acceptor, then the fraction of micelles containing an acceptor is given by

$$[A] \cdot [N]/([\text{Surf}] - [\text{CMC}])$$

where Surf and N stand for the surfactant concentration and micellar aggregation number, respectively. The ratio of intensities of donor and acceptor fluorescence then is given by

$$(I_A/I_D) = (\Phi_D/\Phi_A \eta_T)(\{[\text{Surf}] - [\text{CMC}]\}/[A][N]) \tag{2.65}$$

where Φ_D and Φ_A refer to the fluorescence quantum yields of donor and acceptor, respectively and η_T is the energy transfer efficiency. Koglin et al.[223] have applied this method to singlet–singlet energy transfer from diphenylacetylene to pyrene in SDS micelles and obtained satisfactory values of aggregation number N for SDS micelles ($N \sim 61$).

References

1. J. K. Thomas, *Chem. Rev.* **80**, 283 (1980).
2. K. Kalyanasundaram, *Chem. Soc. Rev.* **7**, 453 (1978).
3. N. J. Turro, M. Grätzel, and A. M. Braun, *Angew. Chem., Int. Ed. Engl.* **19**, 675 (1980).
4. L. A. Singer, *in* "Solution Behaviour of Surfactants" (E. J. Fendler and K. L. Mittal, eds.), Vol. 1, p. 73. Plenum, New York, 1982.

5. K. Kalyanasundaram and J. K. Thomas, *J. Am. Chem. Soc.* **99**, 2039 (1977).
6. B. B. Craig, J. Kirk, and M. A. J. Rodgers, *Chem. Phys. Lett.* **49**, 437 (1977).
7. (a) A. Nakajima, *J. Lumin.* **15**, 277 (1977). (b) A. Nakajima, *Bull. Chem. Soc. Jpn.* **44**, 3272 (1971).
8. (a) P. Lianos and S. Georghiou, *Photochem. Photobiol.* **30**, 355 (1979). (b) P. Lianos and S. Georghiou, *Photochem. Photobiol.* **29**, 843 (1979).
9. V. Glushko, M. S. R. Thaker, and C. D. Karp, *Arch. Biochem. Biophys.* **210**, 33 (1981).
10. D. C. Dong and M. A. Winnik, *Photochem. Photobiol.* **35**, 17 (1982).
11. K. Hara and W. R. Ware, *Chem. Phys.* **51**, 61 (1980).
12. P. Lianos, M. L. Viriot, and R. Zana, *J. Phys. Chem.* **88**, 109 (1984).
13. H. W. Offen and W. D. Turley, *J. Phys. Chem.* **86**, 3501 (1982).
14. (a) A. Nakajima, *Chem. Phys. Lett.* **21**, 200 (1973). (b) A. Nakajima, *Photochem. Photobiol.* **25**, 593 (1977).
15. A. K. Mukhopadhyay and S. Georghiou, *Photochem. Photobiol.* **31**, 407 (1980).
16. K. Kalyananasundaram and J. K. Thomas, *J. Phys. Chem.* **81**, 2176 (1977).
17. (a) A. U. Acuna and J. M. Oton, *J. Lumin.* **20**, 379 (1979). (b) J. M. Oton and A. U. Acuna, *J. Photochem.* **14**, 341 (1980).
18. J. C. Dederen, L. Coosemans, F. C. DeSchryver, and A. V. Dormael, *Photochem. Photobiol.* **30**, 443 (1979).
19. N. J. Turro and T. Okubo, *J. Phys. Chem.* **86**, 159 (1982).
20. C. V. Kumar, S. K. Chattopadhyay, and P. K. Das, *Photochem. Photobiol.* **38**, 141 (1983).
21. K. Muthuramu and V. Ramamurthy, *J. Photochem.* **26**, 57 (1984).
22. K. Y. Law, *Photochem. Photobiol.* **33**, 799 (1981).
23. C. David, E. Szali, and D. Baeyens-Volant, *Ber. Bunsenges. Phys. Chem.* **86**, 710 (1982).
24. M. A. J. Rodgers, *Chem. Phys. Lett.* **78**, 509 (1981).
25. M. A. J. Rodgers, *J. Phys. Chem.* **85**, 3372 (1981).
26. T. Ban, K. Kasatani, M. Kawasaki, and H. Sato, *Photochem. Photobiol.* **37**, 131 (1983).
27. T. Wolff, *Ber. Bunsenges. Phys. Chem.* **85**, 145 (1981).
28. T. Wolff, *J. Colloid Interface Sci.* **83**, 658 (1981).
29. H.-C. Chiang and A. Lukton, *J. Phys. Chem.* **79**, 1935 (1975).
30. R. C. Mast and L. V. Haynes, *J. Colloid Interface Sci.* **53**, 35 (1975).
31. K. S. Birdi, H. N. Singh, and S. V. Dalsager, *J. Phys. Chem.* **83**, 2733 (1979).
32. B. L. Hauenstein, W. J. Dressick, S. L. Buell, J. N. Demas, and B. A. DeGraff, *J. Am. Chem. Soc.* **105**, 4251 (1983).
33. K. Mandal, B. L. Hauenstein, J. N. Demas, and B. A. DeGraff, *J. Phys. Chem.* **87**, 328 (1983).
34. W. J. Dressick, J. N. Demas, and B. N. DeGraff, *J. Photochem.* **24**, 45 (1984).
35. B. L. Hauenstein, W. J. Dressick, T. B. Gilbert, J. N. Demas, and B. A. DeGraff, *J. Phys. Chem.* **88**, 1902 (1984).
36. W. J. Dressick, K. W. Raney, J. N. Demas, and B. A. DeGraff, *Inorg. Chem.* **23**, 875 (1984).
37. W. J. Dressick, B. L. Hauenstein, T. B. Gilbert, J. N. Demas, and B. A. DeGraff, *J. Phys. Chem.* **88**, 3337 (1984).
38. U. Lachish, M. Ottolenghi, and J. Rabani, *J. Am. Chem. Soc.* **99**, 8062 (1977)
39. J. H. Baxendale and M. A. J. Rodgers, *Chem. Phys. Lett.* **72**, 424 (1980).
40. J. H. Baxendale and M. A. J. Rodgers, *J. Phys. Chem.* **86**, 4906 (1982).
41. K. S. Schanze, T. F. Mattox, and D. G. Whitten, *J. Am. Chem. Soc.* **104**, 7133 (1982).
42. R. R. Hautala, N. E. Schore, and N. J. Turto, *J. Am. Chem. Soc.* **95**, 5508 (1973).
43. N. E. Schore and N. J. Turro, *J. Am. Chem. Soc.* **96**, 306 (1974).
44. N. E. Schore and N. J. Turro, *J. Am. Chem. Soc.* **97**, 2488 (1975).
45. N. J. Turro, Y. Tanimoto, and G. Gabor, *Photochem. Photobiol.* **31**, 527 (1980).
46. G. A. Davis, *J. Am. Chem. Soc.* **94**, 5089 (1972).
47. L. K. Patterson and E. Vieil, *J. Phys. Chem.* **77**, 1191 (1973)

48. O. S. Khalil and A. J. Sonnessa, *Mol. Photochem.* **8**, 399 (1977).
49. R. Humphry-Baker and M. Grätzel, *J. Am. Chem. Soc.* **102**, 847 (1980).
50. E. Blatt, K. P. Ghiggino, and W. H. Sawyer, *J.C.S. Faraday I* **77**, 25551 (1981).
51. J. Slavik, *Biochim. Biophys. Acta* **694**, 1 (1982).
52. G. R. Fleming and G. Porter, *Chem. Phys. Lett.* **52**, 228 (1977).
53. H. Dodiuk, H. Kanety, and E. M. Kosower, *J. Phys. Chem.* **83**, 515 (1979).
54. J. H. Fendler, E. J. Fendler, G. A. Infante, P.-S. Shih, and L. K. Patterson, *J. Am. Chem. Soc.* **97**, 89 (1975).
55. K. A. Zachariasse, N. V. Phuc, and B. Kozankiewicz, *J. Phys. Chem.* **85**, 2676 (1981).
56. M. S. Fernandez and P. Fromherz, *J. Phys. Chem.* **81**, 1755 (1977).
57. P. Mukerjee and J. R. Cardinal, *J. Phys. Chem.* **82**, 1620 (1978).
58. N. Funasaki, *J. Phys. Chem.* **83**, 1998 (1979).
59. P. Mukerjee and A. Ray, *J. Phys. Chem.* **70**, 2144 (1966). (a) J. P. Otruba and D. G. Whitten, *J. Am. Chem. Soc.* **105**, 6503 (1983).
60. M. Shinitzky, A.-C. Dianoux, C. Gitler, and G. Weber, *Biochemistry* **10**, 2106 (1971).
61. M. Shinitzky, *Isr. J. Chem.* **12**, 879 (1974).
62. M. Grätzel and J. K. Thomas, *J. Am. Chem. Soc.* **95**, 6885 (1973).
63. M. Chen, M. Grätzel, and J. K. Thomas, *Chem. Phys. Lett.* **24**, 65 (1974).
64. M. Chen, M. Grätzel, and J. K. Thomas, *J. Am. Chem. Soc.* **97**, 2052 (1975). (a) K. Kalyanasundaram, M. Grätzel, and J. K. Thomas, *J. Am. Chem. Soc.* **97**, 3915 (1975).
65. R. C. Dorrance, T. F. Hunter, and J. Philip, *J.C.S. Faraday I* **73**, 89 (1977).
66. N. J. Turro and Y. Tanimoto, *Photochem. Photobiol.* **34**, 157 (1981).
67. M. Aoudia, M. A. J. Rodgers, and W. H. Wade, *J. Colloid Interface Sci.* **101**, 472 (1984).
68. U. K. A. Klein and H. P. Haar, *Chem. Phys. Lett.* **58**, 531 (1978).
69. H. E. Lessing and A. von Zena, *Chem. Phys.* **41**, 395 (1979).
70. S. A. Rice and G. Kenney-Wallace, *Chem. Phys.* **47**, 161 (1980).
71. W. Reed, M. J. Politi, and J. H. Fendler, *J. Am. Chem. Soc.* **103**, 4591 (1981).
72. M. Montal and C. Gitler, *Bioenergetics* **4**, 363 (1973).
73. L. S. Beckman and D. G. Brown, *Biochim. Biophys. Acta* **428**, 720 (1976).
74. A. D. James and B. H. Robinson, *J.C.S. Faraday I* **74**, 10 (1978).
75. U. K. A. Klein and M. Hauser, *Z. Phys. Chem. (Wiesbaden)* **90**, 215 (1974).
76. M. P. Pileni and M. Grätzel, *J. Phys. Chem.* **84**, 2403 (1980).
77. M. Gutman, D. Huppert, E. Pines, and E. Nachliel, *Biochim. Biophys. Acta* **642**, 15 (1981).
78. M. J. Politi and J. H. Fendler, *J. Am. Chem. Soc.* **106**, 265 (1984).
79. U. K. A. Klein and M. Hauser, *Z. Phys. Chem. (Wiesbaden)* **96**, 139 (1975).
80. U. Khauanga, R. J. McDonald, and B. K. Selinger, *Z. Phys. Chem. (Wiesbaden)* **101**, 209 (1976).
81. B. K. Selinger, *Aust. J. Chem.* **30**, 2087 (1977).
82. B. K. Selinger and A. Weller, *Aust. J. Chem.* **30**, 2377 (1977). (a) B. K. Selinger and C. M. Harris, in "Time Resolved Fluorescence Spectroscopy in Biochemistry and Biology," NATO ASI Ser., Vol. A69, p. 735. Plenum, New York, 1983.
83. E. Roelants, E. Geladé, M. van Der Auweraer, Y. Croonen, and F. C. DeSchryver, *J. Colloid Interface Sci.* **96**, 288 (1983).
84. M. Almgren and S. Swarup, *J. Phys. Chem.* **87**, 876 (1983).
85. M. V. Encinas, M. A. Rubio, and E. Lissi, *Photochem. Photobiol.* **37**, 125 (1983).
86. M. V. Encinas and E. A. Lissi, *Photochem. Photobiol.* **37**, 251 (1983).
87. M. V. Encinas, E. Guzman, and E. A. Lissi, *J. Phys. Chem.* **87**, 4770 (1983).
88. Y. Croonen, E. Geladé, M. van der Auweraer, M. van der Zegel, H. Vandendriessche, F. C. DeSchryver, and M. Almgren, *J. Phys. Chem.* **87**, 1426 (1983).
89. M. Almgren and S. Swarup, *J. Colloid Interface Sci.* **91**, 256 (1983).
90. J. E. Lofroth and M. Almgren, *J. Phys. Chem.* **86**, 1632 (1982).
91. M. Almgren and S. Swarup, *J. Phys. Chem.* **86**, 4212 (1982).

92. M. V. Encinas and E. A. Lissi, *Chem. Phys. Lett.* **91**, 55 (1982).
93. M. van der Auweraer, J. C. Dederen, C. Palman-Windel, and F. C. DeSchryver, *J. Am. Chem. Soc.* **104**, 1800 (1982).
94. M. Almgren and J. E. Lofroth, *J. Colloid Interface Sci.* **81**, 486 (1981).
95. M. A. J. Rodgers and J. H. Baxendale, *Chem. Phys. Lett.* **81**, 347 (1981).
96. H. W. Ziemiecki, R. Holland, and W. R. Cherry, *Chem. Phys. Lett.* **73**, 145 (1980).
97. S. M. B. Costa and A. L. Maçanita, *J. Phys. Chem.* **84**, 2408 (1980).
98. N. J. Turro and A. Yekta, *J. Am. Chem. Soc.* **100**, 5951 (1978).
99. A. Henglein and R. Scheerer, *Ber. Bunsenges. Phys. Chem.* **82**, 1112 (1978).
100. G. L. Gaines, *Inorg. Chem.* **19**, 1710 (1980).
101. M. R. Eftink and C. A. Ghiron, *J. Phys. Chem.* **80**, 486 (1974).
102. P. P. Infelta, M. Grätzel, and J. K. Thomas, *J. Phys. Chem.* **78**, 190 (1974).
103. M. W. Geiger and N. J. Turro, *Photochem. Photobiol.* **22**, 273 (1975).
104. N. J. Turro, M. Aikawa, and A. Yekta, *Chem. Phys. Lett.* **64**, 473 (1979).
105. E. A. Lissi, E. B. Abuin, L. Sepulveda, and F. H. Quina, *J. Phys. Chem.* **88**, 81 (1984).
106. M. Gonzalez, J. Vera, E. B. Abuin, and E. A. Lissi, *J. Colloid Interface Sci.* **98**, 152 (1984).
107. E. B. Abuin, E. A. Lisssi, N. Bianchi, L. Miola, and F. H. Quina, *J. Phys. Chem.* **87**, 5166 (1983).
108. L. Miola, R. B. Akaberil, M. F. Ginani, P. B. Filho, V. G. Toscano, and F. H. Quina, *J. Phys. Chem.* **87**, 4417 (1983).
109. T. Wolff, *Ber. Bunsenges. Phys. Chem.* **86**, 1132 (1982).
110. T. Wolff and G. von Bunau, *Ber. Bunsenges. Phys. Chem.* **86**, 225 (1982).
111. S. Shah and M. W. Windsor, *Chem. Phys. Lett.* **92**, 33 (1982).
112. N. J. Turro and Y. Tanimoto, *Photochem. Photobiol.* **34**, 157 (1981).
113. F. Grieser, *Chem. Phys. Lett.* **83**, 59 (1981).
114. S. A. Amire and H. D. Burrows, *J.C.S. Faraday I* **78**, 2033 (1982).
115. H. D. Burrows, S. J. Formoshinho, M. F. J. R. Paiva, and E. J. Rasbum, *J.C.S. Faraday II* **76**, 685 (1980).
116. H. D. Burrows, S. J. Formoshinho, M. Fernanda, and J. R. Paiva, *J. Photochem.* **12**, 285 (1980).
117. N. J. Turro, Y. Tanimoto, and G. Gabor, *Photochem. Photobiol.* **31**, 527 (1980).
118. S. S. Atik, C. L. Kwan, and L. A. Singer, *J. Am. Chem. Soc.* **101**, 5696 (1979).
119. M. Almgren, J. E. Lofroth, and R. Rydholm, *Chem. Phys. Lett.* **63**, 265 (1979).
120. M. Van Bockstaele, J. Gelan, H. Martens, J. Put, J. C. Dederen, N. Boens, and F. C. DeSchryver, *Chem. Phys. Lett.* **58**, 211 (1978).
121. S. S. Atik and L. A. Singer, *J. Am. Chem. Soc.* **100**, 3234 (1978).
122. S. S. Atik and L. A. Singer, *Chem. Phys. Lett.* **59**, 519 (1978).
123. M. A. J. Rodgers and M. F. DeSilva e Wheeler, *Chem. Phys. Lett.* **53**, 165 (1978).
124. K. Kalyanasundaram, M. Grätzel, and J. K. Thomas, *J. Am. Chem. Soc.* **97**, 3915 (1975).
125. M. Grätzel, K. Kalyanasundaram, and J. K. Thomas, *J. Am. Chem. Soc.* **96**, 7869 (1974).
126. H. J. Pownall and L. C. Smith, *Biochemistry* **13**, 2594 (1974).
127. R. R. Hautala, N. E. Schore, and N. J. Turro, *J. Am. Chem. Soc.* **95**, 5508 (1973).
128. J. C. Dederen, M. Vander Anneraer, and F. C. DeDchryer, *Chem. Phys. Lett.* **68**, 451 (1979).
129. T. Nakamura, A. Kira, and M. Imamura, *J. Phys. Chem.* **88**, 3435 (1984).
130. S. Hashimoto and J. K. Thomas, *Chem. Phys. Lett.* **109**, 115 (1984).
131. T. Miyashita, T. Murakata, and M. Matsuda, *J. Phys. Chem.* **87**, 4529 (1983).
132. E. B. Abuin and E. A. Lissi, *J. Colloid Interface Sci.* **93**, 562 (1983).
133. N. Kim-Thnan and J. C. Scaiano, *Chem. Phys. Lett.* **101**, 192 (1983).
134. T. Nakamura, A. Kira, and M. Imamura, *J. Phys. Chem.* **87**, 3122 (1983).
135. H. J. Timpe, G. Israel, H. G. O. Becker, I. R. Gould, and N. J. Turro, *Chem. Phys. Lett.* **99**, 275 (1983).
136. S. Hashimoto and J. K. Thomas, *J. Am. Chem. Soc.* **105**, 5230 (1983).

137. T. Nakamura, A. Kira, and M. Imamura, *J. Phys. Chem.* **86**, 3359 (1982).
138. S. J. Atherton, J. H. Baxendale, and B. M. Hoey, *J.C.S. Faraday I* **78**, 2167 (1982).
139. J. C. Russell and D. G. Whitten, *J. Am. Chem. Soc.* **104**, 5937 (1982).
140. J. C. Dederen, M. Vander Auweraer, and F. C. DeSchryver, *J. Phys. Chem.* **85**, 1198 (1981).
141. J. C. Russell, D. G. Whitten, and A. M. Braun, *J. Am. Chem. Soc.* **103**, 3219 (1981).
142. F. M. Martens and J. W. Verhoeven, *J. Phys. Chem.* **85**, 1773 (1981).
143. H. Ziemiecki and W. R. Cherry, *J. Am. Chem. Soc.* **103**, 4479 (1981).
144. R. H. Schmehl, L. G. Whitesell, and D. G. Whitten, *J. Am. Chem. Soc.* **103**, 3761 (1981).
145. T. K. Foreman, W. M. Sobol, and D. G. Whitten, *J. Am. Chem. Soc.* **103**, 5339 (1981).
146. F. Grieser and R. Tausch-Treml, *J. Am. Chem. Soc.* **102**, 7258 (1980).
147. E. B. Abuin and E. A. Lissi, *J. Phys. Chem.* **84**, 2605 (1980).
148. R. H. Schmehl and D. G. Whitten, *J. Am. Chem. Soc.* **102**, 1938 (1980).
149. M. Von Bockstaele, J. Gelan, H. Martens, J. Put, F. C. DeSchryver, and J. C. Dederen, *Chem. Phys. Lett.* **70**, 605 (1980).
150. M. A. J. Rodgers and M. F. de Silva e Wheeler, *Chem. Phys. Lett.* **53**, 165 (1978).
151. M. Tachiya, *Chem. Phys. Lett.* **33**, 289 (1975).
152. P. P. Infelta, *Chem. Phys. Lett.* **61**, 88 (1979).
153. A. Yekta, M. Aikawa, and N. J. Turro, *Chem. Phys. Lett.* **63**, 543 (1979).
154. P. P. Infelta and M. Grätzel, *J. Chem. Phys.* **70**, 179 (1979).
155. M. Tachiya, *J. Chem. Phys.* **76**, 340 (1982).
156. M. Almgren and J. E. Lofroth, *J. Chem. Phys.* **76**, 2734 (1982).
157. M. Van der Auwerear, J. C. Dederen, E. Geladé, and F. C. DeSchryver, *J. Chem. Phys.* **74**, 1140 (1981).
158. T. F. Hunter, *Chem. Phys. Lett.* **75**, 152 (1980).
159. M. Almgren, G. Gunnarsson, and P. Linse, *Chem. Phys. Lett.* **85**, 451 (1982).
160. M. Almgren, P. Linse, M. Van der Auweraer, F. C. Deschryver, E. Geladé, and Y. Croonen, *J. Phys. Chem.* **88**, 289 (1984).
161. M. Tachiya, *J. Chem. Phys.* **78**, 5282 (1983).
162. P. P. Infelta and M. Grätzel, *J. Chem. Phys.* **78**, 5280 (1983).
163. S. S. Atik, M. Nam, and L. A. Singer, *Chem. Phys. Lett.* **67**, 75 (1979).
164. P. Lianos and J. Lang, *J. Colloid Interface Sci.* **96**, 222 (1983).
165. P. Lianos and R. Zana, *J. Colloid Interface Sci.* **88**, 594 (1982).
166. P. Lianos, J. Lang, and R. Zana, *J. Colloid Interface Sci.* **91**, 276 (1983).
167. A. Malliaris and C. M. Paleos, *J. Colloid Interface Sci.* **101**, 364 (1984).
168. P. Lianos and R. Zana, *Chem. Phys. Lett.* **76**, 62 (1980).
169. P. Lianos and R. Zana, *Chem. Phys. Lett.* **72**, 171 (1980).
170. P. Lianos and R. Zana, *J. Colloid Interface Sci.* **84**, 100 (1981).
171. P. Lianos and R. Zana, *J. Phys. Chem.* **84**, 3339 (1980).
172. P. Lianos and R. Zana, *J. Colloid Interface Sci.* **101**, 587 (1984).
173. M. Almgren, F. Grieser, and J. K. Thomas, *J. Am. Chem. Soc.* **101**, 279 (1979).
174. F. H. Quina and V. G. Toscano, *J. Phys. Chem.* **81**, 1750 (1977).
175. J. Leigh and J. C. Scaiano, *J. Am. Chem. Soc.* **105**, 5652 (1983).
176. F. Greiser, *J. Phys. Chem.* **85**, 928 (1981).
177. T. Förster and B. K. Selinger, *Z. Naturforsch., A* **19**, 38 (1964).
178. S. J. Rehfeld, *J. Colloid Interface Sci.* **34**, 518 (1970).
179. H. J. Pownall and L. C. Smith, *J. Am. Chem. Soc.* **95**, 3136 (1972).
180. R. C. Dorrance and T. F. Hunter, *J.C.S. Faraday I* **68**, 1312 (1972).
181. M. Hauser and U. Klein, *Z. Phys. Chem. (Wiesbaden)* **78**, 32 (1972).
182. M. Hauser and U. Klein, *Acta Phys. Chem.* **19**, 363 (1973).
183. R. C. Dorrance and T. F. Hunter, *J.C.S. Faraday I* **70**, 1572 (1974).
184. U. Khuanga, B. K. Selinger, and R. McDonald, *Aust. J. Chem.* **29**, 1 (1976).

185. K. Kalyanasundaram and J. K. Thomas, *in* "Micellisation, Solubilisation and Microemulsions" (K. L. Mittal, ed.), Vol. 2, p. 569. Plenum, New York, 1977.
186. B. K. Selinger and A. R. Watkins, *Chem. Phys. Lett.* **56**, 99 (1978).
187. M. Aoudia and M. A. J. Rodgers, *J. Am. Chem. Soc.* **101**, 6777 (1979).
188. A. R. Watkins and B. K. Selinger, *Chem. Phys. Lett.* **64**, 250 (1979).
189. D. J. Miller, U. K. A. Klein, and M. Hauser, *Ber. Bunsenges. Phys. Chem.* **84**, 1135 (1980).
190. D. J. Miller, *Ber. Bunsenges. Phys. Chem.* **85**, 337 (1981).
191. B. K. Selinger and A. R. Watkins, *J. Photochem.* **16**, 321 (1981).
192. B. K. Selinger and A. R. Watkins, *J. Photochem.* **20**, 319 (1982).
193. K. A. Zachariasse, *Chem. Phys. Lett.* **57**, 429 (1978).
194. J. Emert, C. Behrens, and M. Goldenberg, *J. Am. Chem. Soc.* **101**, 771 (1979).
195. N. J. Turro, M. Aikawa, and A. Yekta, *J. Am. Chem. Soc.* **101**, 772 (1979).
196. N. J. Turro and T. Okubo, *J. Am. Chem. Soc.* **103**, 7224 (1981).
197. C. N. Henderson, B. K. Selinger, and A. R. Watkins, *J. Photochem.* **16**, 215 (1981).
198. K. Kano, T. Ishibashi, and T. Ogawa, *J. Phys. Chem.* **87**, 3010 (1983).
199. K. A. Zachariasse and W. Kühnle, *Z. Phys. Chem.* (*Wiesbaden*) **101**, 267 (1976).
200. K. A. Zachariasse, W. Kühnle, and A. Weller, *Chem. Phys. Lett.* **59**, 375 (1978).
201. M. J. Snare, P. J. Thistlethwaite, and K. P. Ghiggino, *J. Am. Chem. Soc.* **105**, 3328 (1983).
202. K. A. Zachariasse, G. Duveneck, and R. Busse, *J. Am. Chem. Soc.* **106**, 1045 (1984).
203. H. Masuhara, K. Kaji, and N. Mataga, *Bull. Chem. Soc. Jpn*, **50**, 2084 (1977).
204. B. Katusin-Razem, M. Wong, and J. K. Thomas, *J. Am. Chem. Soc.* **100**, 1679 (1978).
205. Y. Waka, K. Hamamoto, and N. Mataga, *Chem. Phys. Lett.* **53**, 242 (1978).
206. Y. Waka, K. Hamamoto, and N. Mataga, *Chem. Phys. Lett.* **62**, 364 (1979).
207. H. Masuhara, H. Tanabe, and N. Mataga, *Chem. Phys. Lett.* **63**, 273 (1979).
208. Y. Tsutsui, K. Takuma, and T. Matsuo, *Chem. Lett.* p. 617 (1979).
209. Y. Waka, K. Hamamoto, and N. Mataga, *Photochem. Photobiol.* **32**, 27 (1980).
210. S. S. Atik and J. K. Thomas, *J. Am. Chem. Soc.* **101**, 3550 (1981).
211. G. S. Singhal, E. Rabinowitch, J. Hevesi, and V. Srinivasan, *Photochem. Photobiol.* **11**, 537 (1970).
212. M. Almgren, *Photochem. Photobiol.* **15**, 219 (1972).
213. G. A. Kenney-Wallace, J. H. Flint, and S. C. Wallace, *Chem. Phys. Lett.* **32**, 71 (1975).
214. K. Csatorday, E. Lehoczki, and L. Szalay, *Biochim. Biophys. Acta* **376**, 268 (1975).
215. J. R. Escabi-Perez, F. Nome, and J. H. Fendler, *J. Am. Chem. Soc.* **99**, 7749 (1977).
216. G. Rothenberger, P. P. Infelta, and M. Grätzel, *J. Phys. Chem.* **83**, 1871 (1979).
217. M. Almgren, F. Grieser, and J. K. Thomas, *J. Am. Chem. Soc.* **101**, 2021 (1979).
218. Y. Kusumoto and H. Sato, *Chem. Phys. Lett.* **68**, 13 (1979).
219. Y. Usui and A. Gotou, *Photochem. Photobiol.* **29**, 165 (1979).
220. M. D. Hatlee, J. J. Kozak, G. Rothenberger, P. P. Infelta, and M. Grätzel, *J. Phys. Chem.* **84**, 1508 (1980).
221. T. Matsuo, Y. Aso, and K. Kano, *Ber. Bunsenges. Phys. Chem.* **84**, 146 (1980).
222. T. Marszalek, A. Baczynski, W. Orzesko, and A. Rozploch, *Z. Naturforsch., A* **35A**, 85 (1980).
223. P. K. F. Koglin, D. J. Miller, J. Steinwandel, and M. Hauser, *J. Phys. Chem.* **85**, 2363 (1981).
224. K. Mandal and J. N. Demas, *Chem. Phys. Lett.* **84**, 410 (1981).
225. J. C. Scaiano and J. C. Selwyn, *Photochem. Photobiol.* **34**, 29 (1981).
226. T. W. Ebbesen, O. Delgado, A. Valla, M. Giraud, Y. Saito, H. Tachibana, and A. Wada, *Photochem. Photobiol.* **35**, 665 (1982).
227. H. Sato, M. Kawasaki, and K. Kasatani, *J. Phys. Chem.* **87**, 3759 (1983).

Chapter 3

Micellar Photophysics—Triplet-State Reactions

Singlet excited-state reactions described in an earlier chapter take place over a few hundred nanoseconds at most. On these time scales, the micellar aggregates can be considered as intact assemblies (permanent) with the solutes trapped in them. There are, however, several dynamical processes associated with aggregation and solubilization that take place over several hundred microseconds or longer, for example, the exit and entry of solutes in and out of the micelles, exchange of a monomeric surfactant with the micelle, and the complete collapse of the micelle (dissolution). These slower processes are conveniently probed using the triplet excited states.

3.1 Phosphorescence and the Triplet State

Monitoring of the luminescence in the steady-state probably is the simplest way of studying the reactions of an excited-state molecule. Unfortunately, with triplets the observation of phosphorescence at room temperature has always been such a problem (if not a challenge) that the technique of flash photolysis is invariably employed. The rare observance of phosphorescence in solution at room temperature is often attributed to self-quenching and impurity quenching of the triplet state. To facilitate the observance of phosphorescence internal and external heavy atom effects are often introduced to optimize the critical parameters viz., intersystem crossing yields Φ_{isc} and the radiative rate constant k_p. In a few cases under carefully prepared conditions, it is now possible to observe phosphorescence in homogeneous solutions.[1]

In recent years it has been found that introduction of micelles to solubilize the probes leads to some dramatic improvements in our ability to observe phosphorescence in solution at room temperature (in some cases even in *aerated* micellar solutions!).[2-4] A heavy atom can be introduced as a substituent on an aromatic molecule or we can employ a functionalized micellar system with heavy atoms such as Tl^+ or Ag^+ as counterions with comparable efficiency. Thus 1-bromopyrene or 1-bromonaphthalene in aqueous sodium laurylsulfate micelles (or pyrene/naphthalene solubilized in the presence of added Tl^+ or Ag^+ ions) exhibits intense room-temperature phosphorescence. Figure 3.1 illustrates the latter phenomenon for a few aromatic hydrocarbons. The list of aromatic hydrocarbons and heterocyclic and carbocyclic compounds that have been shown to exhibit room-temperature phosphorescence in this manner is growing steadily.[5-15] Analytical potential of the micelle-stabilized room-temperature phosphorescence (MS-RTP) has been exploited in an elegant manner by Cline Love and

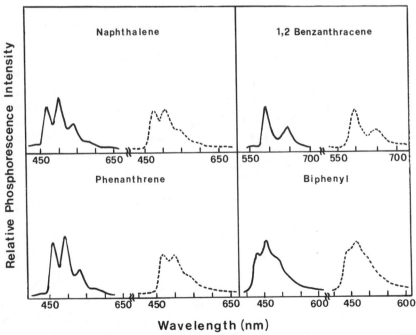

FIG. 3.1. Phosphorescence spectra of aromatic hydrocarbons in 3-methylpentane at 77 K (left) and in room-temperature aqueous micellar solutions of sodium dodecyl sulfate (SDS) containing Tl^+ ions (right): [SDS] = 10^{-2} M and [Tl^+] = 4×10^{-3} M, λ_{excit} = 280 nm for all cases except for 1,2-benzanthracene for which λ_{exit} = 330 nm. (From Kalyanasundaram et al.[2])

her co-workers.[9-14] For molecules such as pyrene, the limits of detection has been estimated to be in parts per billion (ppb), i.e., in the nanomolar range.

There are several factors that can contribute to the observation of MS-RTP. Fluorescence studies described earlier have shown that micelles organize reactants on a molecular level by distributional effects and also provide a protective environment. These effects minimize self-quenching and also quenching by the impurities. In the case of intense phosphorescence observed with nonhalogenated aromatics in SDS micelles in the presence of Tl^+/Ag^+ ions, the micelles increase the proximity of heavy atoms and probes and present locally high concentrations of the heavy atom species. Moderate triplet lifetimes in micellar solutions (microsecond–millesecond range) are also believed to contribute to the sensitivity factor. For any short time integration, molecules with short lifetimes (e.g., 0.5 msec as compared to seconds) recycle through the excited state many times emitting more photons during the same span. In SDS micelles, heavy atoms such as Tl^+ or Ag^+ in the concentration range of 10–20% (Tl^+/Na^+ and Ag^+/Na^+) have been found to be adequate.

The observed phosphorescence lifetimes in aqueous micellar solutions are fairly long, in the range of 0.2–2.0 msec at room temperature. Table 3.1 presents a collection of triplet lifetime data measured via MS-RTP. At low solute concentrations (mean occupancy number per micelle ≪ 1) the phosphorescence decay is invariably exponential. Since the triplet is very much quenched when it is in the aqueous phase and since there are several studies which indicate that solute entry and exit occur on microsecond time scale, it is worthwhile asking what the observed phosphorescence lifetimes represent and whether it is possible to extract the entry and exit rates from an analysis of phosphorescence quenching. There have been a few studies of triplet lifetimes in micellar media from this viewpoint. All models treat the micellar solutions as a two-phase system: an aqueous phase and a micellar phase with the triplet state exhibiting different decay rates in each of these (k_w and k_m). Furthermore, we consider cases of very low $\bar{n}$ and high [M] (where $\bar{n}$ and [M] stand for mean occupancy per micelle and micelle concentration, respectively). The probes exit and enter the micelles with specific rate constants k_- and k_+. In the absence of any secondary quenching, the phosphorescence decay is described by:

$$k_{obs} = 1/\tau = k_m + k_- - [k_+k_-[M]/(k_+[M] + k_w)] \qquad (3.1)$$

Phosphorescence quenching studies (to be discussed later) indicate that for most of the hydrophobic solutes the entry rate constant k_+ is quite high ($\simeq 10^{10} \ M^{-1} \ sec^{-1}$). Hence, the observed lifetime for many species depends primarily on the exit rate k_- and the rate constant for triplet-state deactivation inside the micelle k_m. Which of the above terms predominate depends on the

TABLE 3.1. Micelle-Stabilized Room-Temperature Phosphorescence
Lifetimes of Polynuclear Aromatic Compounds [a]

Solute	Micellar system	τ_{ph} (ref. no) (msec)
1-Bromonaphthalene	SDS	1.15 (4)
4-Bromo-p-terphenyl	SDS	2.15 (4)
1-Chloronaphthalene	SDS	0.19 (5)
1-Chloronaphthalene	CTAB	0.45 (5)
Bromonaphthoyl dodecanoate (BND)	CTAC	0.48 (5)
Pyrene	SDS/Ag$^+$	0.60 (12), 0.75 (3)
Naphthalene	SDS/Ag$^+$	0.31 (12), 0.20 (3)
Biphenyl	SDS/Ag$^+$	0.33 (12)
Pyrene	SDS/Tl$^+$	1.20 (14), 0.93 (7)
Naphthalene	SDS/Tl$^+$	0.70 (4), 0.45 (7)
Biphenyl	SDS/Tl$^+$	1.00 (4), 0.43 (7)
Phenanthrene	SDS/Tl$^+$	0.89 (14)
Chrysene	SDS/Tl$^+$	1.37 (14)
Fluorene	SDS/Tl$^+$	0.42 (14)
Fluoranthrene	SDS/Tl$^+$	2.33 (14)
2,3-Benzofluorene	SDS/Tl$^+$	1.53 (14)
1-Methylnaphthalene	SDS/Tl$^+$	1.0 (4), 0.62 (9)
2-Methylnaphthalene	SDS/Tl$^+$	0.66 (9)
2-Methoxynaphthalene	SDS/Tl$^+$	0.56 (9)
2-Naphthylacetic acid	SDS/Tl$^+$	0.39 (7)
2-Naphthol	SDS/Tl$^+$	0.28 (7)
2-Naphthoic acid	SDS/Tl$^+$	0.32 (7)
1-Naphthylmethylcarbinol	SDS/Tl$^+$	0.69 (7)
4-Biphenylcarboxaldehyde	SDS/Tl$^+$	0.68 (7)
1-Phenylphenol	SDS/Tl$^+$	0.25 (7)
4-Hydroxybiphenyl	SDS/Tl$^+$	0.40 (9)
1-Pyrenebutyric acid	SDS/Tl$^+$	1.10 (7,9)
1-Aminopyrene	SDS/Tl$^+$	1.12 (9)
N-2-Cyanoethyl carbazole	SDS/Tl$^+$	0.13 (10)
N-2-Iodoethyl carbazole	SDS/Tl$^+$	0.02 (10)
N-2-Chloroethyl carbazole	SDS/Tl$^+$	0.02 (10)
Benzo(e)pyrene	SDS/Tl$^+$	0.71 (4)

[a] All values measured in milliseconds.

relative magnitudes of the migration rate (average residence time) versus the
stability of the excited state. In many cases, the exit rate is proportional to
the probe's solubility in the bulk aqueous phase. Thus, the relative ordering
of the phosphorescence lifetimes of different species can be predicted from
their solubilities in water, barring other secondary reactions. Table 3.2
presents partition coefficient data for arenes in micellar solutions determined by solubility measurements and estimates on their exit rates.[4]

TABLE 3.2. Partition Coefficient K Data for Arenes in Micellar Solutions
Determined by Solubility Measurements and Estimates on
the Exit Rates k_{-1}

	Micelles[b]			
	SDS		CTAB	
Arene	$K(M^{-1})$	$k_-(\sec^{-1})$	$K(M^{-1})$	$k_-(\sec^{-1})$
Perylene	1.7×10^7	4.1×10^2	2.7×10^7	2.6×10^2
Anthracene	4×10^5	1.7×10^4	2.6×10^6	3.2×10^3
Pyrene	1.7×10^6	4.1×10^3	1.0×10^7	1.7×10^3
1-Bromonaphthalene	2.1×10^5	3.3×10^4	1.7×10^6	4.1×10^3
Biphenyl	7.3×10^4	9.6×10^4	4.3×10^5	1.6×10^4
1-Methylnaphthalene	7.1×10^4	1×10^5	—	—
Naphthalene	2.5×10^4	2.5×10^5	9.1×10^4	—
p-Xylene	1.6×10^4	4.4×10^5	—	—
Toluene	5.3×10^4	1.3×10^6	—	—
Benzene	1.6×10^3	4.4×10^6	9.3×10^3	7.5×10^5

[a] Data from Almgren et al.[4]

[b] Here k_- has been calculated using $7 \times 10^9 \ M^{-1} \sec^{-1}$ for the entrance rate
k_+ and the relation $k_- = k_+/K$.

3.2   Delayed Fluorescence and Triplet Self-Quenching

Micellar systems loaded with high concentrations of probes (mean
occupancy per micelle $\bar{n} \gg 1$) can also promote self-quenching of triplets.
Thus, under suitable conditions intramicellar annihilation of two triplets
cosolubilized in the same micelle can lead to delayed fluorescence:

$$^3P_m^* + {}^3P_m^* \longrightarrow {}^1P_m^* + P_m \qquad (3.2)$$

Turro and Aikawa, in fact, have demonstrated the occurrence of such
reactions by detecting both the phosphorescence and delayed fluorescence of
1-chloronaphthalene solubilized in anionic and cationic micellar solutions.[5]
The possibility of intermicellar mechanisms (e.g., exit of a triplet probe from
one micelle and entry into another micelle which already has a triplet proble)
has been eliminated by studies of intensity dependence of delayed fluorescence, phosphorescence as a function of $\bar{n}$.

Hunter and Szczepanski[16] have examined triplet lifetimes of pyrene
solubilized in ionic micelles via triplet–triplet absorption over a wide
concentration range to accommodate multiple occupancy of probes. An
attempt has been made with kinetic models to quantitatively describe the
decrease in triplet lifetime as to (a) the influence of the distribution of pyrene
among the micelles (poisson and other types), (b) triplet excimers (quenching
of a triplet by a ground-state pyrene molecule), and (c) T–T annihilation.
The estimated rate constants for pyrene are in expected ranges, except for k_+.

For SDS micelles, the values are $k_- = 18$ sec^{-1}, $k_+ = 1.6 \times 10^5$ M^{-1} sec^{-1}, $k_w = 100$ sec^{-1}, and $k_m = 13$ sec^{-1}.

Rothenberger et al.,[17] in fact, have presented a detailed kinetic statistical model for the analysis of intramicellar T–T annihilation processes. Laser photolysis studies of triplets of 1-bromonaphthalene solubilized in CTAB micelles showed that the temporal behavior of the triplet absorption decay is dependent on both the energy fluence of the laser pulse and the average occupancy $\bar{n}$ of the micelles. At high photon fluxes or $\bar{n}$, we observe a fast initial decay due to T–T annihilation. Kinetic analysis in terms of a model has yielded $k_{T-T} = 1.4 \times 10^7$ sec^{-1} for the intramicellar triplet–triplet annihilation process.

3.3 Phosphorescence and Triplet Quenching: Exit Rates of Solutes

The exit rate of a solubilized arene (hydrophobic probes, in general) from a micelle can be expected to be relatively slow, and, hence, fluorescence techniques are not suited for the study of exit process. However methods based on long-lived phosphorescence can still be used. Earlier we showed that with due care long-lived phosphorescence of a wide variety of aromatic hydrocarbons and heterocyclic compounds can be seen in deaerated solutions. Almgren et al.[4] have proposed a method that utilizes phosphorescent probes to study exit rates of solutes. It is based on a limiting case in which all the quenching of the triplet probe occurs during its residence in the aqueous phase. The kinetic scheme below outlines various processes of interest:

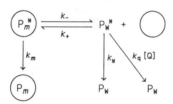

In the above scheme P_m^* and P_w^* refer to the excited-state probe present in the micellar and aqueous phases of the dynamic system. If a quencher Q restricted to the aqueous phase is now added to such a system, quenching will occur exclusively in the aqueous phase during the probe residence in the aqueous phase in competition with the re-entry of the probe into the micelles. No quenching of the excited-state probe occurs in the micelle! With an increase in the quencher concentration, we could expect a decrease in the observed phosphorescence lifetime until the exit rate becomes the controlling step to the quenching. At this point, further increases in the quencher concentration will not change the phosphorescence lifetime. Figure 3.2

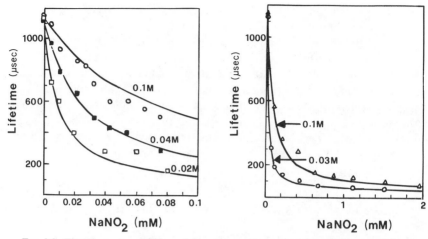

FIG. 3.2. Phosphorescence lifetime of 1-bromonaphthalene as a function of sodium dodecyl sulfate (SDS) and sodium nitrite ($NaNO_2$) concentrations. The solid curves are computer fits using the following parameters: $k_+ = 7 \times 10^9 \ M^{-1} \ \sec^{-1}$, $k_- = 2.5 \times 10^4 \ \sec^{-1}$, $k_q = 5 \times 10^9 \ M^{-1} \ \sec^{-1}$, $k_p = 2.5 \times 10^3 \ \sec^{-1}$, $k_m = 8.7 \times 10^2 \ \sec^{-1}$, and $N = 62$. [Br-Naph] $= 10^{-4} \ M$ in all cases except for 0.02 M SDS for which [Br-Naph] $= 5 \times 10^{-5} \ M$. (From Almgren et al.[4] Copyright 1979 American Chemical Society.)

illustrates such limiting behavior during the NO_2^- quenching of 1-bromo-naphthalene in anionic SDS micelles.

The following kinetic equations can be constructed for this scheme.

$$-d[P_m^*]/dt = (k_-[P_m^*]) + (k_+[M][P_w^*]) + (k_m[P_m^*]) \tag{3.3}$$

$$-d[P_w^*]/dt = (k_q[P_w^*][Q]) + (k_w[P_w^*]) - (k_-[P_m^*]) + (k_+[M][P_w^*]) \tag{3.4}$$

where [M] and [Q] stand for the concentration of the micelles and quencher, respectively. Assuming steady-state conditions for $[P_w^*]$, i.e.,

$$-d[P_w^*]/dt = 0 \tag{3.5}$$

we have

$$[P_w^*] = k_-[P_m^*]/(k_q[Q] + k_w + k_+[M]) \tag{3.6}$$

Substituting Eq. (3.6) into Eq. (3.3) we obtain

$$-\frac{d[P_m^*]}{dt} = (k_-[P_m^*]) - \frac{(k_+[M]k_-[P_m^*])}{(k_q[Q] + k_w + k_+[M])} + (k_m[P_m^*])$$

$$= [P_m^*]\left\{ k_- + k_m - \frac{k_-k_+[M]}{(k_q[Q] + k_w + k_+[M])} \right\} \tag{3.7}$$

Thus, the disappearence of $[P_m^*]$ follows first-order kinetics and the observed rate constant k_{obs} is given by

$$k_{obs} = 1/\tau_{obs} = k_- + k_m - \left(\frac{k_- k_+[M]}{(k_q[Q] + k_w + k_+[M])}\right) \qquad (3.8)$$

Equation (3.8) shows that as the quencher concentration increases, k_{obs} approaches a plateau value of $(k_- + k_m)$. If $k_m \ll k_-$ as in the case of long-lived phosphorescence, the escape rate constant k_- of the probe can be measured.

If the conditions are so chosen that $k_+[M] \gg (k_q[Q] + k_w)$ then Eq. (3.8) can be approximated by

$$k_{obs} = \frac{(k_- k_q + k_m k_q)[Q]}{k_+[M]} + \alpha \qquad (3.9)$$

where

$$\alpha = \frac{k_w k_- + k_w k_m + k_m k_+[M]}{k_+[M]}$$

If $k_- > k_m$, this equation reduces further to

$$k_{obs} = (k_- k_q[Q]/k_+[M]) + \alpha \qquad (3.10)$$

The usefulness of Eq. (3.10) lies in the region where the exit rate k_- cannot be found because it is beyond the time resolution of the equipment. The equilibrium constant can still be found from the slope of k_{obs} versus $[Q]$ provided that k_q and $[M]$ are known. Table 3.3 presents a collection of exit rate constants determined using this method.[4,5,8,18-21]

Mobile Quenchers

For quenchers that partition significantly between the micellar and aqueous phases, depending on the relative mobility of the probe and the quencher molecule, the triplet quenching process will be controlled either by the exit rate of the probe or of the quenchers. In the earlier-described method of Almgren *et al.*, ionic quenchers which are restricted to the aqueous phase were employed. For this case the exit rate of the probe from the micelles controls the quenching process. For quenchers that partition significantly between the micellar and aqueous phases, the quenching will occur both by intramicellar and intermicellar pathways. Intramicellar quenching occurs in micelles where both the probe and quenchers are already present during the production of triplets. This is a very fast process (occurring in times less than a few nanoseconds) with unit efficiency and can be considered as static quenching. By intermicellar exchange processes, the quencher or the probe (depending on their relative mobilities) exits and re-enters several micelles before it finds its reaction partner. For very hydrophobic

TABLE 3.3. Exit Rates Data for Solutes (Probes) in Micelles Determined via the Triplet Quenching Method of Almgren, Grieser, and Thomas

Probe[a]	Micelle/quencher	$K = (k_+/k_-)$ $(\times 10^4 \ M^{-1})$	$k_-(\text{sec}^{-1})$	Reference
1-Bromonaphthalene	SDS/NO_2^-	18	2.5×10^4	4
1-Methylnaphthalene	SDS/NO_2^-	2.5–5.5	$>5 \times 10^4$	4
Naphthalene	SDS/NO_2^-	0.7–1.7	$>5 \times 10^4$	4
Biphenyl	SDS/NO_2^-	2.2–5.4	$>5 \times 10^4$	4
4-Bromo-p-terphenyl	SDS/NO_2^-	—	$<7.7 \times 10^2$	4
BNDA-10	$CTAC/Co(NH_3)_6^{3+}$	1.8	3.2×10^3	8
BNDA-8	$CTAC/Co(NH_3)_6^{3+}$	1.1	1.5×10^4	8
BNDA-5	$CTAC/Co(NH_3)_6^{3+}$	0.26	$>5 \times 10^4$	8
BNDS-10	$SDS/Fe(CN)_6^{3+}$	—	1.4×10^3	8
BNDS-10	$SDS/Fe(CN)_6^{3+}$	—	5.7×10^3	8
BNDS-10	$SDS/Fe(CN)_6^{3+}$	—	0.7×10^3	8
1-Chloronaphthalene	SDS/NO_2^-	—	4.3×10^4	5
1-Chloronaphthalene	$CTAB/Co(NH_3)_6^{3+}$	—	1.9×10^3	5
Acetophenone	SDS/NO_2^-	0.204	7.8×10^6	19,20
Propiophenone	SDS/NO_2^-	0.47	3.0×10^6	19,20
Iso-butyrophenone	SDS/NO_2^-	0.75	1.6×10^6	19,20
p-Methoxyacetophenone	SDS/NO_2^-	0.30	(9.2×10^6)	19,20
Benzophenone	SDS/NO_2^-	(2.6)	(2.0×10^6)	19,20
Xanthone	SDS/NO_2^-	(0.63)	2.7×10^6	19,20

[a] BNDA-8 = 8-(4-Bromo-1-naphthoyl)octyltrimethylammonium bromide, BNDA-10 = 10-(4-Bromo-1-naphthoyl)decyltrimethylammonium bromide, BNDS-10 = 10(4-Bromo-1-naphthoyl)decyl sulfate.

probes this process is again exit rate controlled but with respect to the quencher. Using suitable kinetic models and quantitative analysis, it is possible to extract the exit and entry rate constants of the quencher and estimate those of the probe molecules. Several studies of this type are already available, e.g., quenching of ^{3}bromonaphthalene* by pyrene,[15] ^{3}phenanthrene* by conjugated dienes,[20] ^{3}arylketones* by γ-methylvalerophenone[19] and ^{3}acetone* by biphenyl.[21]

The decay of micelle-solubilized probes can be expected to occur in a pseudo-first-order fashion (k_{expt}) controlled by the concentration of the quenchers in the aqueous phase $[Q_w]$ and the rate of entry of quencher into the micelle k_+:

$$k_{\text{expt}} = k_0 + k_+[Q_w] \tag{3.11}$$

where k_0 is the rate of triplet decay in the absence of the quencher. Using the equilibrium relations between K, k_+, and k_-:

$$K = (k_+/k_-) = [Q_m]/([Q_w][M]) \tag{3.12}$$

Eq. (3.11) can also be written as

$$k_{expt} = k_0 - (k_-[Q_m]/[M]) \qquad (3.13)$$

i.e.,

$$k_{expt} = k_0 - (k_- \bar{n}) \qquad (3.14)$$

where $\bar{n}$ is the mean occupancy number of quenchers per micelle.

For those quenchers which are largely resident in the micellar phase, i.e., $[Q_m] \gg [Q_w]$, then we can replace $[Q_m]$ by the total (or bulk) concentration of the quenchers $[Q_T]$ and Eq. (3.13) reduces to

$$k_{expt} = k_0 - (k_-[Q_T]/[M]) \qquad (3.15)$$

For those systems where the fraction of the quenchers in the aqueous phase cannot be neglected, k_+ and k_- can be evaluated from a study of the kinetics of triplet decay as a function of quencher and surfactant concentrations:

$$[Q_T] = [Q_m] + [Q_w] \qquad (3.16)$$

$$K = ([Q_T] - [Q_w])/([Q_w][M]) \qquad (3.17)$$

and hence

$$\left(\frac{1}{k_{expt} - k_0} \right) = k_0 + \frac{k_+[Q_T]}{1 + K[M]} \qquad (3.18)$$

or

$$k_{expt} = \frac{1}{[Q_T]} \left(\frac{1}{k_+} + \frac{[M]}{k_-} \right) \qquad (3.19)$$

If we call k_m the slopes of the plots of k_{expt} versus $[Q_T]$ [according to Eq. (3.19)], then the values of k_+ and k_- can be obtained from plots of $(1/k_m)$ versus $[M]$:

$$1/k_m = (1/k_+) + ([M]/k_-) \qquad (3.20)$$

In the studies of phosphorescence quenching of 1-bromonaphthalene by pyrene in SDS micelles, Hauser et al.[15] have used slightly different expressions in their analysis. Energy transfer quenching of acetone triplets by biphenyl in the presence of SDS micelles[21] represents yet another variation of the quenching scenarios described earlier: probe triplets are mainly in the micellar phase.

Rothenberger et al.[22] also investigated intramicellar triplet energy transfers (from triplet N-methylphenothiazine to trans-stilbene or naphthalene in CTAB micelles) treating the distribution of donors and acceptors among the micelles to be a static one (no probe exit or re-entry occurs on the time scales of energy transfer quenching). Under such conditions, the donor triplet decay exhibits a fast component corresponding to those triplets quenched and a slow component corresponding to unquenched fraction of the triplets. The quenched fraction (fast component) is adequately represented by a poisson distribution of acceptors among the micelles.

Data presented in Tables 3.2 and 3.3 indicate that the exit rates for nonpolar aromatic hydrocarbons typically fall in the range of $k_- \simeq 10^3 - 10^6$ sec^{-1}. The implications are that these micelle-solubilized arenes are expected to remain with their hosts during the time scales of normal fluorescence ($\leq 10^{-6}$ sec) but should exit at rates comparable to or faster than their phosphorescence decays. Thus, there is some justification for the fluorescence quenching experiments that treat micellar aggregates with the solubilized arenes as intact cages. This also necessitates the quenchers that are located in the aqueous phase to approach the solubilized probes such as pyrene rather than the probes leaving the micelle, a model proposed in the early experiments of fluorescence quenching and which has also been subject to some criticisms.

3.4 Photoionization Processes

Photoionization processes in which the excited state donates an electron to the solvent are analogous to the photoredox reactions. In nonpolar solvents there are no solvation forces to stabilize the electron:

$$P^* \longrightarrow (P^+e^-) \longrightarrow P^+ + e^-_{solv} \qquad (3.21)$$

Recombinative in the geminate pair predominates, and ion yields are quite low. In polar solvents the photoionization yield increases, and the presence of ionic micelles has been found to greatly facilitate this process. Thus, photolysis of aromatic molecules such as pyrene, perylene, and phenothiazine in the micellar media gives high yields of aromatic solute cations and hydrated electrons:[23-38]

$$P \xrightarrow[h\nu]{SDS} P^+ + e^-_{aq} \qquad (3.22)$$

Figure 3.3 presents the spectra of short-lived intermediates formed during the laser photolysis of pyrene in cyclohexane and SDS micelles. It confirms the formation of pyrene cations in high yields in anionic micelles. In anionic SDS micelles, for tetramethylbenzidine, for example, the ratio of solute cations to triplets is 6.0 as compared to 0.17 in methanol.[27] The corresponding values for phenothiazine are 1.7 and 0.16, respectively.[25]

The formation of high yields of ionic products in micellar media is a consequence of the increasing probability of electron escape to the bulk water and effective stabilization of the solute cation in the micelle. In accord with this are the observations that increasing the micellar size causes a decrease in the

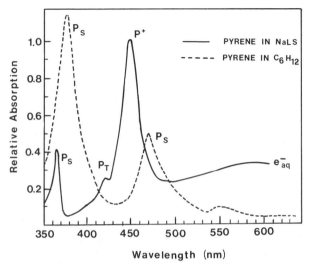

FIG. 3.3. Spectra of short-lived intermediates (triplet P_T, singlet P_S, cation P_+, and hydrated electron e_{aq}^-) formed during the laser photolysis of pyrene in cyclohexane (– – –) and in aqueous SDS micellar solutions. (——). (From Thomas and Piciulo.[30] Copyright 1979 American Chemical Society.)

yields of photoionization. The electrons need to travel longer distances before they are solvated. Relative yields of e_{aq}^- during triphenylene photoionization are 1.0, 0.65, and 0.20 in sodium octyl sulfate, sodium dodecyl sulfate, and sodium cetyl sulfate, respectively.[23]

Aromatic hydrocarbons such as pyrene, perylene, and tetracene photo-ionize by a biphotonic process.[23] For more polar aromatic compounds, e.g., aromatic amines with low ionization potentials such as tetramethylbenzidine (TMB), phenothiazine (PTH), and tetramethyl-p-phenylenediamine (TMPD), the photoionization is mostly monophotonic.[25–31] Mechanistically mono-photonic processes have been pictured to involve electron tunnelling to the unoccupied energy levels of the e^-/e_{aq}^- system. In many instances the photoejected electron in the micelle (dry electron) has excess energy of over 1 V. It can react with cosolubilized solutes prior to solvation or escape the micellar surface to reach the aqueous phase.

Detailed studies have shown that the threshold for photoionization is greatly reduced in micellar systems compared to the gas phase or alkane liquids.[28–31,34,38] Table 3.4 presents data of Thomas and Piciulo[31] on the ionization potential for various solutes in the gas-phase I_g, in tetramethylsilane and in anionic (SDS) micelles. For neutral aromatics solubilized in the inner core, the reduction in micellar media ($\Delta I = I_g - I_s$) is nominal compared to TMS as a solvent. Based on a study of perylene and tetracene in SDS and Triton X-100 micelles, Bernas et al.[34,38] have also reached similar conclusions.

TABLE 3.4. Micellar Effects on the Photoionization Thresholds of Solutes [a]

	TMS			SDS Micelles			
						Photoionization	
Solute	I_g (eV)	I_s (eV)	ΔI (eV)	I_{mi} (eV)	ΔI (eV)	λ_{ex} (nm)	Mechanism
Pyrene	7.55	5.46	2.09	5.4	2.15	347	2-Photon
Perylene	6.92	4.82	2.10	4.55	2.37	347	2-Photon
Tetracene	6.88	4.92	1.96	4.25	2.63	347	2-Photon
Aminoperylene	7.02	4.77	2.25	3.0	4.0	530	1-Photon
PTH	6.96	4.80	2.16	3.0	3.96	347	1-Photon
TMB	6.40	4.52	1.88	3.1	3.30	347	1-Photon

[a] Data from Thomas and Piciulo.[31]

For polar solutes such as TMB or PTH, I_s is lower than I_g by as much as 3.5–4.0 eV. With 3-aminoperylene the lowering in the ionization threshold is so substantial that it photoionizes in SDS micelles even with green light (530 nm) monophotonically, the only known case of this kind.[28,30]

There has been some controversy[28–30] regarding the differences in the photoionization mechanisms of structurally similar molecules: perylene/pyrene (biphotonic) versus aminoperylene (monophotonic). The distinction arises due to the different solubilization sites in the micelles. Polar molecules such as aminoperylene tend to reside closer to the surface as compared to the deeper penetration by hydrophobic solutes.

Generally for the ionization potential in solution, I_s will be reduced from that in the gas phase (I_g) by the polarization energy of the cation (P_+) and the energy of the electron in the solution (V_0):

$$I_s = I_g + P_+ + V_0 \tag{3.23}$$

Electrochemical estimates for V_0 in water (of about -1.6 eV) suggest P_+ to be in the range of 2.5 eV to account for the lowering of the I_g by the micellar system. For the solute cation that is in the micellar interior if the latter is hydrocarbon-like, P_+ can be only about 1.5 eV (alkanes with $\varepsilon \sim 2$). Presumably both P_+ and V_0 are quite different in micellar systems. If we utilises the Born equation to calculate P_+:

$$P_+ = (e^2/2r)[1 - (1/\varepsilon)] \tag{3.24}$$

then the effective dielectric constant for the micellar interior when the cation is located should be much greater than 2, the ε value for hydrocarbon liquids. It is thus of much interest to know the exact solubilization site of the cation in the micelle. Kevan and co-workers are currently studying this question in detail using ESR spin echo techniques.[39–43]

3.5 **Photoredox Reactions and Charge Separation Phenomena**

Photoredox reactions have potential applications in the field of photo-chemical conversion of solar energy. They also play a central role in a wide variety of chemical and biological processes. In photosynthesis, for example, the initial charge separation step and the unidirectional electron transfer that follows continues to intrigue scientists. In homogeneous solvents several systems are known to exhibit efficient light-induced electron transfer, but the endergonic electron transfers are always accompanied by reverse electron transfers in the dark:

$$D + A \stackrel{h\nu}{\rightleftharpoons} [D^+ \cdots A^-] \rightleftharpoons D^+ + A^- \qquad (3.25)$$

Theories of thermal and photochemical redox reactions have matured in recent years to high levels that today we can appreciate much and understand better the various factors which determine the reverse electron transfer rates and the escape yields of redox products from the primary cage.[44-47] However, we are still far away in mimicking this vectorial electron transfer in homogeneous systems (*in vitro*).

With a distinct possibility that lipid membrane, protein architecture, and charged interfaces play a crucial role in achieving net charge separation in photoredox reactions, there has been a growing interest in the studies of model systems composed of such ingredients: micelles, vesicles, liposomes, polyelectrolytes, etc. In this section we consider photoredox processes in micellar media. Micellar systems provide several advantages over homo-geneous solvents. Most of the water-insoluble sensitizers can be used. By keeping the solute-to-micelle ratio fairly low, it is possible to segregate the dye molecules and use fairly a high concentration without encountering efficiency losses due to self-quenching reactions.

Ionic micellar aggregates are characterized by an inner hydrophobic core and a charged interface both of which can play important roles in influencing reaction course and yields. The presence of the micellar interface leads to an inhibition of the quenching reaction of excited-state probes by ionic quenchers if the quencher and the micelles have the same sign of charges. Acceleration of the quenching event occurs in the opposite case by collection of a large number of quencher ions on the surface. Room-temperature phosphorescence and very efficient energy transfers in micelles using heavy atom ions as counterions are manifestations of this sequestering effect. Here we will be concerned with similar effects on the rates and efficiencies of light-induced electron transfer reactions (intramicellar electron transfer in func-tional micelles).

We can anticipate other possible variations of photoredox systems in micellar media. By suitable choice of the donar-acceptor pair (solubility, charges, etc.) it is possible to study electron transfer processes under conditions where the two components are spatially distributed with a charged electrical interface separating the two. The interface can then inhibit or promote reverse electron transfers. Also if we start with electrically neutral molecules (D, A) and produce ions (D^+, A^-) by light absorption, the association of A^- with anionic micelles is destabilized by electrostatic repulsions, and that of D^+ is enhanced. Thus, D^+ and A^- can be spatially separated (a process labelled as a solute-ejection mechanism), and their diffusional re-encounter is inhibited by the micellar double layer. The efficiency of the charge separation process then will depend on the relative rates of A^- ejection from the micelle and the intramicellar back electron transfer from A^- to D^+. The latter is thermodynamically favorable and can, in principle, occur very rapidly. The rate of ejection of A^- from the anionic micelle will depend critically on the degree of hydrophobic interaction of A^- with the micellar aggregates.

Numerous photoredox reactions have been studied in micellar media and the feasibility of several of the above pathways has been unambigously demonstrated.[48-92] Table 3.5 provides a listing of the various photoredox systems that have been examined. Photoreduction of bipyridinium salts (viologens such as methylviologen MV^{2+}) and their surfactant analogs in

$$R—\overset{+}{N}\bigcirc—\bigcirc\overset{+}{N}—R \qquad \text{Viologens (R = Me, } MV^{2+})$$

micellar assemblies is an area which has received particular attention in several laboratories:[48-64]

$$S + RV^{2+} \underset{\text{micelles}}{\overset{h\nu}{\rightleftharpoons}} S^+ + RV^+ \tag{3.26}$$

The reason is that the reduced viologens readily liberate molecular hydrogen (H_2) from water in the presence of suitable redox catalysts such as colloidal Pt:[65-67]

$$2MV^+ + 2H_2O \xrightarrow[\text{catalyst}]{\text{Pt}} 2MV^{2+} + H_2 + 2OH^- \tag{3.27}$$

Sensitized photoreductions involving both oxidative and reductive quenching of excited states have been examined:

$$S^* + Q \rightleftharpoons S^+ + Q^- \qquad \text{(oxidative)} \tag{3.28}$$

$$S^* + Q \rightleftharpoons S^- + Q^+ \qquad \text{(reductive)} \tag{3.29}$$

The majority of these studies have been with some form of functionalized micellar system, viz., those in which an electron donor or acceptor has been made part of the surfactant molecules constituting the micelle. Typical examples are micelles with acceptor counterions such as Cu^{2+} or Eu^{3+} in place of Na^+ ions in sodium laurylsulfate (SDS) or labelled with surfactant

TABLE 3.5. Studies of Photoredox Reactions in Micellar Media

Sensitizer	Quencher	Micelle	Reference
Oxidative quenching/anionic micelles			
RuL_3^{2+}, L = bpy, bpy-C_n, phen, $(\phi)_2$phen,	Viologens (MV^{2+}, PVS, C_nMV^{2+})	SDS, C_{12}—$(EO)_{12}$—OSO_3^-	48,53,54,57, 58,61,62,63
Pyrene	Cu^{2+}	SDS	—
N-MePTH	Eu^{2+}	$M(DS)_n$, M=Zn, Cu	65,66
N-MePTH, TMB	NO_2^-	SDS	76
Co(acac)$_3$	BNAH, DNAH	SDS	86
C_n-PTH, PTH-SO_3^-	Cu^{2+}	$M(DS)_n$, M=Cu, Na..	68
Chlorophyll	Lumiflavin^{2+}	SDS	79
RuL_3^{2+}	Tryptophan	SDS	—
Chlorophyll	Duroquinone	SDS	73
Oxidative quenching/cationic micelles			
RuL_3^{2+}, L = bpy, bpy-C_n	Viologens (C_n-V^{2+})	CTAC	52,54,56,59,60,
RuL_3^{2+}, L = bpy	C_n-MV^{2+}	C_n-MV^{2+}	64
Zn porphyrins (P, TPP)	C_n-MV^{2r}	CTAC	51
Zn porphyrins (P, TPP)	C_n-DQ	CTAC	69,87
N-MePTH	C_n-Cu-crown^{2+}	C_n—Cu—crown^{2+}	67
Oxidative quenching/nonionic micelles			
RuL_3^{2+}, L = bpy	N-MePTH	C_{12}-OEB	70
Chlorophyll a	MV^{2+}	Triton X-100	49
Chlorophyll a	Methyl orange	Triton X-100	72
Zn TPP	Anthraquinone-SO_3^-	Triton X-100	74
Reductive quenching			
Quinones (AQS, DQ)	OH^-	SDS, CTAX, X=OH, Br$^-$	81,83
Quinones (DQ)	N-Ethylcarbazole	SDS, CTAB	84
Thionine, methylene blue	EDTA	SDS	81,83
Methylene blue	Ascorbate	CTAB	83

derivatives of the sensitizers, e.g., long-chain compounds of $Ru(bpy)_3^{2+}$, phenothiazine, or of the acceptors, e.g., long-chain derivatives of viologen salts.

The intrinsic heterogeneous character of the micellar aggregates requires, as in the fluorescence quenching studies, development of novel kinetic concepts in the treatment of photoredox processes. To begin with, the fact has to be taken into account that compartmentalization of reactants occurs as a consequence of their association with the micellar aggregates. This has two main consequences. Unlike in the homogeneous solvent systems, the distribution of solutes occurs statistically over the micellar hosts. Second, the reactive interactions take place between isolated groups of reactant partners contained in the host micelle. Obviously the rate laws of homogeneous solvent systems are no longer valid. As in the case of fluorescence quenching studies (cf.

Chapter 2), the commonly accepted distribution law is poissonian. The probability P_i of finding i solutes in a given micelle is given by

$$P_i = \frac{\bar{n}^i e^{-\bar{n}}}{i!} \qquad (3.30)$$

where $\bar{n}$ is the mean (average) occupancy number of the solutes per micelle.

Redox quenching of an excited state by an electron donor or acceptor in micellar media can occur under a variety of conditions. Both forward and reverse electron transfers can be purely intramicellar in the same native micelle, as between two hydrophobic micelle-associated species. With ionic solutes either one or both the steps occur at the micellar interface with the possibility of some exit and re-entry of the species involved between the micellar and aqueous phases. Irrespective of the mechanism of quenching of the excited state (electron or energy transfer) the analysis of the decay kinetics of the excited state in the presence of quenchers is identical. We have discussed earlier in Section 3.3 kinetic treatments for several quenching scenarios as applied to the triplet quenching. It has been pointed out that with polypyridyl complexes of Ru^{2+} as sensitizers, the lifetime of the charge transfer excited state is very short (often less than a microsecond). Hence, with these metal complexes the exit of the excited probe from the micellar host during the redox quenching event can be safely ignored and the system treated as in the fluorescence quenching of singlet excited states. For long-lived triplet states in some cases there can be the exit and re-entry of the probe among the empty micelles and also onto those which are occupied by the quenchers during the time scales of quenching. Most often this is avoided by careful choice of the sensitizer, viz., choose those whose exit rates are much lower as compared to the triplet lifetime. Here we will consider only the kinetic aspects of reverse electron transfers. Using the technique of laser flash photolysis it is now possible to monitor the kinetics of these electron transfer reactions down to a few hundred picoseconds.

Kinetic Aspects of Reverse Electron Transfer Reactions

As in the singlet excited-state quenching studies we can consider several possible (limiting) scenarios for the reverse electron transfer that follows light-induced electron transfer.

Case A: The simplest case to consider is a pure intramicellar process between donor–acceptor pairs both quantitatively associated with the micelle in the oxidized and reduced forms. (The nature of the association can be purely hydrophobic, electrostatic, or both. The site of the redox event can be in the interior or at the interface. For kinetic analysis these finer details are irrelevant.) The forward quenching process corresponds to the immobile quencher case (case I) described earlier (Section 2.5):

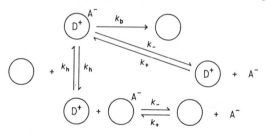

$$\boxed{D \quad A} \;\overset{h\nu}{\underset{}{\rightleftharpoons}}\; \boxed{D^+ \; A^-}$$

The reverse electron transfer is of much kinetic interest. The role of dimensionality in describing kinetics in micelles has been examined recently with several stochastic models. In homogeneous solvents, the reverse electron transfers are usually bimolecular and second order. In micellar systems we can consider them as an intramolecular (to be more precise intramicellar supercage) reaction between the isolated partners. The reverse electron transfer then is expected to follow first-order kinetics.[70]

Case B: In functional micelles with reactive counterions such as Eu^{3+} in anionic micelles, upon reduction the product cation Eu^{2+} can exit the micelle. Consider for example the photoreduction of Cu^{2+} and Eu^{3+} ions by *N*-methylphenothiazine (MPTH) in micelles of $Eu(DS)_3$ and $Cu(DS)_2$ studied by Grätzel and co-workers.[65-67]

$$MPTH + Cu^{2+} \;\overset{h\nu}{\rightleftharpoons}\; MPTH^+ + Cu^+ \tag{3.31}$$

$$MPTH + Eu^{3+} \;\overset{h\nu}{\rightleftharpoons}\; MPTH^+ + Eu^{2+} \tag{3.32}$$

Due to the locally high concentration of the quencher counterions, the triplet quenching reaction is extremely rapid occurring within a short laser pulse (less than 20 nsec). The scheme below outlines possible pathways for the back reactions to occur in addition to the intramicellar back electron transfer in the micellar 'supercage' (a first-order process with rate constant k_b):

The reduced acceptor (A^-) can detach from the micelle (a first-order process with rate constant k_-) to go to the bulk aqueous phase and/or directly transfer (hopping) to a neighbouring micelle (a second-order process with rate constant k_h). Experimental monitoring of the kinetics of the products indicates the presence of two distinct decay domains.

The initial rapid decay is due to the efficient intramicellar back transfer, and this competes with the escape of A^- ions from the native micelle. By a combination of exit and hopping processes, the product A^- eventually will find a micelle that contains D^+ and undergoes intramicellar back transfer. This slower process occurs according to the second-order rate law. Explicit rate expressions are available for the analysis of the decay curves of the photoredox processes. Functionalized micellar versions of this case using a long-chain phenothiazinyl dodecane sulfate (PTHS) and a long-chain polyamine complex of Cu^{2+} (tetraazacyclododecane) have also been examined.[68]

Since divalent cations bind to the anionic micellar surfaces more strongly than monovalent cations, the escape process of the reduced counterion A^- can also be visualized as an exchange process between the micellar counterions (reduced) and the ions present in the bulk aqueous phase. This escape or exchange process can be taken to advantage by scavenging the product cations (Eu^{2+}, Cu^+, etc.) with suitable anions (e.g., $Fe(CN)_6^{3-}$) present in the aqueous bulk:

$$Cu^+ + Fe(CN)_6^{3-} \longrightarrow Cu^{2+} + Fe(CN)_6^{4-} \tag{3.33}$$

Since ferrocyanides are repelled by the anionic micelle, the photoredox system with the bulk anion scavenger provides a means of achieving irreversible light-induced electron transfer.

A close variation of this situation of solute ejection from the micelle following the photoredox event has already been encountered in our discussion of exciplex studies (Section 2.7). Following the electron transfer quenching of pyrene by dimethylaniline in a cationic micelle, the amine cation is expelled from the micelle, and pyrene anion is stabilized:

$$^1Py^* + DMA \xrightarrow[\text{micelles}]{\text{CTAC}} Py_{\text{mic}}^- + DMA^+ \tag{3.34}$$

Using similar principles, a threefold increase in the quantum yields has been achieved during the redox quenching of the $[Os(phen)_2(CO)Cl]^+$ complex excited state by $HgCl_2$ with the introduction of SDS micelles:[88]

($\Phi = 0.20$ in SDS micelles as compared to 0.07 in water):

$$[L_2Os^{III}(CO)(Cl)]^+ + HgCl_2 \overset{h\nu}{\rightleftharpoons} [L_2Os^{III}(CO)(Cl)]^{2+} + HgCl_2^- \tag{3.35}$$

Increased efficiency of charge separation by ejection of one of the photoproducts from the micelle has also been observed[31] in aminoperylene–benzoquinone, tetramethylbenzidine–duroquinone, and chlorophyll–duroquinone systems in anionic micelles.

Case C An inverse situation to the solute ejection from the micelle following the photoredox event has been encountered in the photoreduction of long-chain (surfactant) viologen salts C_n—V^{2+} by cationic sensitizers (such as $Ru(bpy)_3^{2+}$ and zinc porphyrin $ZnTMPyP^{4+}$) in cationic micelles.[48-64] The scheme below outlines the kinetic course of events in the photoredox quenching process:

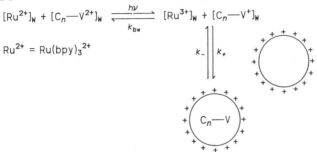

$$[Ru^{2+}]_W + [C_n—V^{2+}]_W \; \underset{k_{bw}}{\overset{h\nu}{\rightleftharpoons}} \; [Ru^{3+}]_W + [C_n—V^{+}]_W$$

$$Ru^{2+} = Ru(bpy)_3^{2+}$$

The initial redox quenching of $Ru(bpy)_3^{2+}$*(indicated in the scheme simply as Ru^{2+}) by the alkyl viologen (C_n—V^{2+}) occurs exclusively in the aqueous phase. Both reagents are very soluble in water and electrostatic repulsions keep the reagents away from the cationic micelles. Alkyl viologens such as C_{14}—V^{2+} show drastic changes in their hydrophobic/hydrophilic properties between their oxidized and reduced forms; C_{14}—V^{2+} is fairly soluble in water (even forms micelles of its own above a certain critical concentration) but the reduced form C_{14}—V^{+} is very insoluble in water and can be readily solubilised in cationic micelles.

In the preceding photoredox process, following the forward electron transfer in the aqueous phase, the product C_n—V^{+} rapidly enters the micelle in competition with the back electron transfer with RuL_3^{3+} in the aqueous phase. The coulombic repulsions between the RuL_3^{3+} and the cationic micellar interface are sufficient to prevent interactions between the oxidized sensitizer and the reduced viologen. Substantial reductions (over 500 fold) in the back electron transfer have been obtained in this manner. Figure 3.4 presents some absorptions decay curves for the reduced form of viologen with and without the CTAC micelles illustrating this charge separation process. Table 3.6 presents a collection of reverse electron transfer rate constants for various viologen–sensitizer systems. (Table 2.16 presents similar data for pyrene–dimethylaniline system). The reverse electron transfer is limited by the exit rate of the reduced viologen from the cationic micelle. The exit rates of surfactant molecules decrease with increasing chain length. Thus k_- values are 1.3×10^9 sec^{-1} for hexyl sulfate as compared to 6×10^4 sec^{-1} for hexadecyl sulfate. In contrast, the micellar entry rates show only minor influence on chain length ($k_+ \simeq 10^9 \; M^{-1} \; sec^{-1}$).

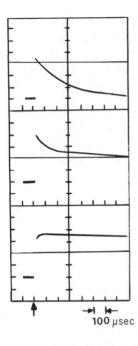

FIG. 3.4. Effect of cetyltrimethylammonium chloride (CTAC) concentration on the decay of reduced functional viologen radical cation monitored at 602 nm. $[C_{14}MV^{2+}] = 5 \times 10^{-4}$ M and $\lambda_{\text{excit}} = 530$ nm. (From Brugger et al.[60] Copyright 1981 American Chemical Society.)

TABLE 3.6. Micellar Effects on the Reverse Electron Transfer in the Photoredox Quenching with Viologen Derivatives

Sensitizer	Viologen	Micelle	k_b(Homogeneous solvent) $(M^{-1} \text{sec}^{-1})$	k_b(Micelle) $(M^{-1} \text{sec}^{-1})$	Reference
$Ru(bpy)_3^{2+}$	C_{14}—MV^{2+}	CTAC	4×10^9	$(0.8-4.0) \times 10^7$	59
$Ru(bpy)_3^{2+}$	C_{14}—MV^{2+}	SDS	4×10^9	10^{10}	59
$Ru(bpy)_3^{2+}$	C_{14}—MV^{2+}	Triton	4×10^9	8×10^8	60
$Ru(bpy)_3^{2+}$	C_{14}—MV^{2+}	C_{14}—MV^{2+}	4×10^9	10^7	60
$ZnTMPyP^{4+}$	C_{14}—MV^{2+}	CTAC	5×10^9	10^7	60
$Ru(bpy)_3^{2+}$	PVS	CTAC	—		61
$Ru(bpy)_3^{2+}$	PVS	SDS	6.5×10^9	9×10^8	62
$Ru(bpy)_3^{2+}$	PVS	TTOES	6.5×10^9	1.7×10^8	62
$Ru(bpy)_3^{2+}-2C_{19}$	PVS	SDS	—	8×10^8	62
$Ru(bpy)_3^{2+}-2C_{19}$	PVS	TTOES	8×10^9	4×10^7	62

Infelta and Brugger[59,60] have presented a detailed kinetic analysis with appropriate rate expressions. The micellar trapping process slows down considerably the reverse electron transfer. It is shown that the decay of the redox products follows a second-order rate law of the form

$$k_{\text{obs}} = k_b/[1 + (k_+[M]/k_-)] = k_b/(1 + K[M]) \qquad (3.36)$$

where k_b is the second-order rate constant for the electron transfer in homogeneous solutions, K the equilibrium constant for the reduced viologen species ($K = k_+/k_-$); and [M] the concentration of the micelles. Estimated K values are 1.0×10^3 M^{-1} and 2.7×10^5 M^{-1} for C_{14},Me—V^{2+} and C_{14}, Me—V^+, respectively.

Analogous experiments demonstrating the beneficial effects of micelles is slowing down the reverse electron transfer using a zwitterionic viologen (propylviologensulfonate, PVS) have also been described:[61,62]

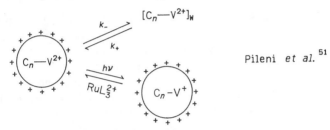

Here both the cationic and anionic micelles are equally effective in distinct contrast to the results obtained for C_{14},Me—V^{2+} in anionic micelles. Other variations of these systems have been reported by Kurihara et al.[64] and Pileni et al.[51] using alkyl viologens in cationic micelles. The schemes below outline the reaction sequences as described by the authors:

Both invoke photoredox reaction between micelle-bound partners with possible exchange of reactants prior to (Pileni et al.) or after (Kurihara et al.) the forward electron transfer. The alkyl viologens used by Kurihara et al. are

analogous to the C_{14},Me—V^{2+}. Hence the experiments can also be interpreted in terms of earlier schemes, viz., photoredox reaction in the aqueous phase followed by the entry of the reduced viologen into the micelle. Features of photoredox quenching of micelle-bound sensitizers [$Ru(bpy)_3^{2+}$ and its surfactant analogs, for example] by alkyl viologens in anionic micelles have been discussed earlier under fluorescence quenching studies (Section 2.5).

The concept of trapping and stabilizing one of the products as a means of achieving net charge separation has also been exploited by Schmehl and Whitten[55] in studies of viologen quenching of palladium porphyrins (methyl and benzyl viologen quenching of $PdTPPS^{4-}$). In aqueous media extensive ion pairing occurs in the ground state and the cage escape yield of the redox products is negligible:

$$[\text{PdTPPS}^{4-} \cdots \text{MV}^{2+}] \xrightleftharpoons{h\nu} [\text{PdTPPS}^{3-} \cdots \text{MV}^{+}] \xrightarrow{\text{SDS}} \text{PdTPPS}^{3-} + \text{MV}^{+} \quad (3.37)$$

Addition of electrolytes slightly improves the yields, but drastic improvements in the lifetimes of redox products are seen on addition of either CTAC or SDS micelles (half-lives upto 100 msec). The same principles have also been applied to the irreversible generation of Ag^0. Photoredox quenching of cyanines and Ru(bpy)_3^{2+} by Ag^+ occurs reversibly in homogeneous solvents (water or acetonitrile) but net irreversible production of Ag^0 is observed upon introduction of simple micelles or crown ether surfactants:[89,90]

$$\text{S}^* + \text{Ag}^+ \xrightarrow{h\nu} \text{S}^+ + \text{Ag}^0 \quad (3.38)$$

3.6 Triplet Energy Transfer: Dynamics of Singlet Oxygen

The chemical reactivity and physical properties of singlet molecular oxygen, O_2^* $(^1\Delta_g)$ have been the subject of detailed investigations in homogeneous and microheterogeneous media, due to their biological importance in photodynamic action and related phenomena. In this section, we will be concerned with two aspects of the singlet oxygen chemistry, viz., its lifetime in solution[93-104] and its role in oxygen-mediated sensitized photo-oxidation processes.[105-115] The interest in studying them in micelles has been that they provide a simple, aqueous, microheterogeneous environment to mimic more complex biological systems, and second, they allow solubilization of a wide variety of sensitizers and singlet oxygen traps that are used in the study of reactions of singlet oxygen.

Lifetime of Singlet Oxygen in Solution

Extensive work has gone into the determination of the natural lifetime in condensed media. The forbidden nature of the $(^3\Sigma_g \rightarrow {}^1\Delta_g)$ transition confers a long lifetime to the $^1\Delta_g$ state in the gas phase (about 2700 secs) but is considerably shortened in condensed media (10^{-3}–10^{-6}-sec range, depending on the nature of the medium). This is due to the effective interactions of the excited oxygen with solvent vibrational modes. Several studies have established that the lifetimes of $^1\text{O}_2^*$ in H_2O and D_2O differ by an order of magnitude, although estimates on the absolute values vary widely.

Until very recently singlet oxygen lifetimes have been determined via excitation energy transfer methods where an appropriate triplet state of a sensitizer transfer its energy to oxygen $(^3\Sigma_g \text{O}_2)$ generating $^1\text{O}_2^*$. The lifetime of $^1\text{O}_2^*$ is determined via its competitive scavenging with trap molecules:

$$\text{S} \xrightarrow{h\nu} {}^1\text{S}^* \longrightarrow {}^3\text{S}^* \longrightarrow \text{S} \quad (3.39)$$

$$^3\text{S}^* + \text{O}_2(^3\Sigma_g) \longrightarrow \text{S} + \text{O}_2^*(^1\Delta_g) \quad (3.40)$$

$$O_2^*(^1\Delta_g) \longrightarrow O_2(^3\Sigma_g) \tag{3.41}$$

$$O_2^*(^1\Delta_g) + Tr \longrightarrow [O_2 \cdots Tr] \tag{3.42}$$

The number of sensitizers and trap molecules found to be useful in this context are numerous. (S = methylene blue, 2-acetonaphthalene, chlorophyll, pyrene, eosin, etc.; Tr = sodium azide, DABCO, DABF, histidine, tryptophan, etc.). Table 3.7 summarizes lifetime data of $^1O_2^*$ determined by Linding and Rodgers[96,101] using these procedures.

The following rate law has been derived for the disappearance of the trap molecule such as DPBF in various ionic and nonionic micellar media:

$$-(d[Tr]/dt) = k_r[Tr][O_2^*] \exp(-k't) \tag{3.43}$$

where [Tr] stands for the micellized trap molecule and k' refers to the composite term $(k_d + k_r[M] + k_q[Q])$ with Q referring to unspecified quenchers which may or may not be in the micellar phase. With the availability of very sensitive infrared response photodiodes and the discovery of luminescence of $^1O_2^*$ occurring at 1.27 μm, it is now possible to directly monitor the excited state decay of $^1O_2^*$:

$$^1O_2^*(^1\Delta_g) \longrightarrow O_2(^3\Sigma_g) + h\nu(1.27 \ \mu m) \tag{3.44}$$

Table 3.7 includes lifetime data measured in this manner. The data are in excellent agreement with those determined earlier using triplet sensitizers (except for Brij 35 for which the recent values are considered more reliable). Lifetime studies in H_2O/D_2O mixtures have yielded extrapolated lifetimes of

TABLE 3.7. Singlet Oxygen Lifetimes and Reactivity in Micellar D_2O Solutions [a]

Micelle[b]	Sensitizer[c]	Emission[d] τ (μsec)	Sensitization τ (μsec)	$k_r(1 \ M^{-1} \ sec^{-1})$ (DPBF)
SDS	ACN	57	53.7	10.9 $\times 10^8$
SDS	MB	—	54.0	10.0 $\times 10^8$
Sodium laurate	ACN	—	48.3	9.84 $\times 10^8$
CTAB	ACN	56	57.4	6.22 $\times 10^8$
CTAB	MB	—	53.7	6.74 $\times 10^8$
Igepal CO-630	ACN	—	24.0	6.28 $\times 10^8$
Igepal CO-660	ACN	—	21.5	6.48 $\times 10^8$
Igepal CO-660	MB	—	23.5	6.63 $\times 10^8$
Brij 35	ACN	52	26.4	6.61 $\times 10^8$
Triton X-100		32		
DDAB (vesicles)		35		

[a] Data of Rodgers.[101]
[b] All surfactant concentrations are 0.1 M, except for sodium laurate = 0.05 M.
[c] ACN = 2-acetonaphthone, MB = methylene blue.
[d] Emission lifetimes measured at 1.27 μm in D_2O-based systems.

$^1O_2^*$ in H_2O as 4.0 μsec and 3.5 μsec in CTAB and Igepal CO-630 micelles, respectively. Direct measurements in neat H_2O also yield $\tau = 4$ μsec. It should be pointed out that earlier reported singlet oxygen lifetimes in micellar media are much shorter and most of these studies apparently are in error.

In addition to the trap molecules we can also add other quenchers of singlet oxygen to the solution, and by the same competitive scavenging methods it is possible to determine the rate constants for the reaction of singlet oxygen with these molecules as well. Table 3.8 presents some rate constant data measured in this manner.

Exit Rates and Distribution Coefficients for Molecular Oxygen

In singlet oxygen studies conditions are often so chosen that $^1O_2^*$ is produced exclusively in the aqueous phase. The $^1O_2^*$ species can decay in the aqueous phase or enter the micelle where it can also decay naturally or be trapped by solubilized monitor (trap) molecules such as diphenylisobenzofuran (DPBF) as shown in the scheme below:

$$O_2(^3\Sigma_g) \xrightarrow[h\nu]{sens.} O_2^*(^1\Delta_g)_w \rightleftharpoons O_2^*(^1\Delta_g)_m$$
$$\downarrow k_w \qquad\qquad\qquad \downarrow k_m$$
$$O_2(^3\Sigma_g)_w \qquad\qquad O_2(^3\Sigma_g)$$

The kinetic processes associated with the above scheme have been examined via direct monitoring of the IR luminescence[100,102] and also by steady-state analysis of data obtained via cw laser irradiation.[97] The distribution coefficients for O_2 K in SDS and CTAB micelles has been estimated as 2.8 and 4.0 M^{-1}, respectively, (k_+, and k_- values are 1×10^8 and 3.7×10^7 sec^{-1},

TABLE 3.8. Quenching Rate Constants $k(M^{-1}\text{ sec}^{-1})$ for the Reaction of Singlet Oxygen ($^1O_2$) in Aqueous and Micellar Media [a]

Quencher[b]	Neat D_2O	SDS (0.1 M)	CTAB (0.1 M)
Sodium azide	5.1×10^8	3.9×10^8	8.2×10^8
Histidine	6.1×10^7	5.9×10^7	5.9×10^7
Tryptophan	6.1×10^7	4.0×10^7	4.2×10^7
Methionine	1.5×10^7	1.8×10^7	1.6×10^7
DABCO		1.4×10^6	1.4×10^6
DPBF		1.0×10^9	6.7×10^8
PMC		4.0×10^8	4.1×10^8
2,3-Dimethyl indole		7.0×10^8	4.4×10^8

[a] From Lindig and Rodgers.[98]
[b] PMC = pentamethylchromol, DPBF = 1,3-diphenylisobenzofuran, and DABCO = 1,4-diazabicyclo[2.2.2]octane.

respectively). A value of $k_w = 2.5 \times 10^5$ sec^{-1} for the lifetime of $^1O_2^*$ in water coupled to the above K estimates yields 20 μsec as the lifetime for singlet oxygen in the interior of the SDS micelles. This is in close agreement with $\tau = 25$ μsec determined in liquid dodecane. The low values of the equilibrium constant K in micelles implies that oxygen partitions between the aqueous and micellar hydrocarbon phase to about the same extent. (It may be recalled that K' values ($= k_+$ [Micelles]/k_-) of 280 and 260 M^{-1} sec^{-1} have been determined in SDS and CTAB micelles via fluorescence quenching methods.[107]

A high value for the partition coefficient K would be consistent with the generally greater solubility of oxygen in organic solvents compared with that in water. If the oxygen concentration in the micellar phase is indeed higher than that in the aqueous phase, then at first glance the observation of increased fluorescence lifetimes for various aromatic molecules upon solubilization even in aerated micellar solutions may appear contradictory. The reason for the increased lifetimes is that under normal aerated conditions (1 atm), the average number of oxygen molecules per micelle is much less than unity. There will be several micelles containing oxygen without probe and, hence, the oxygen concentration in the probe microenvironment is relatively low. Thus, in aggregated systems such as micelles, simple interpretations based on bulk concentrations of quenchers and relative solubility of solutes in neat solvents can be misleading if the statistical distribution effects are not properly taken into account.

Oxygen-Mediated Dye-Sensitized Photooxidations

In principle, singlet oxygen can be produced by a direct process of energy transfer from the excited state of a sensitizer (type I process) or in subsequent dark reactions of primary electron transfer product superoxide. (type II):

$$\text{Type I:} \qquad S^* + O_2 \longrightarrow S + {}^1O_2^* \tag{3.45}$$

$$\text{Type II:} \qquad S^* + O_2 \longrightarrow S^+ + O_2^- \longrightarrow S + {}^1O_2^* \tag{3.46}$$

The fundamental questions posed in sensitized photooxidations are the occurrence of type I or II process as the source of singlet oxygen and the role of the microheterogeneous environment in promoting photooxidations.

Table 3.9 presents a listing of oxgen-mediated photosensitized oxidation processes studied in micellar media, intermediates observed, and the mechanisms proposed. Among various systems studied porphyrin-sensitized photooxidation of various substrates and their own (self-sensitized) oxidations are of much interest in photobiology. Often the same photoproducts are observed as in homogeneous solvents but with a different distribution ratio. Though most of the photooxidations were originally thought to proceed solely by a singlet oxygen pathway, there is now increasing evidence pointing out the importance of electron transfer/radical pathways.

TABLE 3.9. Oxygen-Mediated Photosensitized Oxidation of Compounds in Aqueous Micellar Solution

Micelle	Sensitizer	Substrate oxidized	Intermediate and mechanisms	Reference
CTAB, Triton X-100	Hematoporphyrin	L-Tryptophan and tryptamine	Mixed (Electron transfer/radical mechanism)	110
SDS	Hematoporphyrin	L-Tryptophan and tryptamine	$^1O_2$ pathway minor	111
CTAB, SDS, Triton X-100	Chlorpromazine, promazine, anthracene furosemide	2,5-Dimethyl furan and acrylamide	$^1O_2$ (for anthracene) electron transfers (others)	
SDS, DTAC, CTAC	Methylene blue	9,10-Dimethyl anthracene	$^1O_2$	107
SDS	(10-X)-Phenothiazine (X = methyl, aceyl)	Arylsulfides and (Ph)$_3$P	$^1O_2$	114
SDS, DTAB	Protoporphyrin IX	Protoporphyrin IX	$^1O_2$, electron transfer	113
SDS, CTAB	N-methyldiphenylamine	N-Methyldiphenylamine	$^1O_2$	109
Triton X-100	Chlorophyll a, pheophytin	Chlorophyll a pheophytin	$^1O_2$	112

References

1. N. J. Turro, K. C. Liu, M. F. Chow, and P. Lee, *Photochem. Photobiol.* **27**, 523 (1977)
2. K. Kalyanasundaram, F. Grieser, and J. K. Thomas, *Chem. Phys. Lett.* **51**, 501 (1977).
3. R. Humphry-Baker, Y. Moroi, and M. Grätzel, *Chem. Phys. Lett.* **58**, 207 (1978).
4. M. Almgren, F. Grieser, and J. K. Thomas, *J. Am. Chem. Soc.* **101**, 279 (1979).
5. N. J. Turro and M. Aikawa, *J. Am. Chem. Soc.* **102**, 4886 (1980).
6. L. J. Cline Love, M. Skrilec, and J. G. Habarta, *Anal. Chem.* **52**, 754 (1980).
7. M. Skrilec and L. J. Cline Love, *Anal. Chem.* **52**, 1559 (1980).
8. J. D. Bolt and N. J. Turro, *J. Phys. Chem.* **85**, 4029 (1981).
9. L. J. Cline Love and M. Skrilec, *in* "Solution Behaviour of Surfactans" (K. L. Mittal and B. Lindman, eds.), Vol. 3. Plenum, New York, 1981.
10. M. Skrilec and L. J. Cline Love, *J. Phys. Chem.* **85**, 2047 (1981).
11. L. J. Cline Love and M. Skrilec, *Anal. Chem.* **53**, 2103 (1981).
12. L. J. Cline Love, J. G. Habarta, and M. Skrilec, *Anal. Chem.* **53**, 437 (1981).
13. R. A. Femia and L. J. Cline Love, *Anal. Chem.* **56**, 327 (1984).
14. R. J. Woods, S. Scypinski, and L. J. Cline Love, *Anal. Chem.* **56**, 1395 (1984).
15. K. Gläsle, U. K. A. Klein, and M. Hauser, *J. Mol. Struct.* **84**, 353 (1982).
16. T. F. Hunter and A. J. Szczepanski, *J. Phys. Chem.* **88**, 1231 (1984).
17. G. Rothenberger, P. P. Infelta, and M. Grätzel, *J. Phys. Chem.* **85**, 1850 (1981).
18. J. C. Scaiano and J. C. Selwyn, *Can. J. Chem.* **59**, 2368 (1981).
19. J. C. Scaiano and J. C. Selwyn, *Photochem. Photobiol.* **34**, 29 (1981).
20. J. C. Selwyn and J. C. Scaiano, *Can. J. Chem.* **59**, 663 (1981).
21. J. C. Leigh and J. C. Scaiano, *J. Am. Chem. Soc.* **105**, 5652 (1983).
22. G. Rothenberger, P. P. Infelta, and M. Grätzel, *J. Phys. Chem.* **83**, 1871 (1979).
23. S. C. Wallace, M. Grätzel, and J. K. Thomas, *Chem. Phys. Lett.* **23**, 359 (1973).
24. M. Grätzel and J. K. Thomas, *J. Phys. Chem.* **78**, 2248 (1974).
25. S. A. Alkaitis, G. Beck, and M. Grätzel, *J. Am. Chem. Soc.* **97**, 5723 (1975).
26. S. A. Alkaitis, M. Grätzel, and A. Henglein, *Ber. Bunsenges. Phys. Chem.* **79**, 541 (1975).
27. S. A. Alkaitis and M. Grätzel, *J. Am. Chem. Soc.* **98**, 3549 (1976).
28. J. K. Thomas and P. Piciulo, *J. Am. Chem. Soc.* **100**, 3239 (1978).
29. G. E. Hall, *J. Am. Chem. Soc.* **100**, 8260 (1978).
30. J. K. Thomas and P. Piciulo, *in* "Interfacial Photoprocesses: Energy Conversion and Synthesis" (M. S. Wrighton, ed.), Adv. Chem. Ser. No. 184, p. 97. Am. Chem. Soc., Washington, D.C., 1980.
31. J. K. Thomas and P. Piciulo, *J. Am. Chem. Soc.* **101**, 2502 (1979).
32. H. Bauer and G. Reske, *J. Photochem.* **9**, 43 (1978).
33. R. Humphry-Baker, A. M. Braun, and M. Grätzel, *Helv. Chim. Acta* **64**, 2036 (1981).
34. A. Bernas, D. Grand, S. Hautecloque, and A. Chambaudet, *J. Phys. Chem.* **85**, 3684 (1981).
35. S. C. Wallace, G. E. Hall, and G. Kenney-Wallace, *Chem Phys.* **49**, 279 (1980).
36. J. P. Chauvet, R. Viovy, M. Bazin, R. Santus, and L. K. Patterson, *Chem. Phys. Lett.* **86**, 135 (1982).
37. S. J. Atherton, *J. Phys. Chem.* **88**, 2840 (1984).
38. D. Grand, S. Hautecloque, A. Bernas, and A. Petit, *J. Phys. Chem.* **87**, 5236 (1983).
39. P. A. Narayana, A. S. W. Li, and L. Kevan, *J. Am. Chem. Soc.* **103**, 3603 (1981); **104**, 6502 (1982).
40. B. L. Bales and L. Kevan, *J. Phys. Chem.* **86**, 3836 (1982).
41. P. A. Narayana and L. Kevan, *Photochem. Photobiol.* **37**, 105 (1983).
42. A. Plonka and L. Kevan, *J. Chem. Phys.* **80**, 5023 (1984).
43. E. Szajdzinska-Pietek, R. Maldonado, L. Kevan, and R. R. M. Jones, *J. Am. Chem. Soc.* **106**, 4675 (1984).
44. T. J. Meyer, *Prog. Inorg. Chem.* **30**, 389 (1983).

45. N. Sutin and C. Creutz, *Pure Appl. Chem.* **52**, 2717 (1980).
46. V. Balzani and F. Scandola, *in* "Energy Resources through Photochemistry and Catalysis" (M. Grätzel, ed.), Chap. 1. Academic Press, New York, 1983.
47. V. Balzani, F. Bolletta, M. T. Gandolfi, and M. Maestri, *Top. Curr. Chem.* **75**, 64 (1978).
48. K. Kalyanasundaram, *J.C.S. Chem. Commun.* p. 627 (1978).
49. K. Kalyanasundaram and G. Porter, *Proc. R. Soc. London, Ser. A* **364**, 29 (1978).
50. T. Tsutsui, K. Takuma, T. Nishijima, and T. Matsuo, *Chem. Lett.* p. 617 (1979).
51. M. P. Pileni, A. M. Braun, and M. Grätzel, *Photochem. Photobiol.* **31**, 423 (1980).
52. P. A. Brugger and M. Grätzel, *J. Am. Chem. Soc.* **102**, 2461 (1980).
53. M. A. J. Rodgers and J. C. Becker, *J. Phys. Chem.* **84**, 2762 (1980).
54. R. H. Schmehl and D. G. Whitten, *J. Am. Chem. Soc.* **102**, 1938 (1980).
55. R. H. Schmehl and D. G. Whitten, *J. Phys. Chem.* **85**, 3473 (1981).
56. T. Matsuo, T. Sakamoto, K. Takuma, K. Sakura, and T. Ohsaka, *J. Phys. Chem.* **85**, 1277 (1981).
57. T. K. Foreman, W. M. Sobol, and D. G. Whitten, *J. Am. Chem. Soc.* **103**, 5333 (1981).
58. R. H. Schmehl, L. G. Whitesell, and D. G. Whitten, *J. Am. Chem. Soc.* **103**, 3761 (1981).
59. P. P. Infelta and P.-A. Brugger, *Chem. Phys. Lett.* **82**, 462 (1981).
60. P.-A. Brugger, P. P. Infelta, A. M. Braun, and M. Grätzel, *J. Am. Chem. Soc.* **103**, 320 (1981).
61. T. Nakamura, T. Kurihara, and T. Matsuo, *J. Phys. Chem.* **86**, 4368 (1982).
62. P.-A. Brugger, M. Grätzel, T. Gaurr, and G. McLendon, *J. Phys. Chem.* **86**, 944 (1982).
63. T. Miyashita, T. Murakata, and M. Matsuda, *J. Phys. Chem.* **87**, 4529 (1983).
64. K. Kurihara, P. Tundo, and J. H. Fendler, *J. Phys. Chem.* **87**, 3777 (1983).
64a. J. Kiwi, K. Kalyanasundaram, and M. Grätzel *Struct. Bonding (Berlin)* **49**, 37 (1982).
64b. A. Harriman and M. A. West, eds., "Photogeneration of Hydrogen." Academic Press, New York, 1982.
64c. M. Grätzel, ed., "Energy Resources through Photochemistry and Catalysis." Academic Press, New York, 1983.
65. Y. Moroi, A. M. Braun, and M. Grätzel, *J. Am. Chem. Soc.* **101**, 567 (1979).
66. Y. Moroi, P. P. Infelta, and M. Grätzel, *J. Am. Chem. Soc.* **101**, 573 (1979).
67. R. Humphry-Baker, Y. Moroi, M. Grätzel, and E. Pelizetti, *J. Am. Chem. Soc.* **102**, 3689 (1980).
68. Y. Moroi, *Bull. Chem. Soc. Jp.* **54**, 3265 (1981).
69. M. P. Pileni and M. Grätzel, *J. Phys. Chem.* **84**, 1822 (1980).
70. M. Maestri, P. P. Infelta, and M. Grätzel, *J. Chem. Phys.* **69**, 1522 (1978).
71. D. Meisel, M. S. Matheson, and J. Rabani, *J. Am. Chem. Soc.* **100**, 117 (1978).
72. P. Massini and G. Voorn, *Biochim. Biophys. Acta* **153**, 589 (1968).
73. C. Wolff and M. Grätzel, *Chem. Phys. Lett.* **52**, 542 (1977).
74. K. Kano, K. Takuma, T. Ikeda, D. Nakajima, Y. Tsutsui, and T. Matsuo, *Photochem. Photobiol.* **29**, 695 (1978).
75. L. Putna, G. Reske, and H. Schmidt, *Photochem. Photobiol.* **30**, 723 (1979).
76. A. J. Frank and M. Grätzel, *Inorg. Chem.* **21**, 3834 (1982).
77. T. Nakamura, A. Kira, and M. Imamura. *J. Phys. Chem.* **88**, 3435 (1984).
78. T. Nakamura, A. Kira, and M. Imamura, *Bull. Chem. Soc. Jpn.* **57**, 2033 (1984).
79. A. Yoshimura and S. Kata, *Bull, Chem. Soc. Jp.* **53**, 1877 (1980).
80. K. Kano and T. Matsuo, *Bull. Chem. Soc. Jp.* **47**, 2836 (1974).
81. Y. Usui, S. Kodera, and Y. Nishida, *Chem. Lett.* p. 1329 (1976).
82. M. Seno, K. Kousaka, and H. Kise, *Bull. Chem. Soc. Jp.* **52**, 2970 (1979).
83. Y. Usui and K. Saga, *Bull. Chem. Soc. Jpn.* **55**, 3302 (1982).
84. Y. Yamaguchi, T. Miyashita, and M. Matsuda, *J. Phys. Chem.* **85**, 1369 (1981).
85. H. Inone and M. Hida, *Bull. Chem. Soc. Jpn.* **55**, 1880 (1982).
86. K. Okubo, K. Yamashita, and S. Sakaki, *J.C.S. Chem. Commun.* p. 787 (1984).

87. M. P. Pileni and M. Grätzel, *J. Phys. Chem.* **84**, 2402 (1980).
88. W. J. Dressick, K. W. Raney, J. N. Demas, and B. A. De Graff, *Inorg. Chem.* **23**, 875 (1984).
89. R. Humphry-Baker, M. Grätzel, P. Tundo, and E. Pelizetti, *Angew. Chem. Int. Ed. Engl.* **18**, 630 (1979).
90. K. Chandrasekaran, T. K. Foreman, and D. G. Whitten, *Nouv. J. Chim.* **5**, 275 (1981).
91. A. J. Frank, M. Grätzel, A. Henglein, and E. Janata, *Int. J. Chem. Kinet.* **8**, 817 (1976).
92. A. J. Frank, M. Grätzel, and J. J. Kozak, *J. Am. Chem. Soc.* **98**, 3317 (1977).
93. A. A. Gorman, G. Lovering, and M. A. J. Rodgers, *Photochem. Photobiol.* **23**, 399 (1976).
94. A. A. Gorman and M. A. J. Rodgers, *Chem. Phys. Lett.* **55**, 52 (1978).
95. I. B. C. Matheson, J. Lee, and A. D. King, *Chem. Phys. Lett.* **55**, 49 (1978).
96. B. A. Lindig and M. A. J. Rodgers, *J. Phys. Chem.* **83**, 1683 (1979).
97. I. B. C. Matheson and R. Massoudi, *J. Am. Chem. Soc.* **102**, 1942 (1980).
98. B. A. Lindig and M. A. J. Rodgers, *Photochem. Photobiol.* **33**, 627 (1981).
99. I. B. C. Matheson and M. A. J. Rodgers, *J. Phys. Chem.* **86**, 884 (1982).
100. P. C. Lee and M. A. J. Rodgers, *J. Phys. Chem.* **87**, 4894 (1983).
101. M. A. J. Rodgers, *Photochem. Photobiol.* **37**, 99 (1983).
102. M. A. J. Rodgers and P. C. Lee, *J. Phys. Chem.* **88**, 3480 (1984).
103. N. J. Turro, M. Aikawa, and A. Yekta, *Chem. Phys. Lett.* **64**, 473 (1979).
104. N. Miyoshi and G. Tomita, *Z. Naturforsch.*, B **33B**, 622 (1978).
105. Y. Usui, M. Tsukada, and H. Nakamura, *Bull. Chem. Soc. Jpn.* **51**, 379 (1978).
106. I. Kraljic, N. Barboy, and J. P. Leicknam, *Photochem. Photobiol.* **30**, 631 (1979).
107. O. Bagno, J. C. Soulignac, and J. Joussot-Dubien, *Photochem. Photobiol.* **29**, 1079 (1979).
108. C. Sconfienza, A. Van de Vorst, and S. Jori, *Photochem. Photobiol.* **31**, 351 (1980).
109. N. Roessler and T. Wolff, *Photochem. Photobiol.* **31**, 547 (1980).
110. E. Rossi, A. Van de Vorst, and G. Jori, *Photochem. Photobiol.* **34**, 447 (1981).
111. D. E. Moore and C. D. Burt, *Photochem. Photobiol.* **34**, 431 (1981).
112. J. P. Chauvet, F. Villain, and R. Viovy, *Photochem. Photobiol.* **34**, 557 (1981).
113. G. S. Cox, M. Krieg, and D. G. Whitten, *J. Am. Chem. Soc.* **104**, 6930 (1982).
114. M. C. Hovey, *J. Am. Chem. Soc.* **104**, 4196 (1982).
115. M. Krieg and D. G. Whitten, *J. Am. Chem. Soc.* **106**, 2477 (1984).

Chapter 4

Micellar Photochemistry and Photoreactions

Chapters 2 and 3 have been concerned with micellar effects on various photophysical processes. It was shown that micellar aggregates in addition to providing a medium to dissolve (solubilize) a wide variety of organic molecules, do alter significantly the excited-state chemistry and reactivity for the singlets and triplets. It is of interest then to ask whether micellar effects can be observed in the overall products of photochemical reactions as well. In this chapter we address ourselves to micellar control of various organic photochemical reactions. Photofragmentation, photocycloaddition, and photo-induced H-abstraction are some of the photochemical processes where the nature and the yields of photoproducts depend critically on the nature of the chromophore environment and its relative mobility there.[1-3] Excited-state quenching studies have shown that dynamic solubilization of probes in micelles in effect leads to their transient trapping in environments quite different from those of the bulk solution. Photochemical studies have shown that such solubilization leads to pronounced differences in the product distribution and even their regioselectivity!

4.1 Photolysis of Dibenzylketones: Micellar Cage Effect, Isotope Enrichment, and Magnetic Field Effects

Practical applications of micellar cage effects are best illustrated with the elegant studies of Turro *et al.* on the photolysis of ketones in micellar solutions.[4-19] Consider for example, the photolysis of dibenzylketone (DBK). In homogeneous solvents such as benzene, photolysis of DBK leads to

122

quantitative decarbonylation, yielding CO and 1,2-diphenylethane (DPE)

$$Ph—CH_2—CO—CH_2—Ph \xrightarrow{h\nu} Ph—CH_2—CH_2—Ph + CO \qquad (4.1)$$
$$(DBK) \qquad\qquad\qquad (DPE)$$

Photolysis of DBK in aqueous micellar solutions however shows dramatic differences. The quantum yield for the disappearance of DBK (Φ_{-DBK}) is significantly reduced compared to homogeneous solvents ($\Phi_{-DBK} \simeq 0.7$ in benzene versus 0.3 in 0.05 M CTAC micelles). Furthermore, careful examination of the photolysis products showed substantial isotope enrichment in the regenerated DBK. The regeneration of DBK and consequent isotope enrichment

$$DBK \xrightarrow[micelle]{h\nu} DPE + CO + \quad DBK \qquad\qquad (4.2)$$
$$(\text{enriched in } {}^{13}C)$$

are strongly dependent on the presence of a magnetic nuclei such as ^{13}C, substituents on the DBK, and applied (external) magnetic fields. Detailed analysis of the products under a wide variety of conditions along with the elegant use of CIDNP techniques have enabled Turro and co-workers, to establish a reaction sequence that satisfactorily accounts for the observed yields, and isotope and magnetic field effects in micellar media. Figure 4.1 presents the overall reaction scheme.

Disappearance in homogeneous and micellar media. Excitation of DBK to its singlet excited state is followed by rapid radiationless transition to a lower-lying (n, π*) triplet state:

$$DBK \longrightarrow {}^1DBK^* \longrightarrow {}^3DBK^* \qquad\qquad (4.3)$$

The latter undergoes homolytic cleavage to produce a geminate triplet radical pair in a solvent cage:

$$^3DBK \longrightarrow {}^3\overline{Ph—CH_2—CO \quad \overset{\cdot}{C}H_2—Ph} \qquad\qquad (4.4)$$
$$(^3RP)$$

(denoted as ^3RP in reaction (4.4) with a bar over the pair to indicate a solvent cage). In homogeneous, nonviscous solvents, the ^3RP separates quantitatively into free radicals:

$$^3\overline{Ph—CH_2—CO \quad \overset{\cdot}{C}H_2—Ph} \longrightarrow {}^3Ph—CH_2—CO\cdot + \overset{\cdot}{C}H_2—Ph \qquad (4.5)$$

The phenylacetyl radical (Ph—CH$_2$—CO ·) subsequently undergoes decarbonylation, and the resulting free benzyl radicals combine to form diphenylethane (DPE):

$$^3Ph—CH_2—CO\cdot \longrightarrow {}^3Ph—\overset{\cdot}{C}H_2 + CO \qquad\qquad (4.6)$$

$$2(Ph—CH_2\cdot) \longrightarrow Ph—CH_2—CH_2—Ph \qquad\qquad (4.7)$$
$$(DPE)$$

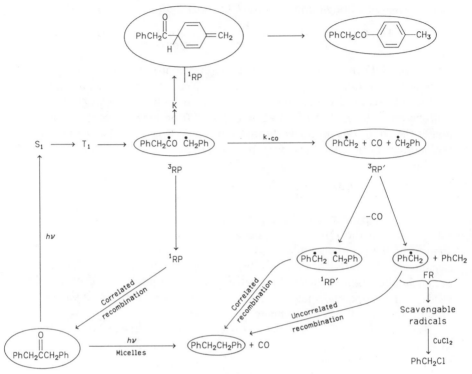

FIG. 4.1. Schematic representation of various reaction sequences that occur during the photolysis of dibenzylketone in aqueous micellar media. (From Turro.[5])

The production of free radicals via reaction (4.5) occurs very rapidly $(10^{-10} - 10^{-11}$ sec) in these solvents. For electron-spin correlated radical pairs such as ^3RP, there are other competitive processes which are available:

$$\underset{(^3RP)}{^3\overline{Ph-CH_2-CO\ \ \overset{.}{C}H_2-Ph}} \xrightarrow{\text{ISC}} \underset{(^1RP)}{^1\overline{Ph-CH_2-CO\ \ \overset{.}{C}H_2-Ph}} \quad (4.8)$$

$$\underset{(^3RP)}{^3\overline{Ph-CH_2-CO\ \ \overset{.}{C}H_2-Ph}} \longrightarrow \underset{(^3RP')}{^3\overline{Ph-CH_2 +CO +CH_2-Ph}} \quad (4.9)$$

The triplet radical pair can undergo a change in spin angular momentum (intersystem crossing) to form a singlet radical pair (^1RP) or decarbonylate to form another spin-correlated benzyl radical pair ($^3RP'$) [reactions (4.8) and (4.9)]. The singlet radical pair can undergo recombination to regenerate DBK or an isomer of DBK, 1-phenyl-4-methylacetophenone (PMAP):

$$\overline{\text{Ph—CH}_2\text{—CO \quad CH}_2\text{—Ph}} \xrightarrow{\hspace{1cm}}$$
$$(^1\text{RP})$$

$$\text{Ph—CH}_2\text{—CO—CH}_2\text{—Ph} + \text{Ph—CH}_2\text{—CO—C}_6\text{H}_4\text{—CH}_3 \quad (4.10)$$
$$\text{(DBK)} \qquad\qquad\qquad \text{(PMAP)}$$

In homogeneous solvents, the formation of free radicals occurs very rapidly on time scales too short to allow hyperfine interaction, electronic Zeeman, and other effects to influence the intersystem crossing step [reaction (4.8)]. The spin correlations of the triplet radical pair must be preserved for periods of 10^{-9} sec for these effects to operate. Thus, for competitive steps (4.8) and (4.9) to operate, we need to preserve the ^{3}RP in a supercage environment which will (1) allow the fragments of ^{3}RP to remain geminate without diffusing apart too far from each other, yet allow the fragments to execute diffusional and rotational motion that will decrease the electron exchange energy to appropriate values for hyperfine interactions to operate and (2) encourage the ^{1}RP to undergo recombination reaction (4.10).

By their unique dynamic association and solubilization processes, micellar systems can amply fulfill both of these requirements. The average time solutes remain captured in the micelle (residence time) is long enough (10^{-3}–10^{-6} sec) for various spin exchange mechanisms to operate and also for the ^{1}RP to undergo the recombination reaction. The micellar size is also large enough for the ^{3}RP to move around tens of angstroms and loose the excess energy without loss of spin correlations. The micellar cage can thus promote efficiency of cage reactions by providing slow escape of radicals. Enhanced radical recombination also will decrease the quantum yield for the overall reaction as is observed experimentally. With regard to DBK photolysis, in homogeneous solvents the extent of cage reaction is negligible but is substantial in micellar solutions (about 30% for the recombination of geminate benzyl radicals). The cage effect numbers are best understood in terms of photolysis products of unsymmetrical dibenzylketones:

$$\text{PhCH}_2\text{COCH}_2\text{Ph}' \xrightarrow[-\text{CO}]{h\nu} \text{PhCH}_2\text{CH}_2\text{Ph} + \text{PhCH}_2\text{CH}_2\text{Ph}' + \text{Ph}'\text{CH}_2\text{CH}_2\text{Ph}' \quad (4.11)$$
$$\text{(ACOB)} \qquad\qquad \text{(AA)} \qquad\quad \text{(AB)} \qquad\qquad \text{(BB)}$$

The cage effect is defined by the product ratio

$$\text{cage effect} = [\text{AB} - (\text{AA} + \text{BB})]/(\text{AA} + \text{AB} + \text{BB}) \qquad (4.12)$$

The ratio of products AA : AB : BB would be 0 : 100 : 0 if there were 100% cage effect and 25 : 50 : 25 if there were 0% cage effect.

As with the primary caged ^{3}RP, the secondary radical pair ^{3}RP' generated via reaction (4.9) can undergo either intersystem crossing to generate ^{1}RP' (which can also undergo rapid combination to yield DPE in a cage reaction) or separate to give random free radicals (scavengable with suitable free radical traps):

$$^3\overline{Ph-CH_2\cdot + CO + CH_2-Ph}\quad \begin{cases} \xrightarrow{\text{(a)}}\quad Ph-CH_2\cdot \xrightarrow{X} Ph-CH_2-X \\ \qquad\qquad\qquad \text{(random free radical)} \\ \xrightarrow{\text{(b)}}\quad {}^1\overline{Ph-CH_2\ CH_2-Ph}\ \longrightarrow DPE \end{cases}$$

$$(^3RP')\qquad\qquad\qquad\qquad\qquad\qquad (^1RP') \qquad\qquad (4.13)$$

In summary, we can distinguish two reaction sequences (A and B) coupling various reaction steps:

Sequence A: DBK $\longrightarrow$ S_1^* $\longrightarrow$ T_1^* $\longrightarrow$ ^3RP $\xrightarrow{\text{ISC}}$ ^1RP $\longrightarrow$ DBK/PMAP

$$(4.14)$$

Sequence B: DBK $\longrightarrow$ S_1^* $\longrightarrow$ T_1^* $\longrightarrow$ ^3RP $\xrightarrow{-CO}$ $^3RP'$ $\longrightarrow$ ^1RP $\longrightarrow$ DPE

$$\downarrow$$
$$Ph-CH_2\cdot \longrightarrow Ph-CH_2X \qquad (4.15)$$

The elaborate description of reaction sequences also serves to explain the origin of observed isotope enrichment and magnetic field effects. The intersystem crossing step [reaction (4.8)] of ^3RP to ^1RP in the micellar supercage can occur via coupling of the electronic spins of the radical pair to an external magnetic field (such as provided by a laboratory magnet) or by an internal magnetic field (such as that provided by a nearly magnetic nuclei or by the orbital motion of the electron). Presence of a ^{13}C nucleus in a radical pair plays essentially the role of an internal magnetic field by hyperfine interactions. The hyperfine interactions (hfi) will promote ISC via spin flipping $(T_+ \rightarrow S$ or $T_- \rightarrow S)$ or by the spin rephasing $(T_0 \rightarrow S)$ mechanisms. The fact that a given ^3RP that contains a ^{13}C nucleus will enjoy a faster ISC than an ^1RP implies that DBK will be regenerated more efficiently from such pairs and that PMAP will be formed more efficiently from such pairs. In the extreme case, ^3RP containing ^{13}C will undergo only recombination reactions (sequence A) whereas ^3RP containing only ^{12}C nuclei will undergo escape reactions (i.e., decarbonylation (sequence B). Thus, one effect of the nuclear spins on the reaction of DBK in micelles will be that the combination products (DBK and DMAP) will be enriched in ^{13}C.

There are other effects of nuclear spins as well: the ^{13}C enrichment will be different at various distinguishable atoms of DBK and PMAP, and the extent of enrichment at each distinguishable atom will be field dependent, decreasing on the application of a high laboratory magnetic field. All these effects have been confirmed by experiment. For example, starting with a DBK sample enriched with 48% ^{13}C, after photolysis to 91% conversion in CTAC micelles, the recovered DBK was found to be enriched in ^{13}C to 62%. For photolysis under similar conditions but in the presence of magnetic field at $H = 15$ KG,

TABLE 4.1. Micellar Cage and External Magnetic Field Effects on the Quantum Yields for Photolysis of Dibenzyl Ketones in 0.05 M CTAC at Room Temperature[a]

			Cage effect[c]	
Ketone	Medium	DBK[b]	0 G	13 kG
DBK (natural abundance)	Benzene	0.72	0	0
DBK (natural abundance)	0.05 M CTAC	0.30	31 ± 1	16 ± 1
DBK-^{13}C	0.05 M CTAC	0.22	31 ± 1	16 ± 1
DBK-2,2'-^{13}C	0.05 M CTAC	0.25	46 ± 3	22 ± 2
DBK-2,2'-^{2}H$_4$	0.05 M CTAC	0.32	29 ± 1	14 ± 1
DBK-^{2}H$_{10}$	0.05 M CTAC	0.32	28 ± 1	
4-Me-DBK	0.05 M CTAC	0.23	52 ± 1	$31 \pm 1(25)$
4-Cl-DBK	0.05 M CTAC	0.21	52 ± 1	30
4-Br-DBK	0.05 M CTAC	0.11	70 ± 4	70(58)
4,4'-di-Me-DBK	0.05 M CTAC	0.16	59 ± 3	31 ± 2
4,4'-di-t-Bu-TBK	0.05 M CTAC	0.13	95 ± 5	76 ± 5

[a] Data from Turro and Weed.[15]

[b] Values for disappearance of DBK in the presence or absence of CuCl$_2$ as a scavanger.

[c] Cage effect refers to the efficiency of combination of geminate benzyl radical pairs.

the enrichment is limited to 55% ^{13}C. The ^{13}C enrichment actually showed a maximum at a magnetic field strength of 200–300 G. This is qualitatively understood in terms of Zeeman's inhibition of ^{1}H hfi ISC and ^{13}C hfi ISC of radical pairs. (At sufficiently high magnetic fields, ISC may be determined by the difference in g factors of the radical pairs, Δg and not by the hfi.) Table 4.1 gives representative data on the quantum yields, micellar cage effects, isotope enrichment, and magnetic field effects in the photolysis of various DBK derivatives in aqueous micellar solutions at room temperature.

4.2 Dynamic Aspects of Micellar Cage and Magnetic Field Effects: Laser Flash Photolysis Studies

A basic concept in the earlier interpretations of DBK photochemistry in micelles is that the triplet radical pair (^{3}RP) survives in the micellar interior for a long period of time (10^{-7} sec) to undergo intersystem crossing forming the ^{1}RP, which itself undergoes recombination, regenerating the DBK. These reactions occur in competition with the exit reactions which separate the spin-correlated radical pairs. There have been several laser flash photolysis studies in recent years that are directed towards measurements of absolute rates for radical pair intersystem crossing, free-radical exit as well as magnetic field effects on these rates and yields of escaping radicals. Several aliphatic and

TABLE 4.2. Triplet Lifetimes of Ketones in Aqueous Micellar
Solutions at 25°C

Ketone[a]	Micelle	τ_T (μsec)	Reference
Acetophonone	SDS (0.23 M)	3.85	24
Propiophenone	SDS (0.23 M)	3.80	24
Isobutyrophenone	SDS (0.23 M)	(17.0)	24
p-Methoxy acetophenone	SDS (0.23 M)	19.0	24
Xanthone	SDS (0.23 M)	7.65	24
Benzophenone(BP)	SDS (0.23 M)	0.36	30
Benzophenone	CTAC (0.05 M)	0.31	30
Benzophenone	CTAB (0.05 M)	0.017	30
Acetone	SDS (0.03 M)	1.63	32
Acetone	SDS (0.05 M)	1.43	32
Acetone	SDS (0.10 M)	1.08	32
Acetone	CTAC (0.01 M)	1.70	32
Acetone	CTAC (0.05 M)	1.24	32
Acetone	35(0.02 M)	1.00	32
Benzophenone(BP)	SPFO (0.1 M)	6.0	22
Benzophenone	SDS	0.34	37
Benzophenone	DTAC	0.28	37
BP$^-$	SDS	0.39	37
BP$^-$	DTAC	0.285	37
BP$^+$	DTAC	0.21	37
BP$^+$	SDS	0.165	37
BP—C$_{12}$	SDS	0.375	37

[a]BP$^-$ = Ph—CO—⟨◯⟩—CH$_2$SO$_3^-$Na$^+$,

BP—C$_{12}$ = Ph—CO—⟨◯⟩—(CH$_2$)$_{11}$—CH$_3$,

SPFO = sodium perfluorooctanoate,

BP$^+$ = Ph—CO—⟨◯⟩—CH$_2$—$\overset{+}{N}$Me$_3$Cl$^-$

aromatic ketones and quinones have been examined.[20-42] Table 4.2 summarizes data on the triplet lifetimes of several ketones in aqueous media that have been measured in this context. Results for a few selected systems are elaborated in the following.

4.2.1 Laser Photolysis of Dibenzyl Ketones

Let us consider laser flash excitation of DBK to give the triplet excited state ^{3}DBK* via the singlet state ^{1}DBK*. The triplet state is very labile and undergoes homolytic cleavage to form the spin-correlated triplet radical pair of the phenylacetyl and benzyl radicals. The triplet radical pair can undergo intersystem crossing forming the ^{1}RP or decarbonylate to form the triplet radical pair of benzyl radicals. From measurements of fluorescence lifetime, the quenching of ^{3}DBK* and trapping of the benzyl and phenylacetyl radicals, the lifetime of ^{1}DBK*, ^{3}DBK*, and phenylacetyl radicals have been estimated to be 3.6 nsec, about 100 psec and about 10 nsec, respectively. Thus, for 15–30 nsec laser pulse excitation reactions (4.3)–(4.6) all occur rapidly (decarbonylation complete in $\leqslant 50$ nsec) and the early transient is that of a benzyl radical (characterized by its absorption maximum at 318–320 nm) and not that of phenylacetyl radicals. In homogeneous solvents, the decay of benzyl radicals follows second-order kinetics in the same time range of 50 nsec–50 μsec.

In micellar SDS solutions the benzyl radical decay exhibits biphasic character ($k_{fast} \sim 2.5 \times 10^6$ sec^{-1} and $k_{slow} \sim 2.1 \times 10^4$ sec^{-1} at 0 G). Scavenging experiments with traps such as CuCl$_2$ had no effect on both the fast and slow components indicating that the species responsible for these are associated with the micelles. Magnetic fields exert effects on both the fast and slow components. For magnetic fields less than 700 G, the decay rate of the fast component is reduced and the absorbance of the slow one increased. The fast component has been assigned to the decay of spin-correlated benzyl radical pair [through reactions (4.9) and (4.13a)] and the slow component to the decay of escaping benzyl radical (2 PhCH$_2$·). The term *escaping* indicates that these are relaxed radicals where the spin memory between the radical pairs has been lost. Since reaction (4.13a) is considered to be faster than the reaction (4.13b), the fast component corresponds to the intersystem crossing of triplet benzyl radical pairs (^{3}RP′ → ^{1}RP′).

We can take the transient ratios [I(400 nsec)/I(0 nsec)] in the SDS CTAC solutions to represent the amount of escaping benzyl radicals normalized to the initial amount of radicals produced by laser excitation. The presence of the magnetic field increases this ratio quite rapidly below 150 G and only slowly above 150 G. The ratio for SDS (CTAC) solutions increases by 10% (20%) at 150 G and by 29% (31%) at 700 G compared to that observed at zero field. In n-hexane solutions, the transient absorbance at the same wavelength shows neither the fast component nor any magnetic field effects on the yield and decay of the benzyl radicals.

4.2.2 Laser Photolysis of Benzophenones

Production of triplets of benzophenone in micellar media is accompanied by a rapid photoreduction step leading to generation of isolated radical pairs whose behavior resembles that of biradicals. This is because surfactants such as SDS or CTAC function both as micelle-forming agents and also as good

H-donors (like alcohols) for the photoreduction of benzophenone. The following scheme outlines various reactions that follow light excitation of BP:

$$BP \xrightarrow{h\nu} {}^1BP* \longrightarrow {}^3BP* \tag{4.16}$$

$${}^3BP* + RH \longrightarrow {}^3(K \cdot \cdot R) \tag{4.17}$$

$${}^3(K \cdot \cdot R) \longrightarrow {}^1(K \cdot \cdot R) \tag{4.18}$$

$${}^{3,1}(K \cdot \cdot R) \longrightarrow K \cdot + R \cdot \tag{4.19}$$

$${}^1(K \cdot \cdot R) \longrightarrow K{-}R, BP + RH \tag{4.20}$$

Here ${}^3BP*$ abstracts an H atom from a detergent molecule RH, forming the triplet radical pair of benzophenone ketyl radical and an alkyl R within a micelle [reactions (4.16) and (4.17)]. The rest of the steps in the scheme are similar to that of dibenzyl ketones: conversion of the triplet radical pair to the singlet radical pair by the influence of hyperfine interaction and electronic Zeeman mechanisms, separation of the correlated radicals in the triplet and singlet pairs to escaping radicals, and combination reactions of radicals in the singlet pair generating BP and some products. The yield of the escaping radicals from the singlet pair is considered to be much lower than that from the triplet pair because of the parallel path way of reactions (4.19) and (4.20) for the singlet pair.

From a detailed kinetic analysis of the transient decay curves at several wavelengths Scaiano[30] has noted that around 525 nm (where we normally monitor the triplet excited state) there is substantial ketyl absorptions. A monitoring wavelength of 600 nm has been recommended for ${}^3BP*$. By deconvolution of the triplet decay (taken at 600 nm) from the composite curves at 540 nm, it is possible to generate the growth and decay curves of pure ketyl radical K. Such an analysis has shown that ketyl radicals are formed in good yields but that the majority of radical pairs decay during overlapping with the triplet state process. To overcome this problem, we can add a good H-donor such as 1,4-hexadiene so that the triplet-state decay becomes fast enough to occur within the laser pusee, leaving mainly ketyl radical absorptions at the end of the laser pulse.

Examination of the ketyl radical decay curves at 540 nm in the presence of 1,4-hexadiene reveals a rapidly decaying initial component (over in about 2 μsec) leaving some long-lived absorptions. (The latter corresponds to the fraction of radical pairs that underwent the escape process). Thus, the ratio of the end (or plateau) to initial absorption is a measure of the extent of the escape process. In the presence of external applied magnetic field ($\geqslant 2000$ G) the fraction of escape radicals increases substantially without affecting much the rate constant of the radical decay (cf. Figure 4.2). Table 4.3 provides experimental data of Scaiano and Lougnot[37] on the percentage exit of ketyl radicals during the photolysis of benzophenone and several of its derivatives with and without the external H-atom donor 1,4-hexadiene. Addition of 1,4-

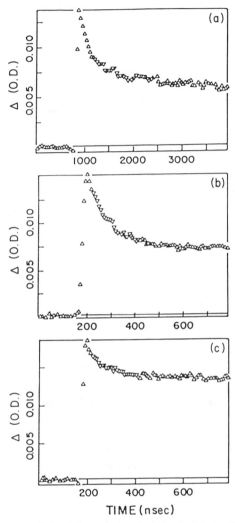

FIG. 4.2. Decay traces recorded at 540 nm during the laser photolysis of benzophenone in surfactant solutions in the presence of sufficient 1,4-cyclohexadiene to quench the triplet state
quantitative. (a) CTAC (0.043 M), $H = 0\,G$; (b) SDS (0.22 M), $H = 0\,G$; SDS (0.22 M),
$H = 2000$ G. Temperature = 27°C. (From J. C. Scaiano, *et al.*[30] Copyright 1982 American
Chemical Society.)

hexadiene in addition of simplying the transient decay monitoring, also
provides a highly mobile radical in the radical pair (percent exit is much higher
in all cases). The apparent decay rate constant of the radical pair (k_{app}) is given
by:

$$k_{app} = k_- + k_{ISC} \qquad\qquad (4.21)$$

TABLE 4.3. Exit Behavior of Radical Pairs of Benzophenone and Derivatives during Photolysis in Aqueous Micellar Solutions at 300 K Determined by Laser Flash Photolysis[a]

		% exit $(R \cdot)$		% Exit (with 1,4-hexadiene)	
Ketone[b]	Micelle	$H = 0$ G	$H = 2000$ G	$H = 0$ G	$H = 2000$ G
BP	SDS	7	60	49	82
BP	DTAC	25	78	55	84
BP$^-$	DTAC	22	51	53	81
BP$^-$	SDS	45	62	59	78
BP$^+$	SDS	6	41	49	70
BP$^+$	DTAC	61	81	59	82
BP—C$_{12}$	SDS	2	14	33	60

[a] From Scaiano and Lougnot.[37]

[b] BP = benzophenone, BP$^-$ = ⟨◯⟩—CO—⟨◯⟩—CH$_2$SO$_3^-$(Na$^+$),

BP$^+$ = ⟨◯⟩—CO—⟨◯⟩—(CH$_2$)—N$^+$(CH$_3$)$_3$(Cl$^-$).

For the benzophenone–ketyl–cyclohexadienyl pairs in CTAC micelles, the evaluated rate constants are $k_{ISC} = 1.7 \times 10^6$ sec^{-1} and $k_- = 1.6 \times 10^6$ sec^{-1}. For the SDS micelles, the corresponding values are $k_{ISC} = 5.8 \times 10^6$ sec^{-1} and $k_- = 4.4 \times 10^6$ sec$^-$, respectively. Since the ketyl exit is a minor process in the presence of cyclohexadine (as indicated by their % exit), the k_- values represent essentially the exit rate of cyclohexadienyl radicals from the micelles. In their independant studies Sakaguchi et al.[20,21,27] have obtained very similar effects on the influence of magnetic fields of the radical exit rates and the yields. Their estimates of the yields of escaping radicals, however, are much lower than those of Scaiano and co-workers.

4.3   Intra- and Intermolecular Photoinduced H-Abstraction Reactions

4.3.1   Norrish Type II Photoreactions[44–46]

The photochemistry of dibenzyl ketones and related compounds in micelles discussed earlier represents the so-called Norrish Type I reaction (photofragmentation/cleavage). The (n, π*) triplet states of carbonyl compounds are known to abstract H atoms from a variety of substrates including relatively unactivated hydrogens of saturated hydrocarbons. These γ-hydrogen abstraction reactions of ketones (known as Norrish Type II processes) have also been studied in micellar media for two main reasons: (1) It is known that yields of type II products are extremely sensitive to the medium polarity. On going from a polar to nonpolar medium, the state-switching occurs in the triplet from (n, π^*) to (π, π^*) and the efficiency of the type II processes also decreases. Hence the solvent sensitivity of type II photoreactions can serve as a probe for solute location in micelles. (2) Extensive photophysical studies in homogeneous solvents have shown that the type II process is intramolecular in nature involving formation of a 1,4-diradical. The H-atom abstraction occurs via a six-numbered (cyclic) transition state:

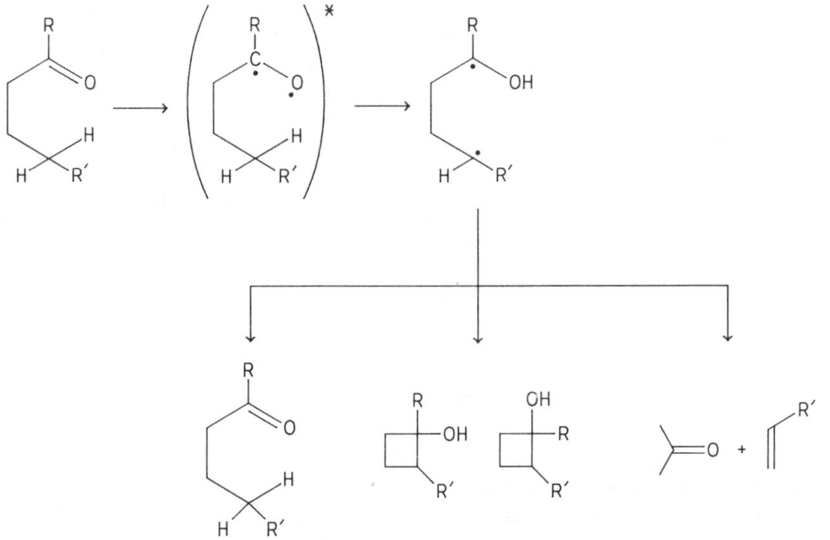

The microheterogeneous environment of the micellar structures can have a pronounced effect in affecting the ease of formation of this crucial intermediate. There can also be cage effects on the types of products formed. With flexible long-chain surfactant ketones, depending on their orientation in the organized assemblies, the selectivity for H-abstraction can be different.

The feasibility of using type II processes as probes was first examined by Turro et al.[44] with simple ketones such as valerophenone and octanophenone in CTAC micelles. Whitten et al.[45,46] subsequently explored further by using long-chain surfactant ketones. Phenyl ketones as indicated in the scheme can yield both cleavage products (acetophenone and alkene) and cyclization products (stereoisomeric cyclobutanols). Quantum yield data for type II processes measured under various conditions are collected in Table 4.4. Upon incorporation into cationic micelles, Φ_{II} for the valerophenone remains unaltered while that of acetophenone drops to 0.7. The ratio of stereoisomeric cyclobutanols also follows the same trend. These results are consistent with a

TABLE 4.4. Quantum Efficiency of Norrish Type II Photoprocesses of Ketones in Micellar Media[a]

Ketone	Homogeneous solvent		Micelles	
	Solvent	ϕ_{II}	Surfactant	ϕ_{II}
Valerophenone	t-Butanol	1.0	CTAC	1.06
	Benzene	0.33	CTAC + Eu^{3+}(5 × 10^{-3} M)	1.02
Octanophenone	t-Butanol	1.0	CTAC	0.71
	Benzene	0.29	SDS	0.76
			SDS + Eu^{3+}(2 × 10^{-4} M)	0.75
Ketoacid I[b]	t-Butanol	1.02	SDS	0.81
	Benzene	0.27	SDS + Eu^{3+}(2 × 10^{-4} M)	0.66
			CTAC	0.72
			DODAC vesicles	0.21
			Assemblies	10^{-3}
Ketoacid II[b]			SDS	0.84
			SDS + heptanol (+0.06 M)	0.66
			SDS(+2 × 10^{-4} M)Eu^{3+}	0.76

[a] Data taken from Turro et al.[44] and Winkle et al.[46]

[b] Ketoacids I and II: $n = 14(I); n = 10(II)$.

model where the ketones (or the intermediate biradical) spend at least some time in a polar environment (Stern layer) of the micelle. The biradical lifetime, 92 ± 10 nsec for γ-methylvalerophenone in SDS micelles measured by Scaiano and Selwyn,[47] is comparable with those obtained in methanol or benzyl alcohol, thus confirming the above proposition.

The threefold increase in the quantum yields of type II products in the photolysis of long-chain (surfactant) ketoacids as compared to that in benzene imply a hydropholic nonviscous environment for the chromophore. (UV absorption spectral data are also in accord with this.) There are several modes of solubilization that we can visualize, depending on the model of the ionic micelle. In the early idealized Hartely model with an inner, organized hydrophobic core, the ketone molecule has to assume a bent configuration such that both the carboxylate and the carbonyl functions are located at the interface. Alternatively in the recent porous micelle model of Menger (with extensive penetration of water into the interior of micelle) the ketone could simply be located in the interior. In larger organized assemblies such as DODAC surfactant vesicles and monolayers more severe geometrical constraints are placed on the molecule to align themselves with the hydrocarbon chains. With less flexibility to assume a cyclic six-membered transition state, the efficiency of type II process decreases in DODAC vesicles and nonexistent in the monolayer assemblies.

4.3.2 Intermolecular Photoinduced Hydrogen Abstraction Reactions

In the presence of alkanes benzophenone derivatives undergo inter-molecular H-abstraction reactions as well. The rapid photoreduction of benzophenone in micellar media with the surfactant chains themselves acting as H-atom donors was mentioned already in earlier discussions on the laser photolysis studies of these compounds. Breslow and co-workers[48] have investigated the photoreduction of ionic benzophenones (alkyl carboxylates and trimethylammonium derivatives) in micelles as conformational probes. Photolysis of the solubilized compounds in various micelles was followed by degradative analysis to determine the distribution of the functionalization positions. The scheme shown below illustrates the methodology:

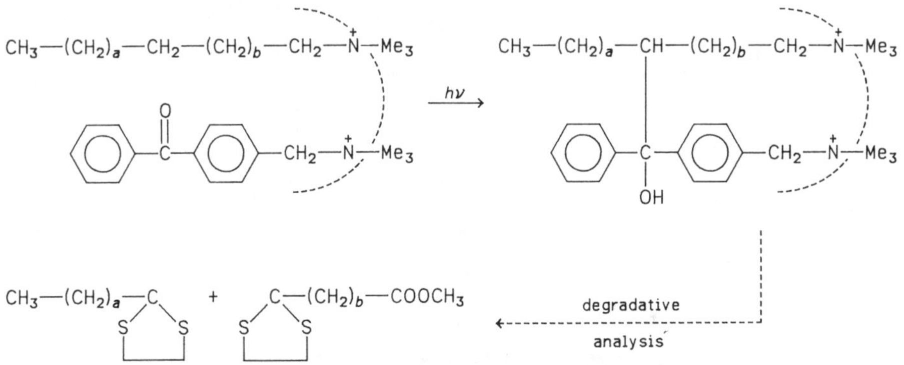

$$CH_3-(CH_2)_a-CH_2-(CH_2)_b-CH_2-\overset{+}{N}-Me_3 \qquad \xrightarrow{h\nu} \qquad CH_3-(CH_2)_a-CH-(CH_2)_b-CH_2-\overset{+}{N}-Me_3$$

Benzophenone trimethylammonium derivative $\longrightarrow$ hydroxyl-substituted product (with OH at the functionalized carbon)

$$CH_3-(CH_2)_a-C\underset{S}{\overset{S}{\diagdown}} \quad + \quad \underset{S}{\overset{S}{\diagup}}C-(CH_2)_b-COOCH_3 \qquad \xleftarrow{\text{degradative analysis}}$$

In anionic and cationic micelles, the attack occurs over the entire chain (C_{5-15}) and the random distribution suggests extensive coiling and folding of the detergent chains. With the CTAB–benzophenone carboxylate, there is more selective attack at the end of the chain C_{15} presumably due to ion-pair-type interactions (as confirmed by similar process even below CMC). The selectivity decreases as the alkyl chain length of the probe is increased.

4.4 Micellar Effects of Photoreactions

4.4.1 Photofragmentation Reactions

Earlier discussions on the photochemistry of dibenzylketones in micelles have shown that, by providing a supercage in dynamic solubilization processes, micellar systems can promote recombination processes in photo-fragmentation reactions. Turro and co-workers have demonstrated this effect on the product distribution in the photolysis of several other alkyl ketones.[49–51] Schemes below summarise the results for benzyl phenylacetate, deoxybenzoin, and 2,4-diphenylpentan-3-ones.

Benzylphenylacetate[49]

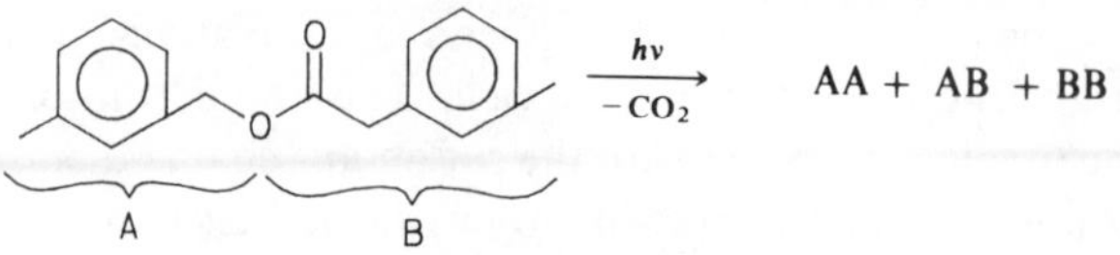

$$\xrightarrow[-CO_2]{h\nu}\ AA + AB + BB$$

i-Propanol	1%	5%	1%
CTAC	1%	50%	1%

Deoxybenzoin[50]

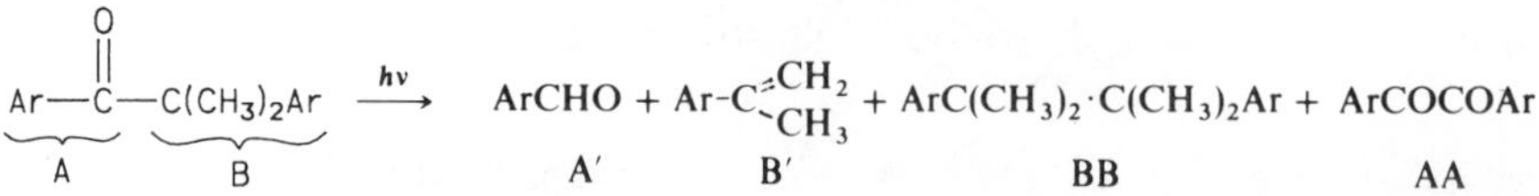

$$\xrightarrow{h\nu}\ ArCHO + Ar-C\overset{CH_2}{\underset{CH_3}{\cdot}} + ArC(CH_3)_2\cdot C(CH_3)_2Ar + ArCOCOAr$$

	A′	B′	BB	AA
Benzene	15%	3.4%	21%	8%
				ArCH(CH₃)₂
CTAC	30%	3.0%	2%	1%

2,4-Diphenylpentan-3-ones[51]

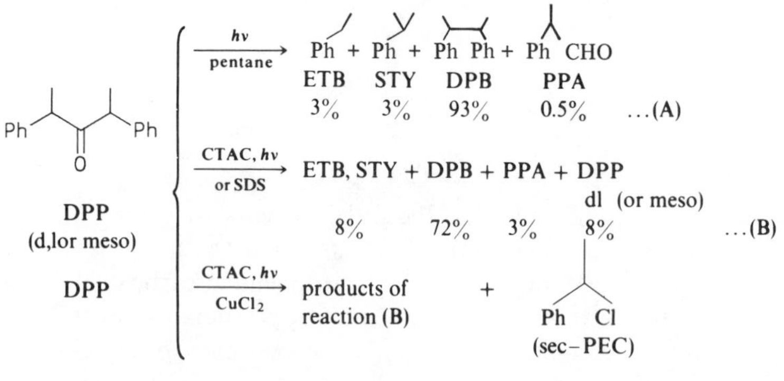

$$\xrightarrow[pentane]{h\nu}$$ Ph + Ph + Ph Ph + Ph CHO
ETB   STY   DPB   PPA
3%   3%   93%   0.5%   …(A)

$$\xrightarrow[or\ SDS]{CTAC, h\nu}$$ ETB, STY + DPB + PPA + DPP
                                                                        dl (or meso)
                                   8%   72%   3%   8%   …(B)

DPP

$$\xrightarrow[CuCl_2]{CTAC, h\nu}$$ products of   +   [structure]
                             reaction (B)   Ph   Cl
                                                          (sec–PEC)

4.4.2 Photocycloaddition Reactions

Photocycloaddition (and photodimerization) reactions are another class of organic photoreactions in which the product distribution and their stereoselectivity are sensitive to the polarity of the medium. Micellar media with appropriate choice of conditions can provide different microenvironments, help orient the substrate molecules, and even control their local concentrations. In photocycloaddition all these can affect their efficiencies, the product distribution, and possibly even the regioselectivity of the photoproducts. de Mayo, Ramamurthy, and others[52–66] have demonstrated such novel micellar effects.

Photodimerization of 3-*n*-butylcyclopent-2-en-1-one occurs very efficiently in aqueous micellar solutions of potassium dodecanoate with the regioselectivity of the addition reversed as compared with the reaction in homogeneous solution. As shown in the scheme below, in the micellar environment the head-to-head dimer is formed almost exclusively, while in benzene, the head to tail dimer is the major product:

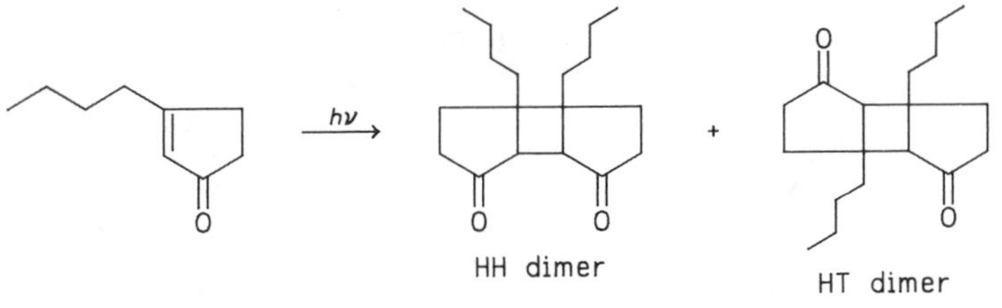

HH dimer

HT dimer

Benzene	98%	2%	(100%)
Potassium Dodecanoate	9%	91%	(1-6%)

In the presence of added cyclohexane, depending on its concentration, we also observes the formation of a cycloaddition product along with the above dimers:

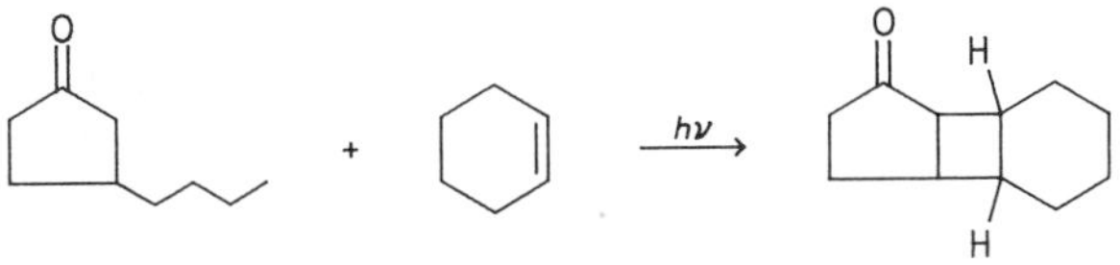

Detailed studies as a function of occupancy number of these reactants in the micelle have established several features of the reaction. The cycloaddition giving the mixed adduct or dimers proceeds selectively with respect to mixed adduct formation if the cyclohexane–enone ratio is 2:1 or greater. A more hindred 1,2-dimethylhexane required a ratio of 4:1 for the same selectivity. Also increasing the enone and alkane concentrations together (for example 0.5 enone, 0.5 alkene per micelle to 2 enone, 2 alkene per micelle) did not greatly alter the observed photodimer to adduct ratio. The latter result illustrates one feature of the micellar solubilization, viz, they serve not only to isolate enones molecules from one another but can concentrate the enone and alkene to locally high levels.

The regiospecificity of the products in micelles is quite dramatic and more substantial than is obtainable in solvents of higher polarity. It is pertinent to ask whether this is due to some orienting effects of the micelle. By careful examination of the regiospecificity of the adducts formed in the photocyclo-addition of the enones with a series of alkenes (1-hexane, 1-octane, and vinylacetate) de Mayo *et al.* have confirmed the origin of those effects to be in the mode of solubilization of the reactant partners in the micelle: the enone is oriented in the micelle with the polar carbonyl group at the micelle–water interface and the hydrophobic butyl tail in the interior. The hydrophobic alkene likewise is incorporated in the nonpolar interior. The close proximity of the partners with the required orientation would account for the high efficiency as well as the regiospecificity of the photoadducts.

The following schemes outline similar examples of the micellar effects controlling the regiospecificity of photocycloaddition products:

Isophorone[64]

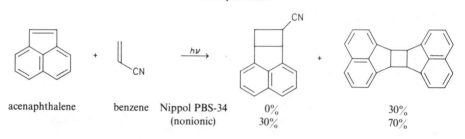

isophorone	syn-HT	anti-HT	anti-HH
C$_6$H$_{12}$	90%	(anti + traces syn)	10%
CH$_3$OH	20%		80%
SDS	5%		95%
(0.02–0.1 M)			

Acenaphthalene[66]

acenaphthalene	benzene	Nippol PBS-34	0%	30%
		(nonionic)	30%	70%

9-Methylanthracene[61]

9-Methylanthracene	HH dimer	HT dimer

Solvent	HT/HH
cyclohexane	1.0
benzene	1.5
methanol	2.0
SDS	1.7–4.0
CTAC	1.7–2.0
CTAB	1.5–2.5

Acenaphthalene[60,65]

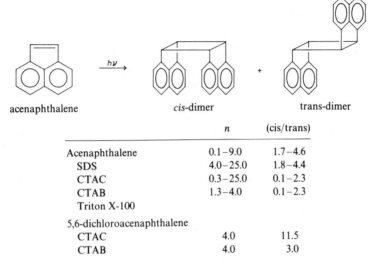

| acenaphthalene | | cis-dimer | | trans-dimer |

	n	(cis/trans)
Acenaphthalene	0.1–9.0	1.7–4.6
SDS	4.0–25.0	1.8–4.4
CTAC	0.3–25.0	0.1–2.3
CTAB	1.3–4.0	0.1–2.3
Triton X-100		
5,6-dichloroacenaphthalene		
CTAC	4.0	11.5
CTAB	4.0	3.0

Coumarin[57,59]

Coumarin   syn-HT   anti-HT

2-Substituted Naphthalene[58]

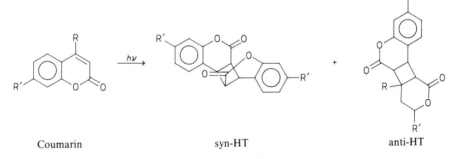

2-substituted naphthalene	trans-dimer	cis-dimer	rearranged ketone (from cis)	cage product (from cis)
R = H benzene	80%	—	—	—
$R_2 = OCH_3$ SDS			60%	—
CTAB			30%	—
CTAC			71%	—

4.4.3 Other Photoreactions in Micelles

There have been several reports on other photoreactions in micelles.[67-74] In a study of the micellar effects on the photobleaching of rhodopsin, Nakanishi et al.[67] found that the yields of cis- and trans-retinals depended on the nature of the surfactants used. The all-trans isomer was the only product in Triton X-100. In micelles of ALO (lauryl/tetradecyl dimethylamine-N-oxide), CTAB, and sodium deoxycholate, relatively large amounts of cis-retinals (9 and 13-retinals) were obtained along with the trans isomer. Photooxidation of ferrocene by CCl_4 and other organic oxidants is normally observable only upon photolysis of the CTTS (charge-transfer-to-solvent) bands around 300 nm but is induced by visible light when the reactants are solubilized in the micelles.[68] Benzene is the only product observable during the photolysis of diphenylmercury in n-hexane or methanol but in SDS micelles biphenyl is formed to as much as 1–5% of the converted products.[69] Photohydrolysis reactions of substrates which are poorly soluble in water, e.g., m-methoxybenzylacetates can be performed readily in micellar media.[70] No special micellar effects, however, are observable.

Giannotti et al.[71] have used the unique hydrophobic environment of solubilized metalloporphyrins in micelles to observe light-induced formation of a reversible dioxygen complex of CoTPP in room temperature solutions. Fouassier and co-workers[72,73] have made a comparative study of the photoinitiation processes in bulk and micellar polymerization. High rates of photopolymerization obtained in the micelles have been rationalized as due to the role of micellar assemblies in providing a low probability of having several radicals simultaneously in the same particle and to prevent the termination of growing polymer chains by radical coupling.

Scaiano and co-workers[74] had partial success in duplicating by photochemical sensitization methods the remarkable micellar effects of influencing the product distribution in thermal dediazoniation reactions. Dediazoniation of p-diazoniobenzyldimethyl-n-alkylammonium bromide leads to >95% of the corresponding phenol in nonmicellar solutions while in micelles the product consists mainly (>95%) of the corresponding aryl bromide. The latter product arises due to the efficient trapping of aryl cations by the locally high concentration of bromide ions. The behavior of the aryldiazonium salts upon direct triplet photosensitized and electron-transfer photosensitized pathways is summarized by

Phenol is formed almost exclusively on direct photolysis in both aqueous homogeneous and DTAB micellar solutions. Presumably due to the electrostatic repulsions of the micellar surface charge, all of the diazonium salts stay in the aqueous phase. However with micelle-bound sensitizers such as phenanthrene, there is a dramatic increase in the yield of PhBr in the sensitized photolysis though the increase is not as pronounced as in thermal reactions. The ready displacement of chloride ions by CN^- and the availability of a locally high concentration of counterions have been exploited by Hautala and Letsinger[75] to achieve high quantum efficiency in photoinduced substitution reaction of 4-methoxy-1-nitronaphthalene with cyanide ions.

References

1. N. J. Turro, "Modern Molecular Photochemistry." Benjamin, New York, 1981.
2. D. O. Cowan and R. L. Drisko, "Elements of organic Photochemistry." Plenum, New York, 1981.
3. W. L. Horspool, "Organic Photochemistry." Wiley (Interscience), New York, 1981.
4. N. J. Turro, *Proc. Natl. Acad. Sci. U.S.A.* **80**, 609 (1983).
5. N. J. Turro, *Pure Appl. Chem.* **53**, 259 (1981).
6. N. J. Turro and B. Kreutler, *Acc. Chem. Res.* **13**, 369 (1980).
7. N. J. Turro and B. Kreutler, *J. Am. Chem. Soc.* **100**, 7432 (1978).
8. R. S. Hutton, H. D. Roth, B. Kreutler, W. R. Cherry, and N. J. Turro, *J. Am. Chem. Soc.* **101**, 2227 (1979).
9. N. J. Turro, B. Kreutler, and D. R. Anderson, *J. Am. Chem. Soc.* **101**, 7435 (1979).
10. B. Kreutler and N. J. Turro, *Chem. Phys. Lett.* **70**, 270 (1980).
11. N. J. Turro, D. R. Anderson, and B. Kreutler, *Tetrahedron Lett.* **21**, 3 (1980).
12. G. F. Lehr and N. J. Turro, *Tetrahedron* **37**, 3411 (1981).
13. N. J. Turro, M.-F. Chow, C.-J. Chung, and B. Kreutler, *J. Am. Chem. Soc.* **103**, 3886 (1981).
14. N. J. Turro, D. R. Anderson, M.-F. Chow, C.-J. Chung, and B. Kreutler, *J. Am. Chem. Soc.* **103**, 3892 (1981).
15. N. J. Turro, M.-F. Chow, C.-J. Chung, Y. Tanimoto, and G. C. Weed, *J. Am. Chem. Soc.* **103**, 4574 (1981).
16. N. J. Turro, C.-J. Chung, G. Jones, and W. G. Becker, *J. Phys. Chem.* **86**, 3677 (1982).
17. N. J. Turro, M. B. Zimmt, and I. R. Gould, *J. Am. Chem. Soc.* **105**, 6347 (1983).
18. N. J. Turro and G. C. Weed, *J. Am. Chem. Soc.* **105**, 1861 (1983).
19. M. B. Zimmt, C. Doubleday, and N. J. Turro, *J. Am. Chem. Soc.* **106**, 3363 (1984).
20. Y. Sakaguchi, S. Nagakura, and H. Hayashi, *Chem. Phys. Lett.* **72**, 420 (1980).
21. Y. Sakaguchi, S. Nagakura, A. Minoh, and H. Hayashi, *Chem. Phys. Lett.* **82**, 213 (1981).
22. A. M. Braun, M. Krieg, N. J. Turro, M. Aikawa, I. R. Gould, G. A. Graf, and P. C. Lee, *J. Am. Chem. Soc.* **103**, 7312 (1981).
23. J. C. Scaiano and E. B. Abuin, *Chem. Phys. Lett.* **81**, 209 (1981).
24. J. C. Scaiano, and J. C. Selwyn, *Can. J. Chem.* **59**, 2368 (1981).
25. Y. Tanimoto and M. Itoh, *Chem. Phys. Lett.* **83**, 626 (1981).
26. Y. Sakaguchi and H. Hayashi, *Chem. Phys. Lett.* **87**, 539 (1982).
27. Y. Sakaguchi, H. Hayashi, and S. Nagakura, *J. Phys. Chem.* **86**, 3177 (1982).
28. D. J. Lougnot, P. Jacques, and J.-P. Fouassier, *J. Photochem.* **19**, 59 (1982).
29. N. J. Turro and P.-C. Lee, *J. Phys. Chem.* **86**, 3367 (1982).

30. J. C. Scaiano, E. B. Abuin, and L. C. Stewart, *J. Am. Chem. Soc.* **104**, 5673 (1982).
31. Y. Tanimoto, M. Takashima, and M. Itoh, *Chem. Phys. Lett.* **100**, 442 (1983).
32. W. J. Leigh and J. C. Scaiano, *J. Am. Chem. Soc.* **105**, 5652 (1983).
33. W. J. Leigh and J. C. Scaiano, *Chem. Phys. Lett.* **96**, 423 (1983).
34. Y. Tanimoto, H. Udagawa, Y. Katsuda, and M. Itoh, *J. Phys. Chem.* **87**, 3976 (1983).
35. Y. Tanimoto, H. Udagawa, and M. Itoh, *J. Phys. Chem.* **87**, 724 (1983).
36. M. C. Thurnauer and D. Meisel, *J. Am. Chem. Soc.* **105**, 3729 (1983).
37. J. C. Scaiano and D. J. Lougnot, *J. Phys. Chem.* **88**, 3379 (1984).
38. D. J. Lougnot and J. C. Scaiano, *J. Photochem.* **26**, 119 (1984).
39. Y. Sakaguchi and H. Hayashi, *Chem. Phys. Lett.* **106**, 420 (1984).
40. Y. Sakaguchi, H. Hayashi, H. Murai, and Y. J. I'Haya, *Chem. Phys. Lett.* **110**, 275 (1984).
41. Y. Sakaguchi and H. Hayashi, *J. Phys. Chem.* **88**, 1437 (1984).
42. Y. Tanimoto, K. Shimizu, and M. Itoh, *Photochem. Photobiol.* **39**, 511 (1984).
43. H. Hayashi and S. Nagakura, *Bull. Chem. Soc. Jpn.* **57**, 322 (1984).
44. N. J. Turro, K.-C. Liu, and M.-F. Chow, *Photochem. Photobiol.* **26**, 413 (1977).
45. P. R. Worsham, D. W. Eaker, and D. G. Whitten, *J. Am. Chem. Soc.* **100**, 7091 (1978).
46. J. R. Winkle, P. R. Worsham, K. S. Schanze, and D. G. Whitten, *J. Am. Chem. Soc.* **105**, 3951 (1973).
47. J. C. Scaiano and J. Selwyn, *Photochem. Photobiol.* **34**, 29 (1981).
48. R. Breslow, S. Kitabatake, and J. Rothbard, *J. Am. Chem. Soc.* **100**, 8156 (1978).
49. N. J. Turro and W. R. Cherry, *J. Am. Chem. Soc.* **100**, 7431 (1978).
50. N. J. Turro and J. Mattay, *J. Am. Chem. Soc.* **103**, 4200 (1981).
51. B. H. Baretz and N. J. Turro, *J. Am. Chem. Soc.* **105**, 1309 (1983).
52. K.-H. Lee and P. de Mayo, *J.C.S. Chem. Commun.* p. 493 (1979).
53. K.-H. Lee and P. de Mayo, *Photochem. Photobiol.* **31**, 311 (1980).
54. P. de Mayo and L. K. Sydnes, *J.C.S. Chem. Commun.* p. 994 (1980).
55. N. Berenjian, P. de Mayo, M.-E. Sturgeon, L. K. Sydnes, and A. C. Weedon, *Can. J. Chem.* **60**, 425 (1982).
56. K. Muthuramu and V. Ramamurthy, *J. Org. Chem.* **47**, 3976 (1982).
57. K. Muthuramu, N. Ramnath, and V. Ramamurthy, *J. Org. Chem.* **48**, 1872 (1983).
58. V. Ramesh and V. Ramamurthy, *J. Org. Chem.* **49**, 536 (1984).
59. N. Ramnath and V. Ramamurthy, *J. Org. Chem.* **49**, 2827 (1984).
60. V. Ramesh and V. Ramamurthy, *J. Photochem.* **24**, 395 (1984).
61. T. Wolff, N. Müller, and G. Von Bünau, *J. Photochem.* **22**, 61 (1983).
62. N. Müller, T. Wolff, and G. Von Bünau, *J. Photochem.* **24**, 37 (1984).
63. T. Wolff and G. Von Bünau, *Ber. Bunsenges. Phys. Chem.* **88**, 1098 (1984).
64. R. Farques, M.-T. Maurette, E. Oliveros, M. Riveriere, and A. Lattes, *Nouv. J. Chim.* **3**, 487 (1979).
65. Y. Nakamura, Y. Imakura, T. Kato, and Y. Morita, *J.C.S. Chem. Commun.* p. 887 (1977).
66. Y. Nakamura, Y. Imakura, and Y. Morita, *Chem. Lett.* p. 965 (1978).
67. W. H. Waddell, A. P. Yudd, and K. Nakanishi, *J. Am. Chem. Soc.* **98**, 238 (1976).
68. D. M. Papsun, J. K. Thomas, and J. A. Labinger, *J. Organomet. Chem.* **208**, C36 (1981).
69. T. Wolff, *J. Photochem.* **18**, 269 (1982).
70. T. Wolff, *J. Photochem.* **18**, 285 (1982).
71. D. A. Lerner, F. R. Richiero, C. Giannotti, and P. Maillard, *J. Photochem* **18**, 193 (1982).
72. J. P. Fouassier and D. Riveriere, *Polym. Chem.* **3**, 29 (1983).
73. J. P. Fouassier and D. J. Lougnot, *Polym. Chem.* **3**, 79 (1983).
74. J. C. Scaiano, N. Kim-Thuan, and W. J. Leigh, *J. Photochem.* **24**, 79 (1984).
75. R. R. Hautala and R. L. Letsinger, *J. Org. Chem.* **36**, 3762 (1971).

Chapter 5

Photoprocesses in Reversed Micelles and in Microemulsions

5.1 General Features of Reversed Micelles and Microemulsions

The association of surfactant molecules to form micellar aggregates is not restricted to aqueous solutions. Micellar aggregates exist in a variety of nonpolar solvents as well. Physical studies of these systems are fairly recent in origin.[1-6] The aggregation occurs in an inverted manner with the polar headgroups collected together in an inner core and the hydrocarbon tails spread out into the nonpolar oil phase. Hence, these aggregates carry the name *reversed or inverted micelles.* The physicochemical properties of these inverted micelles are often as markedly different from those of the normal micelles in aqueous solutions that they are treated separately. A distinguishing feature of these inverted micelles is their ability to solubilize fairly large amounts of water in the inner polar core. The ability to control the size of the water pool and the properties of the water present therein itself is such a unique feature that a good majority of studies in reversed micelles are devoted to the characterization and exploitation of various chemical reactions in these water-pools or pockets.

Closely related to these systems are the microemulsions, which are clear isotropic solution phases composed of three or four components: a surfactant, a hydrocarbon and water, with or without a cosurfactant, usually a long-chain alcohol.[7-12] In a very simplistic picture, these systems can be described as small oil droplets included in larger micellar aggregates which are in turn dispersed in the bulk aqueous phase. While the normal micelles are small aggregates with radii in the range of 10–20 Å, microemulsions are micellelike aggregates of larger dimensions, typically 200–500 Å. Depending on the nature of the constituents and their concentrations, a microemulsion can be of oil-in-water type (O/W microemulsions) or water-in-oil type (W/O emulsions). While the binary systems of surfactants dispersed in a nonpolar solvent are clearly distinguishable as inverted micelles with a third component (water) solubilized in the inner core of these inverted micelles, the ternary system can be considered as a water-in-oil microemulsion as well. In fact, isotropic solutions of Aerosol OT surfactant–alkane–water are synonymously referred to as inverted micelles with solubilized water by some and as water-in-oil microemulsions by others. Actually, in the absence of some well-defined guidelines, currently there is some confusion prevailing in this area of classification and nomenclature of ternary and quarternary systems. For this reason we consider inverted micelles with and without water and microemulsions together in this chapter. In recent years light scattering[13-17c] and small angle neutron scattering[18-19] studies have been profitably employed to characterize the microemulsion systems.

Inverted micelles formed by surfactant Aerosol OT [sodium bis(2-ethylhexyl)sulfosuccinate] in alkanes such as heptane or isooctane and those of dodecyl ammonium propionate (DAP) in cyclohexane or benzene are two of the widely studied, fairly well characterized systems of this kind.[13-19] Aerosol OT, in particular, is of special interest due to its ability to solubilize relatively large amounts of water in a wide variety of nonpolar solvents. The aggregation process is fairly well-characterized with respect to size and shape

TABLE 5.1. Light Scattering Data on the Dependence of Micellar (r_{mic}) and Waterpool Radii (r_{wp}) with the Water Content in Micellar Systems[a]

AOT–isooctane–water			AOT–toluene–water		
w_0[b]	r_{mic} (Å)	r_{wp} (Å)	w_0	r_{mic} (Å)	r_{wp} (Å)
0	15	—	3.36	16.9	7.9
3.7	21	9.7	3.85	18.5	9.5
5.5	23	12.1	4.78	19.2	10.2
7.4	26	14.7	5.83	20.7	11.9
9.2	28	17.6	7.50	22.5	13.5
11.1	32	21.0	8.34	23.0	14.0
18.5	43	32.3	8.97	24.5	15.5
29.6	64	52.6	9.56	4.8	15.8
51.8	111	100.3			

[a] Data for AOT–isooctane–water from Zulauf and Eicke[13] and for AOT–toluene–water from Day et al.[16]

[b] w_0 = molar ratio of [H$_2$O]/[AOT].

at various levels of water addition. Table 5.1 presents data on the micellar and waterpool radii (r_{mic} and r_{wp}) at various levels of water content in isooctane ⁻nd toluene media, determined by light scattering methods. The aggregation

$$^-O_3S-\underset{\underset{CH_2}{|}}{CH}\overset{\diagup CO-O-(CH_2)_4-CH(C_2H_5)-CH_3}{\diagdown CO-O-(CH_2)_4-CH(C_2H_5)-CH_3}\qquad \text{Aerosol OT}$$

takes place in a manner where the aliphatic tails of the surfactant molecules extend into the bulk organic phase and the polar headgroups and the sodium counterions reside in the aqueous core where they are hydrated by the water molecules. The size of the water pools in the center is controlled by the surfactant-to-water ratio $\omega = [H_2O]/[AOT]$ and is very much independent of the nature of the nonpolar solvent and the concentration of the surfactant.

In alkanes such as heptane or decane, the AOT molecules are completely associated in uniformly size assemblies, each consisting of about 23 AOT molecules. The structure of such an aggregate is highly asymmetric and may be best represented by a rounded cylinder with a rod length of 33.4 Å and a cylinder diameter of 23.9 Å. The degree of asymmetry is highly reduced in the presence of water molecules which form a spherical pool in the micellar center. It was pointed out that the main focus of all studies in inverted micelles has been in the characterization of the waterpool (size, polarity, viscosity, etc.) and reactivity of the solutes dissolved in them. The water pool is an interesting model candidate to mimic the water pockets that are often found in various bioaggregates such as proteins, membranes, and mitochondria. Several physical studies have clearly shown that the water core's properties change as a function of the water content, especially at very low water levels. At very low water content, most of the water molecules are utilized to form the hydration sphere of the surfactant head groups or of counterions. At higher water content, there are more free water molecules so that water properties can resemble that of the ordinary/bulk water. From a photochemist's point of view, the interests are how photophysical and photochemical methods can be used to probe this microenvironment and the transition process in the properties of water molecules as a function of water content in Aerosol OT and similar surfactant aggregates.

In three-component systems composed of a surfactant, water, and oil (a long-chain hydrocarbon), we observe different phases depending on the relative concentrations of these components: an oil-in-water (O/W) microemulsion in the high-water content region, an water-in-oil (W/O) microemulsion in the high-oil region and a surfactant phase at the intermediate ranges. With ionic surfactants we often use a fourth component called a cosurfactant, usually a long-chain alcohol. Since a mixture of oil and water separates into two phases in the absence of surfactants, the latter is known as an emulsifier.

Knowledge of the surfactant–oil–water phase diagram is usually necessary in order to be able to select a particular microemulsion composition so that it possesses the necessary stability as well as the desired properties (cf. Fig. 1.4). For four-component systems, we often use a pseudo-three-component phase diagram with one apex of the equilateral triangle to represent a fixed ratio of two components: surfactant/cosurfactant (emulsifier) or surfactant/water.

Microemulsions, thus, are isotropic, monodisperse, and kinetically stable phases. They are clear (transparent or at least translucent) fluids containing a high dispersed phase volume fraction and, hence, an enormous oil–water interfacial area. From earlier discussions, the interrelationship between the inverted micelles and the microemulsions is also obvious. There is wide interest in the characterization and exploitation of novel chemical reactions in the isotropic phases in the high oil and water regions of the phase diagram, namely, in W/O and O/W microemulsions. There are many practical applications of these systems in the oil and detergent industry, especially in the former for usage in tertiary oil recovery from petroleum crudes.

Exchange of Solutes among the Waterpools

There is an increasing realization that solutes present in the waterpools of reversed micelles exchange freely with other waterpools through intermicellar collisions. Most of these have come from rapid mixing or stopped flow kinetic studies of various reactions in reversed micelles. The possibility of such an exchange was first pointed out by Menger et al. in 1973[20] when they observed hydrolysis of an ester to occur on mixing of two Aerosol OT–octane–water solutions, one containing imidazole and the other the ester. In similar experiments, Eicke and co-workers[21,22] used the enhanced fluorescence of Tb^{3+} on complexation with hydroxyphenylacetic acid (HPA) as a probe. Exchange of solutes was monitored during the mixing of two Aerosol OT–isooctane–water solutions, one containing water-solubilized Tb^{3+} ions and the other water-solubilized HPA. The enhanced fluorescence was observed as rapidly as the mixing of two solutions. A transient dimer model for the coalescing water droplets was proposed with the opening of a water channel at the point of collision. The exchange of solutes has been pictured to take place through diffusion through these channels during the collision time $(10^{-10}–10^{-11}$ sec).

From a stopped-flow study of the complexation of Ni^{2+} by murexide, Robinson et al.[15,17] have deduced that the water droplet collision and solute exchange processes are rapid. The rate of collision was assumed to be diffusion controlled $(10^{10} \ M^{-1} \ sec^{-1})$. This would imply breakdown of droplets occurring in about 10^{-8} sec. Thus, solute exchange between the waterpools appears to be a rapid process occurring in a few microseconds at the most. As we shall see later, there have been a few direct estimates of exchange rate constants in this range using fluorescence quenching methods.

5.2 Dynamics of Photoprocesses in Inverse Micelles

5.2.1 Fluorescence Probe Analysis

The nature of the aqueous core of inverted micelles formed by Aerosol OT in alkane–water mixtures has been examined by Thomas and co-workers using the medium-dependent fluorescence of 8-anilino-1-naphthalene-sulfonate (ANS).[23-25] The emission properties were found to be extremely sensitive to the size of the solubilized water cluster. As illustrated in Fig. 5.1, with increasing water content there is an enormous decrease in the fluorescence yield and lifetime, accompanied by red shifts in the emission maximum. The changes, all indicative of increasing polarity of the water core, are very pronounced at very low water content (AOT/water weight ratios of less than 1% to 3% in Aerosol OT–heptane solutions). Interestingly even at the largest water content studied (6 wt.% in 3% AOT), the observed emission maximum and quantum yield are considerably different from those of bulk water.

The pronounced changes at very low water content are interpreted as arising from an involvement of the water molecules in the solvation of sodium counterions, up to a water/Na^+ molar ratio of 6 or about 0.7 wt.% in 3% AOT. The continued observance of the effects even when $H_2O/Na^+ = 14$ suggest that the water molecules not involved in hydration are still involved in

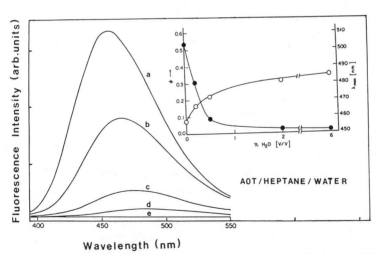

FIG. 5.1. Influence of water addition on the fluorescence maximum and quantum yields of 8-anilino-1-naphthalenesulfonate (ANS) (10^{-6} M) in Aerosol OT/heptane/water solutions. Water content (a) 0, (b) 0.2, (c) 0.5, (d) 2, and (e) 6%. *Insert*: variation in the emission maximum (λ_{max}, nm) and quantum yield (ϕ) as a function of water content for ANS fluorescence. (From Wong *et al.*[23] Copyright 1976 American Chemical Society.)

some bonding interactions (hydrogen bonding or other ion–dipole inter-actions) with the sulfonato or carboxyl group of AOT molecules present in the cavity. NMR relaxation measurements (^{1}H and ^{23}Na) by Wong *et al.*[26] have confirmed similar behavior at low and medium water content. In the largest waterpool (6%) it has been estimated that about 28% of sodium ions are dissociated from the sulfosuccinate group.

Hermann and Schelly[27] have utilized several surfactant derivatives of (C_8, C_{12}, and C_{14}) to detect the aggregation of different alkylammonium carboxylates in pure benzene, cyclohexane, CCl_4, and of Aerosol OT in benzene. In addition to the determination of critical micelle concentration using fluorescence intensities, in a few cases aggregation numbers and association constants have been determined by vapor-pressure osmometry. The onset of aggregation in alkylammonium carboxylates is gradual leading to small aggregation numbers ($n > 3$). With Aerosol OT, the onset is sharper, and aggregation numbers are much larger ($n \geqslant 15$).

The relative intensities of the vibronic bands in the pyrene monomer fluorescence (Ham effect) have also been used to identify the location of pyrene and its derivatives in the inner aqueous core or in the outer nonpolar phase.[28,29] In the absence of water, pyrene sulfonic acid (PSA) is in a hydrocarbon environment in the inverted micelles of cetyldimethylbenzyl-ammonium halides (CDBA) or dodecyldimethylbenzylammonium halides (DDBA) in benzene. However, the addition of water leads to the probe experiencing a more polar, water-disturbed environment. Presumably the probe is located in the interface region with the sulfonato groups lined up with the surfactant groups and the pyrenyl moiety placed in the hydrocarbon phase.

5.2.2 Depolarization of Fluorescence

Monitoring the depolarization of solubilized probe fluorescence often allows the determination of the microviscosity of probe environment. In the inverted micelles of AOT, the fluorescence of rhodamine B is strongly polarised in the absence of water, indicating a very rigid structure of the inner core.[24] Increasing the size of the waterpool decreases the effect. The changes are very pronounced at low water content as with the ANS fluorescence. The calculated rotational correlation times are in the range of few hundred pico-seconds at $[H_2O/Na^+] < 6$ as compared to 3 psec in bulk water. Fig. 5.2 shows the variation of mean anisotropy $\bar{r}$ and rotational correlation times for the two probes, perylene tetracarboxylate and ANS, at various levels of water content in Aerosol OT inverted micelles.

Microviscosity values in the range of 40–70 cP have been determined for the aqueous core of the inverted micelles formed by dodecylammonium

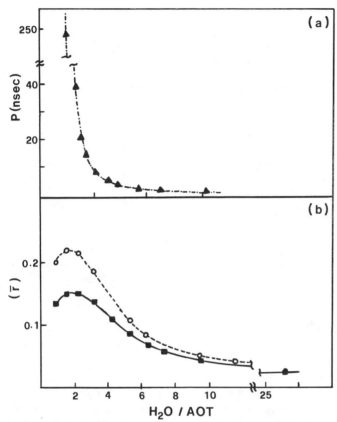

FIG. 5.2. (a) Variation in the internal rotational relaxation time P of perylene carboxylate and (b) mean anisotropy $\bar{r}$ of 2,6-toluidinylnaphthalenesulfonate (TNS) in Aerosol OT/heptane/water solutions. (From Valeur and Keh[31] and Zinsli.[32] Copyright 1979 American Chemical Society.)

propionate(DAP) in cyclohexane at low water content using the polarization of pyranine fluorescence.[30] In nonpolar solvents, each DAP is hydrated by four water molecules. Hence, at water concentrations below this molar ratio of water to surfactant, the inner aqueous core is very viscous.

The effect of water on the volume of the inverted micelles of Aerosol OT has been investigated by Valeur and Keh[14,31] using the fluorescence polarization of perylene tetracarboxylate. Two rotational correlation times have been identified, one due to the micellar rotation and the other due to the probe rotation. The results have confirmed that the micellar volume increases with increasing water content. A value of 3.4 nm has been estimated for the hydrodynamic radius of AOT inverse micelles at $\omega = 11$, a value which

TABLE 5.2. Data on the Microviscosity of the Boundary
Layer (Head Group Region) and on the
Rotational Correlation Times for Dye and
Micelle Rotation Determined via Fluorescence
Anisotropy Measurements in Inverted Micelles
of Aerosol OT–Isooctane–Water [a]

	1-N fluorescence	TNS fluorescence	
w_0	$\eta_{\text{boundary layer}}$ (cP)	$\tau_{\text{rot}}^{\text{micelle}}$ (nsec)	$\tau_{\text{rot}}^{\text{dye}}$ (nsec)
3.7	195 ± 20	5.0	8.4
5.55	85 ± 15	6.5	7.6
7.4	28 ± 8	9.4	4.0
9.25	24 + 4	11.8	3.6
11.1	16 ± 4	17.6	2.9
18.5	8 ± 3	42.7	2.6
29.6	5 ± 2	140.6	2.1
51.8	4 ± 1	733.7	2.1

[a] Data of Zinsli.[32] 1-N = 1-aminonaphthalenesulfonic acid
and TNS = 2-(N-tetradecyl)aminonaphthalenesulfonic acid.

compares well that obtained by photon correlation spectroscopy by Zulauf
and Eicke.[13] For an increase in the ratio ω from 3 to 11, the rotational correlation time decreases by an order of magnitude. In a related work, Zinsli[32]
has examined the fluorescence spectra and time-resolved emission anisotropy
of ANS and TNS in the waterpools of Aerosol OT in isooctane. Table 5.2
presents derived data on the microviscosity for the headgroup region and
rotational correlation times. These studies, in addition to confirming the
presence of the probe in the inner waterpools of increasing size, also
demonstrate the variation of the core fluidity with the water content.

5.2.3 Ground- and Excited-State Acid–Base Equilibria

Given the observation that the properties of water in the aqueous core
differ significantly from that of bulk water, especially at low water content
(concentrations below that necessary to hydrate the surfactant head groups
and counterions), the acidity of the waterpool and proton transfer reactions in
it merit detailed study in their own right. In addition they help in understanding of acid–base catalysis of enzymatic processes in the waterpool.

The efficiency of proton transfer within the waterpools of inverted micelles
of Aerosol OT and cetyltrimethylammonium chloride in $CHCl_3$ or heptane
has been investigated using pyranine (8-hydroxy-1,3,6-pyrenetrisulfonate)
as a probe.[33–35] The scheme below outlines the proton transfer reactions of

interest in the aqueous phase:

$$\text{POH}^* \;+\; \text{H}_2\text{O} \;\underset{pK_a = 0.5}{\overset{-\text{H}^+}{\rightleftharpoons}}\; \text{PO}^{-*} \;+\; \text{H}_3\text{O}^+$$

Emission Maximum: 450–460 nm   510 nm

$h\nu \qquad\qquad h\nu$

$$\text{POH} \;+\; \text{H}_2\text{O} \;\underset{pK_a = 7.2}{\overset{-\text{H}^+}{\rightleftharpoons}}\; \text{PO}^- \;+\; \text{H}_3\text{O}^+$$

Absorption Maximum: 405–410 nm   450 nm

In the inverted micelles with a low water content, i.e., $\omega = 0\text{–}3$ for [H_2O]/[AOT], only the emission from the unionized species POH* was observed. Further addition of water caused POH* emission to decrease with a concomittant increase in PO$^-$* emission. The evolution of the emission spectra revealed that the conversion of POH* to PO$^-$* in the excited state becomes more and more competitive with the fluorescence emission as the water content increases.

To explain the anamolous behavior, Kondo et al.[34] have invoked a biphasic model proposed earlier by Zinsli.[32] In this model two types of water molecules are envisaged: type I bound to the ionic head groups and type II unbound (free) water molecules. The anamolous behavior has been attributed to dye interactions with type I water molecules. Since the involvement of water molecules in the hydration of the surfactant head groups or counterions is now fairly well established, there is no real need to invoke a biphasic model to explain the properties of water in the waterpools of inverted micelles.

The rates of deprotonation and reprotonation (k_1 and k_{-1}) of pyranine has been measured as a function of water content in the inverted micelles of Aerosol OT in heptane.[35] With increasing water content, k_1 increases and k_{-1} decreases and at [H_2O]/[AOT] $\geqslant 12$ both reach values comparable to those observed in bulk, ordinary water, as shown in Fig. 5.3. Values for water activity have been derived using the kinetic constants for proton ejection and Gutman's relation:

$$k_1 = k_1^0(a_{\text{H}_2\text{O}})^n \tag{5.1}$$

where k_1^0 is the rate constant in bulk water and n has the value of 6.9. The water activity decreases rapidly as the ratio [H_2O]/[AOT] decreases below 12.

Using nanosecond time-resolved fluorimetry, the protonation of the singlet excited state of pyrene-1-carboxylate in the waterpool interfaces of (DAP/benzene/water) inverted micelles has been found to be ultrafast ($k_{\text{P}*} = 2 \times 10^{12}$ M^{-1} sec^{-1} as compared to 7.7×10^{10} M^{-1} sec^{-1} in ordinary water).[36] Such ultrafast, nondiffusional proton transfers can occur only if the probe is in the immediate vicinity of the proton donor in the aqueous phase.

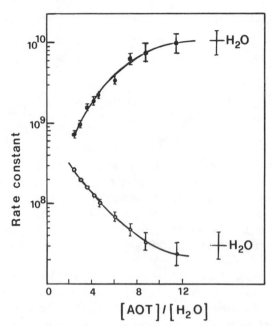

FIG. 5.3. Variation in the rate constants of deprotonation (k_1) and back recombination (k'_{-1}) for pyranine as a function of content in Aerosol OT/heptane/water solutions. (From Bardez et al.[35] Copyright 1984 American Chemical Society.)

Hence, it is likely that protonation occurs in the hydration shell of the carboxylate head group itself.

5.2.4 Fluorescence Quenching

Fluorescence quenching studies in inverse micelles are best discussed in terms of two limiting cases, based on whether or not the probe is associated with the micelle.[37-42] In all cases, at low quencher concentration, the luminescence decay of the probes exhibits a single exponential.

Case I: Probe not associated with the micelle. Neutral arenes such as pyrene or 1-methylnaphthalene due to their hydrophobic character are solubilized in the outer hydrocarbon phase. With these probes with increasing concentration of the quencher in the aqueous phase, the decay curves are still single exponential. The apparent quenching rate constant k_q^{mic} can be obtained from the Stern–Volmer equation. The probes in general feel a time-averaged quencher concentration. The measured k_q values, in general, decrease with increasing water concentration. With increasing water content, the aggregation number and micellar size increase. The micelles become more structural with hydrocarbon chains coming close to each other, and the probability of probe–quencher encounters decreases. Due to the hydration of the surfactant head groups or counterions, the mobility of the quencher molecules in the waterpool is rather inefficient until the maximum hydration number is reached (e.g., six water molecules per AOT or four per DAP surfactant) and increases further on. The following cases illustrate behavior conforming to these situations: Cu^{2+}, and I^- quenching of pyrene in Aerosol OT/heptane/water[38,39] and I^- quenching of 1-methylnaphthalene in DAP/cyclohexane/water.[40]

Case II: Probe associated with the micelle. This case corresponds to probes that usually reside at the interface region, e.g., pyrenesulfonic acid (PSA), pyrenetetrasulfonate(PTSA), napthaleneacetic acid(NAA) and $Ru(bpy)_3^{2+}$. Here the probe emission decay becomes nonexponential at high quencher concentrations. In early experiments, Rodgers and Becker[41] analyzed the quenching of $Ru(bpy)_3^{2+}$ by methylviologen (MV^{2+}) in Aerosol OT/heptane/water system in terms of two exponentials. It is now widely accepted that for ionic solutes present in the waterpools, there is a rapid exchange of solutes between the various waterpools. So we can treat the fluorescence quenching in inverted micelles in terms of a general four-parameter equation along lines similar to those employed in normal micelles.

Let us consider the case where, as in normal micelles the probe does not leave the micelle during its excited state lifetime. Various processes that can contribute to the decay of the probe emission are the following: (1) transition of a quencher from the free water volume (unbound state) to micellar head groups (bound state) described by rate constant k_+; (2) transition of a bound quencher to the free water volume described by the rate constant k_-; (3) exchange of quenchers between the waterpools by intermicellar collisions represented by the rate constant k_{ex} and (4) fluorescence quenching process itself, described by the first-order rate constant k_q:

The processes described are exactly the same as considered earlier in the general schemes for fluorescence quenching in normal micelles (cf. Section 2.4). Hence, we can derive the following expression:

$$F(t) = A_1 \exp\{-A_2 t - A_3[1 - \exp(-A_4 t)]\} \qquad (5.2)$$

On the assumption that $k_q \gg (k_- + k_{ex}[M])$, the parameters A_1, A_2, A_3, A_4

have the following meaning:

$$A_1 = 0 \tag{5.3}$$

$$A_2 = k_0 + \alpha[Q_t], \qquad \text{where} \quad \alpha = \frac{k_+ + k_{ex}K[M]}{1 + K[M]} \tag{5.4}$$

$$A_3 = B[Q_t], \qquad \text{where} \quad \beta = \left(\frac{1}{K} + [M]\right)^{-1} \tag{5.5a}$$

$$A_4 = k_q \tag{5.5b}$$

As in the case of normal micelles, [M] and K stand for the micelle concentration and the partition coefficient, respectively:

$$K = k_+/k_- \tag{5.6}$$

$$[M] = ([Surf] - [CMC])/\text{Aggregation number} \tag{5.7}$$

It is obvious that the processes represented by k_+ and k_- are meaningful only if the water concentration is higher than the minimal amount needed to hydrate the polar head groups or counterions. For water concentrations less than this amount, the preceding equations reduce to

$$A_2 = k_0 + k_{ex}[Q_t] \tag{5.8}$$

$$A_3 = [Q_t]/[M] = \bar{n}, \qquad \text{the mean occupancy} \tag{5.9}$$
$$\text{number of the quencher}$$

$$A_4 = k_q \tag{5.10}$$

Atik and Thomas[38,39] have examined the quenching of pyrenetetrasulfonate fluorescence by Fremy's salt along these lines and have determined $k_{ex} = 1.3 \times 10^7 \ M^{-1} \ \text{sec}^{-1}$ for the quencher exchange in AOT/heptane/water system. Various additives have been found to affect the solute exchange rate. In particular, the effect of benzyl alcohol is notable. On addition of 0.3 M benzyl alcohol, k_{ex} increases to $3.3 \times 10^8 \ M^{-1} \ \text{sec}^{-1}$. These authors have also investigated the quenching of Ru(bpy)$_3^{2+}$ by methylviologen (MV^{2+}) and Fe(CN)$_6^{3-}$. From an analysis similar to that of normal micelles, the distribution of quenchers among the waterpools was found to be better described by a poisson distribution rather than by a geometric distribution

$$P_i = (\bar{n}^i e^{-n})/i! \qquad \text{(poisson)} \tag{5.11}$$

$$G_i = \bar{n}^i/(1 + \bar{n})^{i+1} \qquad \text{(geometric)} \tag{5.12}$$

From an analysis of the fluorescence quenching of naphthaleneacetic acid (NAA) by nitrate and iodide in reverse micelles of DAP in cyclohexane, Geladé and DeSchryver[40] have determined k_{ex} values of 1×10^9 and $8 \times 10^8 \ M^{-1} \ \text{sec}^{-1}$ for NO$_3^-$ and I$^-$, respectively.

5.2.5 Aspects of Excimers and Exciplexes

Excimers

Hunter and Younis[29] have examined the features of excimer formation for pyrenesulfonic acid (PSA) in reversed micelles composed of didodecyldimethylammonium bromide (DDAB)–benzene–water. The excimer of PSA is much less stable in this medium. The energy difference ($v_{0,0}^{mon} - v_{max}^{excimer}$) is only 3800 cm^{-1}, unaffected by water addition, as compared to > 5400 cm^{-1} for pyrene and PSA in water. This is due to the fact that in inverse micelles, the coulombic attractions of the counterions and of water necessitate an orientation of the excimer where the SO$_3^-$ groups are much closer to each other:

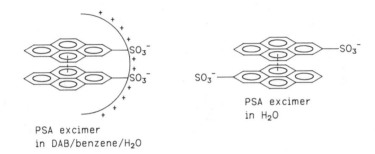

PSA excimer
in DAB/benzene/H$_2$O

PSA excimer
in H$_2$O

In homogeneous solvents they can be placed on opposite ends of the paired molecules.

Attempts to quantitatively interpret the relative intensity ratios of the monomer to excimer fluorescence with increasing probe concentration as due to a statistical distribution of probes among the micelles had only limited success. Consequently it has been proposed that models for excimer formation in inverse micelles must assume that PSA molecules are capable of intermicellar migration in the probe excited-state lifetime. In the light of earlier discussion on the rapid exchange of ionic solutes with other waterpools and their implications in fluorescence quenching, such a possibility of probe exchange is not unreasonable. In the limit of very fast exchange, distributional effects will play only a minor role and the system approximates a homogeneous solvent.

The concentration-dependent behavior of dodecylammonium pyrene butyrate in the excimer formation has also been examined in ethanol and benzene.[43] It is known that simple alkylammonium carboxylates associate in nonpolar solvents to form small aggregates containing 5–10 monomers. For DAPB, the mean aggregation number is less than 3. Consequently the data have been analyzed in terms of an indefinite stepwise association model:

$$\text{monomer} \rightleftharpoons \text{dimer} \rightleftharpoons \text{trimer} \rightleftharpoons \ldots \rightleftharpoons n\text{-mer} \qquad (5.13)$$

Exciplexes

The features and efficiencies of exciplex formation in reversed micellar systems composed of DAP–cyclohexane–water have been explored by DeSchryver and co-workers using several naphthalene derivatives as probes and triethylamine, *m*-dicyanobenzene, and dimethylaniline as quenchers:[44,45]

$$^1Naph^* + Q \rightleftharpoons {}^1[Naph \cdots Q]^* \longrightarrow Naph + Q \qquad (5.14)$$

Their data on the features of exciplexes formed during the triethylamine quenching of several napthalene carboxylic acids are presented in Table 5.3. Compared to homogeneous solvents, the quantum yield for exciplex emission is quite small for all the probes in the reversed micellar system. The difference in the exciplex yields between the two media decreases with increasing chain length and is smallest for 1-methylnaphthalene. While the exciplex emission maximum is identical for 1-methylnaphthalene in both media, for others there is a bathochromic shift. For naphthaleneacetic acid (NAA), the fluorescence quenching rate (K_{sv}) is 4–5 times smaller than that in neat cyclohexane but is 2–3 times larger for others. These data are readily understood in terms of different average localization sites for various probes in the reversed micelles. Hydrophobic probes such as 1-methylnapthalene reside mainly in the bulk apolar phase and the probes with polar head groups are bound to the micelle. For probes attached to a long chain $(Naph—(CH_2)_n—COOH)$, the number of methylene groups determines the average distance between the chromophore and the waterpool. The Φ_E also reflect the radical ion formation as well. For the PSA–DMA pair, only quenching without exciplex

TABLE 5.3. Features of Exciplexes Formed during the Fluorescence Quenching of Various Naphthalene Derivatives by Triethylamine in Neat Cyclohexane and in Reversed Micelles (R.M.) Composed of Dodecylammonium pripionate–Cyclohexane–Water [a,b]

Property	Medium	Probe[c]				
		NAA	NBA	NHA	NDA	1-MeN
$\phi_{ex}(\times 10^{-2})$	C_6H_{12}	7.11	7.57	5.5	4.64	5.5
$\phi_{ex}(\times 10^{-2})$	R.M.	n.a.	0.41	0.87	1.43	3.1
$\lambda_{max}^{exciplex}$ (nm)	C_6H_{12}	429	410	408	407	408
$\lambda_{max}^{excipler}$ (nm)	R.M.	n.a.	427	423	413	408
$K_{sv}(M^{-1})$	C_6H_{12}	91.1	24.4	20.9	20.7	24.8
$K_{sv}(M^{-1})$	R.M.	25	60.4	73.2	65.9	66.4

[a] From Gélade *et al.*[44]
[b] $[H_2O]/[DAP] = 1.325$.
[c] NAA = 2-(1-naphthyl)acetic acid; NBA = 4-(1-naphthyl)butyric acid; NHA = 6-(1-naphthyl)hexanoic acid; NDA = 12-(1-naphthyl)dodecanoic acid, and 1-MeN = 1-methylnapthalene.

formation was observed in accordance with a model where the probe is localized near the waterpool (PSA has limited solubility in cyclohexane).

5.2.6 Aspects of Energy Transfer Processes

Transfer of excitation energy from an arene singlet excited state to rare earth ions such as Tb^{3+} generally occurs over short distances via electron exchange mechanism proposed by Dexter:

$$\text{Arene}^* + TB^{3+} \longrightarrow \text{Arene} + Tb^{3+*} \tag{5.15}$$

In homogeneous solutions, reactions such as (5.15) do not occur due to their low efficiency as compared to other deactivating modes of the arene excited state. Earlier we saw that in micelles, with their ability to organize reactants at a microscopic level it is possible to bring the donor–acceptor pair closer. For example, neutral arenes are placed in the inner core and the acceptor ions at the interface region. Similarly in reversed micelles we can achieve such novel transfers by placing the rare earth ions in the inner waterpool and the arene derivatives at the interface region.

Transfer of the excitation energy between arenecarboxylic acids [pyrene-butyric acid (PBA)[30] and naphthalene acetic acid (NAA)[45]] and Tb^{3+} have been demonstrated in inverse micellar systems of DAP/cyclohexane/water. The process is very efficient especially at low water content when the distance between the donor–acceptor pair is indeed small ($\leqslant 5$ Å). Representative results are presented in Fig. 5.4. In reversed micelles, efficient transfer is obtained at acceptor concentrations an order of magnitude less than that necessary in ionic micelles such as sodium dodecyl sulfate (SDS). Presumably the inner core solubilization in the inversed micelles is superior in bringing together the reactants as compared to the interface solubilization in normal micelles.

Studies using a series of napthalene carboxylic acids, it has been shown that energy transfer efficiency decreases exponentially with the number of $-CH_2-$ units separating the chromophore and the polar head groups bound to the waterpool (cf. insert in Fig. 5.4). For energy transfer by an exchange mechanism this would indicate that methylene chains of the probe and possibly of the detergent as well are most of the time in a folded conformation. Increasing the amount of solubilized water enlarges the water-pool size. This in turn changes the location site of the donor and acceptor species with respect to each other and the efficiency decreases.

Singlet–singlet energy transfer between species dissolved in the bulk nonpolar phase and acceptors present in the inner waterpool has also been the subject of an investigation.[46] An efficient S—S energy transfer has been observed in the toluene–sodium salicylate and fluoranthrene–rose bengal pairs in reversed micellar systems of Manoxol OT/cyclohexane/water.

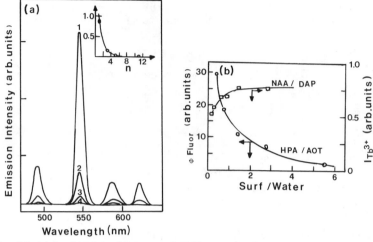

FIG. 5.4. (a) Sensitized luminescence of Tb^{3+} in the presence of napthalene carboxylic acids Naph—(CH$_2$)$_n$—COOH in the waterpools of DAP inverse micelles in cyclohexane: $\lambda_{exit} = 283$ nm. Insert shows the dependence of emission intensity at 546 nm as a function of methylene units between the naphthalene chromophore and the carboxylic acid group. (b) Influence of amount of water added (shown as the ratio [Surf]/[H$_2$O]) on the energy transfer between naphthalene acetic acid (NAA) and Tb^{3+} in DAP/cyclohexane/water compared with energy transfer between hydroxyphenyl acetic acid (HPA) and Tb^{3+} ions in Aerosol OT/heptane/water system. (From Geladé and DeSchryver.[45] Copyright 1984 American Chemical Society.)

(Manoxol OT = dioctylester of sulfosuccinic acid). It has been suggested that the energy transfer occurs by a long-range Förster-type mechanism. The decay of donor fluorescence is exponential. The rate constant increases with increase in acceptor concentration, eventually leveling off in such a way suggestive of multiple occupancy of the micelles by the acceptor molecules.

Quantitative studies have shown that as a result of the rapid exchange of solutes in intermicellar collisions during the time scales of energy transfer, as initial statistical (poisson) distribution of solutes had no influence on the transfer kinetics. Kinetic analysis analogous to that in homogeneous solutions (using total concentrations) is quite adequate in explaining the observed results.

5.2.7 Dynamics of Solvation, Reactivity of Electrons, and Photoionization

The production, solvation, and reactivity of electrons have been studied in the waterpools of inverted micelles using the techniques of flash photolytic ionization and pulse radiolysis techniques.[25,47–51] Studies in Aerosol OT/heptane/water system have revealed that hydrated electron (e_{aq}^{-}) production is observable only above a certain critical size of the waterpool (i.e., above a water content of 1% v/v). The reactivity of e_{aq}^- also depends on the level of water content.[47,49,25]

In the largest size waterpool studied (6% v/v in 3% AOT), the optical absorption spectrum of the solvated electron shows a maximum around 720 nm, similar to that observed in pure water. Decreasing the water content causes a decrease in the yield of e_{aq}^- and a concomittant blue shift in the absorption maximum. In the small water bubbles, the lifetime of the hydrated electron is also significantly shorter (less than 100 nsec). The absorption spectrum of e_{aq}^- in the waterpool is relatively sharper indicating a better ordering of the water molecules in the polar cavities of reversed micelles. Competitive scavenging studies of e_{aq}^- in the water bubbles have shown that the efficiency of electron trapping increases with increasing size of the waterpools.

Measurement of rates of electron attachment to reversed micelles of Aerosol OT/isooctane/water has yielded rates below the diffusion controlled limit at molar ratios $\omega = [H_2O]/[AOT]$ less than 12. At mole ratios of 37 or more (micellar radius > 100 Å), the rates increase to diffusion-controlled limits ($10^{15}\ M^{-1}\ sec^{-1}$). The results clearly indicate that free or non-AOT-bound water molecules are necessary for efficient electron attachment. Laser photolysis studies of electrons produced via photoionization of solutes such as phenothiazine, pyranine, and pyrene derivatives[50] have yielded similar results on the nature of the spectra and reactivity of hydrated electrons. Using the hydrated electron produced by pulse radiolysis as a probe and the nitrate ion as a scavenger, Pileni et al.[48,51] have explored various distribution models to determine the waterpool radii of AOT inverted micelles. Poisson statistics gave results comparable to those obtained via neutron or light scattering.

5.2.8 Photoredox Processes

A major limitation for the practical application of light-induced electron transfer reactions in homogeneous solvents is the fast (diffusion-controlled) reverse electron transfer reaction between the redox products

$$D + A \underset{hv}{\rightleftharpoons} D^+ + A^- \qquad (5.16)$$

The reverse reaction rapidly degrades all the chemical potentials stored in these products. In an earlier chapter (Section 3.5) several cases were discussed where the structural features and solute organization in normal micelles were utilized to achieve net charge separation in photoredox processes. In reversed micelles containing waterpools, the presence of an inner aqueous core, an outer hydrocarbon phase and a charged interface all can provide similar benefits in controlled electron transfer reactions.

Calvin and co-workers[52-54] have demonstrated efficient charge separation during the sensitized electron transfer from EDTA to hexadecylviologen in reversed micelles composed of DAP/toluene/water:

$$\text{EDTA} + \text{HV}^{2+} \xrightarrow[\text{Ru(bpy)}_3^{2+}]{hv, \text{ rev. micelles}} \text{EDTA}_{ox} + \text{HV}^+ \qquad (5.17)$$

The scheme below outlines the overall sequence of reactions. The donor EDTA is confined to the inner aqueous phase while the redox partners Ru(bpy)_3^{2+} and HV^{2+} are located at the hydrocarbon–water interface. A mediating acceptor 1-benzylnicotinamide (BNA^+) is used to mediate the

EDTA → Ru^{2+} —hv→ Ru^{2+*} HV^{2+} ←   BNA$^+$ ←   Dye

EDTA$_{ox}$ ←   — Ru^{3+} ←   → HV$^+$ → BNA   → Dye$_{red}$

waterpool   interface   oil phase (DAP/toluene)

transfer of electrons from the reduced viologen radical to the terminal acceptor, Butter Yellow (4-dimethylaminoazobenzene) that undergoes irreversible reduction.

Using the same phase transfer principle, efficient sensitized photoreduction of several water-soluble viologens using thiophenol as a donor has also been demonstrated by these authors in a complimentary situation (cf. scheme below). The donor thiophenol being amphiphilic is located at the interface, but its oxidized product Ph—S—S—Ph is expected to be extracted into the hydrocarbon phase:

V   S* ←hv— S ←   → Ph—S—S—Ph

V$^-$ ←   → S$^+$ — Ph—SH

waterpool   interface   oil phase (DAP/toluene)

The quenching of the singlet and triplet excited states of MgTPP (magnesium tetraphenylporphyrin) by viologen derivatives [MV^{2+}, PVS, and $(\text{C}_8)_2\text{V}^{2+}$] has been examined in AOT/isooctane inverted micelles as a function of water content.[55] Static quenching is the only process observed with the singlets. Photoinduced electron transfer to viologen occurs from the triplet, and the process is dependent on the water content. Kinetic analysis using poisson statistics along the lines used in fluorescence quenching has been made and rate constants for electron transfer and intermicellar exchange of quenchers have been deduced.

5.2.9 Singlet Oxygen Generation and Its Reactivity

From a biochemical point of view, the distribution of oxygen between the waterpools and hydrophobic alkane phase of the inversed micelles are of much interest. Rodgers et al. have made some studies in this area.[56–58] As with the other reactions occurring in the waterpool, the rate constant for the quenching of singlet oxygen $^1O_2^*$ by azide and tryptophan shows a strong dependence on the amount of water content in Aerosol OT/heptane/water system. Using a kinetic model elaborated earlier in our discussions of singlet oxygen chemistry in normal micelles (Section 3.5), a partition coefficient value of $K_{eq} = 0.11$ has been derived ($k_+ = 2.5 \times 10^5 \text{ sec}^{-1}$ and $k_- = 2.7 \times 10^4 \text{ sec}^{-1}$).

Oxygen-mediated photooxidation reactions of various substrates in reversed micelles formed by DAP in cyclohexane have been investigated in a series of papers by Miyoshi and Tomita.[59-61] Fluorescein and thiazine dyes have been used as sensitizers and diphenylisobenzofuran (DPBF) as the singlet oxygen monitor. Evidence has been obtained for the participation of both type I (radical) and type II (singlet oxygen) mechanisms:

$$S^* + O_2 \longrightarrow S^+ + O_2^- \quad \text{(type I)} \tag{5.18}$$

$$S^* + O_2 \longrightarrow S + {}^1O_2^* \quad \text{(type II)} \tag{5.19}$$

Self-sensitized photooxidation of dye molecules such as 8-anilino-1-naphthalenesulfonic acid in aerated DAP/cyclohexane solutions also have been shown to involve singlet oxygen. CCl_4 enhances the oxidation remarkably.

5.3 Photoprocesses in Microemulsions (Oil-in-Water Type)

Given the information that ionic surfactants aggregate in aqueous solution to form normal micelles and that these normal micelles can solubilize fairly large quantities of long-chain alcohols (cosurfactant) and hydrocarbons (oil) to form larger swollen micelles, it is tempting to visualize the transition of normal micelles to oil-in-water (O/W) microemulsions as proceeding through swollen micelles with or without changes in the overall shape of the micelles:

$$\text{surfactant} \xrightarrow{\text{water}} \begin{array}{c}\text{normal}\\\text{micelles}\end{array} \xrightarrow{\text{alcohols}} \begin{array}{c}\text{swollen}\\\text{micelles}\end{array} \xrightarrow{\text{oil}} \begin{array}{c}\text{oil-in-water}\\\text{microemulsions}\end{array}$$

Currently, a growing number of studies employ fluorescence techniques to obtain information on the nature (aggregation number, size, polarity, viscosity, etc.) and dynamics of the aggregates present at each stage.

5.3.1 Fluorescence Probe Analysis

A versatile chromophore for fluorescence studies is pyrene. Several of its properties have found applications in many areas: probe environment from relative intensities of vibronic bands of monomer fluorescence, mobility by fluorescence quenching, and aggregation number by excimer formation to cite a few. Thomas and co-workers have made some early studies applying these methods to probe O/W and W/O microemulsions.[62-65] Table 5.4 presents a

TABLE 5.4. Rate Constants for Intrawaterpool Quenching k and for Interwaterpool Exchange k_{ex} of Quenchers in the Fluorescence Quenching Reactions in Water-in-Oil Microemulsion Systems [a]

Probe	Quencher	w_0	$\bar{n}$	$k_0(sec^{-1})$ $(\times 10^6)$	$k_q(sec^{-1})$ $(\times 10^6)$	$k_{ex}(M^{-1}\,sec^{-1})$ $(\times 10^8)$	r_{wp} (Å)	Aggregation number
				Potassium oleate/hexadecane/hexanol/water[65]				
$Ru(bpy)_3^{2+}$	$Fe(CN)_6^{3-}$	16	0.52	2.0	2.5	2.2		
$Ru(bpy)_3^{2+}$	$Fe(CN)_6^{3-}$	31	1.00	1.75	0.45	2.8		
$Ru(bpy)_3^{2+}$	$Fe(CN)_6^{3-}$	53	2.05	1.50	0.07	1.1		
				Potassium oleate/dodecane/pentanol/water[65]				
$Ru(bpy)_3^{2+}$	$Fe(CN)_6^{3-}$	8.5	0.31	1.50	25	1.7	13	
$Ru(bpy)_3^{2+}$	$Fe(CN)_6^{3-}$	13	0.87	1.52	15	0.95	17	
$Ru(bpy)_3^{2+}$	$Fe(CN)_6^{3-}$	21	2.2	1.54	6.0	1.1	28	
				CTAB/dodecane/hexanol/water[64]				
$Ru(bpy)_3^{2+}$	MV^{2+}	20.3	0.80	1.75	20		32	225
$Ru(bpy)_3^{2+}$	MV^{2+}	40.6	0.82	1.75	7.5		54	548
$Ru(bpy)_3^{2+}$	MV^{2+}	60.9	3.70	1.72	0.15		81	1250
				Potassium oleate/hexanol/benzene/water[65]				
$Ru(bpy)_3^{2+}$	$Fe(CN)_6^{3-}$	23.5			10	0.28	26	
$Ru(bpy)_3^{2+}$	$Fe(CN)_6^{3-}$	49.5			0.75	0.28	50	

[a] w_0 = molar ratio of $(H_2O)/(AOT)$; $\bar{n}$ = mean occupation number of quencher per waterpool, k_0 = decay rate constant for the unquenched probe* r_{wp} = radius of the waterpool.

comparison of data on the fluorescence and its quenching for pyrene and pyrene butyric acid (PBA) in normal micelles, alcohol-swollen micelles, and O/W microemulsions of anionic and cationic types. During the gradual evolution of the O/W microemulsion from normal micelles, the environment of pyrene becomes more and more hydrophobic. Microviscosity measurements via fluorescence polarization measurements on diphenylhexatriene (DPH) or naphthylamine (NPN) indicate a very fluid interior ($\leqslant 3$–5 cP) for the O/W microemulsions. In general, O/W microemulsions are pictured to consist of an oil droplet center coated with a mixed monolayer of surfactant and cosurfactant molecules with their ionic or polar groups directed outward in contact with the aqueous phase. Photophysical data presented in the table are consistent with this description.

In addition to O/W microemulsions, characteristics of waterpools in W/O microemulsions have also been probed by methods analogous to those used with inverted micelles containing water. Table 5.5 presents a collection of data

TABLE 5.5. Comparison of the Fluorescence and Fluorescence Quenching Properties in Normal Micelles, Alcohol-Swollen Micelles, and in O/W Microemulsions

Property	Ionic micelle (S/H_2O)	Swollen micelles ($S/Co\text{-}S/H_2O$)	O/W microemulsion ($S/Co\text{-}S/Oil/H_2O$)
S = Potassium oleate, Co-S = hexanol, Oil = hexadecane[62]			
Py* fluor. rel. int. (I/III)	1.01		0.71
Microviscosity (DPH Fl. Pol.)			<3 cP
$k_q(\text{Py*} + CH_2I_2)(M^{-1}\,sec^{-1})(\times 10^{10})$	8.0		23.0
$k_q(\text{Py*} + Tl^+)(M^{-1}\,sec^{-1})(\times 10^{10})$	2.0		3.4
$k_q(\text{PBA*} + Tl^+)(M^{-1}\,sec^{-1})(\times 10^{9})$	8.5		3.
S = SDS, Co-S = pentanol, Oil = dodecane[63]			
Py* fluor. rel. int. (I/III)	1.15	0.98	0.85
Microviscosity (DPH Fl. Pol.)	52 cP	0–2 cP	4–5 cP
$k_q(\text{Py*} + Tl^+)(M^{-1}\,sec^{-1})(\times 10^{9})$	7.0	9.7	0.8
$k_q(\text{PBA*} + Tl^+)(M^{-1}\,sec^{-1})(\times 10^{9})$	5.0	2.6	3.0
$[O_2], M(\times 10^{-3})$	1.4		2.5
S = CTAB, Co-S = Hexanol, Oil = Dodecane[64]			
Py* fluor. rel. int. (I/III)	1.30	1.14	0.88
$k_q(\text{Py*} + I^-)(M^{-1}\,sec^{-1})(\times 10^{10})$	5.2	1.60	0.72
$k_q(\text{PSA*} + I^-)(M^{-1}\,sec^{-1})(\times 10^{10})$	13	9.3	7.9
$[O_2], M(\times 10^{-3})$	0.25	0.44	1.50
$k_q(\text{Py*} + Py)(M^{-1}\,sec^{-1})(\times 10^{7})$	0.50	1.40	0.80
Aggregation number N	75	33	135

on the rate constants for the intrawaterpool quenching and for interwaterpool exchange of quenchers in the reactions of $Ru(bpy)_3^{2+*}$ in a few W/O microemulsion systems. A point to note is that in these systems the quenching efficiency is a complex function of the average number of quenchers per waterpool and the encounter frequency between the excited probe and the quenchers in a waterpool (k_q). At fixed surfactant concentration but with increasing water content, the concentration of waterpools decreases as a result of increases in the size of the waterpools and k_q decreases.

An interesting application of the fluorescence probe analysis has been in unraveling the origin of significant differences in the properties of W/O microemulsions made with pentanol as a cosurfactant compared to those prepared with hexanol. Conductance of pentanol system, for example, is much larger than that of their hexanol counterpart. The difference has been attributed to the presence of true W/O microemulsions with hexanol and only molecularly dispersed solutions in the systems composed with pentanol. Photophysical studies [involving quenching of $Ru(bpy)_3^{2+*}$ by $Fe(CN)_6^{3-}$ to determine the size of the waterpools] have established that discrete waterpools exist in both systems but that the exchange of ions between the waterpools is much more rapid in pentanol system compared to that with hexanol!

Zana and co-workers have applied fluorescence probe techniques to monitor changes in the aggregation number during the transition of normal micelles to microemulsions by successive addition of co-surfactants (medium- and long-chain alcohols) and oils (hydrocarbons).[66-69] The determination of aggregation number utilizes an analysis of the fluorescence decay curves of pyrene monomers at various probe concentrations. Quenching of pyrene monomer excited states by ground-state pyrene molecules is known to lead to the formation of excimers (Eq. 5.20):

$$^1Py^* + Py \rightleftharpoons {}^1[Py\cdot\cdot Py]^* \longrightarrow 2Py \tag{5.20}$$

In earlier discussions on micelles (Section 2.6) we outlined a method for the determination of aggregation number N from an analysis of fluorescence decay curve according to.

$$\ln[I(t)/I(0)] = -k_0 t + \bar{n}[\exp(-k_q t) - 1] \tag{5.21}$$

where k_0 is the rate constant for the unquenched pyrene monomer fluorescence, $\bar{n}$, the mean occupancy number of the probe per aggregate and k_q, the effective first order rate constant for intraaggregate excimer formation.

Figure 5.5 illustrates some of the changes observed in the aggregation number of sodium dodecyl sulfate micelles upon successive addition of alcohols and various alkanes determined using the above procedures. In the absence of additives normal micelles of SDS grow only at surfactant

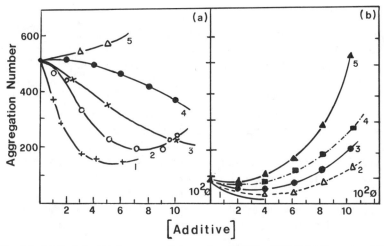

FIG. 5.5. (a) Variation in the aggregation number N of the SDS ($0.5\ M$)–pentanol ($1\ M$) with volume fraction ϕ of various alkane additives: (1) n-hexadecane, (2) n-dodecane, (3) n-decane, (4) n-octane, and (5) n-hexane. (From Lianos et al.[66] Copyright 1982 American Chemical Society.) (b) Variation in the aggregation number N of sodium dodecyl sulfate (SDS) micelles upon addition of 1 M 1-pentanol. [SDS] = 0.1 M(1), 0.2 M(2), 0.3 M(3), 0.4 M(4), and 0.5 M(5). (From Lianos et al.[67] Copyright 1982 American Chemical Society.)

concentrations above 0.3 M. This increase however is relatively small even at 0.5 M. Addition of cosurfactant 1-pentanol, for example, leads to a growth (swelling of the micelles), the effect being very pronounced at high SDS concentrations. For a given surfactant concentration, the relative increase is in the order: 1-pentanol > 3-methyl-butanol $\geqslant$ 2-pentanol. For subsequent addition of alkane, further changes in N (increase or decrease if any) depend on their chemical structure (linear or acylic, chain length, etc.). For linear-chain alkanes the decrease in N is very pronounced for chain length $n \geqslant 12$.

Along with the aggregation number, the rate constant for intermolecular formation of excimers and microviscosity η for the probe environment has also been determined using an intramolecular excimer-forming probe, dipyrenylpropane. The relative intensity changes of the monomer to excimer fluorescence referenced to a calibration curve determined in a suitable solvent mixture (hexadecane–paraffin oil mixture, for example) provided the η values. The scope and limitation of the method has been discussed earlier (Section 2.6). The measured η value for SDS normal micelles is about 21 cP, and it drops significantly to values in the range of 5 cP on addition of alcohols and alkanes. The very fluid nature of the inner core oil droplets has also been noted from fluorescence polarization measurements using DPH, NPN or 2-methylanthracene as probes.[70,71] From an overall consideration of the changes observed in these properties and geometric factors, models for shape changes in the transition of normal micelles to O/W microemulsions have been deduced.

5.3.2 Photoredox Reactions

Apart from the characterization of microemulsions using photophysical methods, there is a growing interest to use these systems in photoredox reactions of potential application in the field of photochemical conversion of solar energy.[72-82] In addition to providing distinct hydrophobic and hydrophilic domains with an associated interface, there are other advantages of microemulsions over micelles. The larger dimensions allow utilization of larger molecules. By varying the composition of the microemulsion, the size of the oil/water droplets in the core can be varied systematically. Table 5.6 provides a listing of various photoredox systems that have been examined in different anionic, cationic, and neutral microemulsion systems. The focus of our discussions would be on the kinetics of forward and reverse electron transfers and the efficiency of charge separation as they differ from that observed in homogeneous solvents and other forms of microheterogeneous systems:

$$S^* + D \rightleftharpoons S^- + D^+ \quad \text{(reductive quenching)} \qquad (5.22)$$

$$S^* + A \rightleftharpoons S^+ + A^- \quad \text{(oxidative quenching)} \qquad (5.23)$$

Mackay and co-workers were the first to explore the utility of microemulsions as media for photochemical reactions using sensitizers such as chlorophyll a.[73-75] An attractive feature of these systems is the inefficient

TABLE 5.6. Photoredox Reactions Studied in Microemulsion Media

Sensitizer	Electron donor/acceptor	Microemulsion	Reference
Chlorophyll a	Ascorbate, duroquinone	SCS/1-pentanol/mineral oil/water	77
Chlorophyll a	Ascorbate, methyl red	SCS/1-pentanol/mineral oil/water	74
Chlorophyll a and Pheophytin a	Ascorbate, methyl red	(a) SCS/1-pentanol/ mineral oil/water	73
		(b) CTAB/1-butanol/ hexadecane/water	73
		(c) Brij 96/1-butanol hexadecane/water	73
Duroquinone	Diphenylamine	(a) SCS/1-pentanol	76
N-Me PTH	Methyl viologen	hexadecane/water	
		(b) potassium oleate/cyclohexanol benzene/water	
Chlorophyll a	Methyl viologen	SCS/1-pentanol/hexadecane/water	77
Pyrene, PBA, PSA	Dimethylaniline		80
Duroquinone	Dimethylnaphthalene, TTF	SCS/1-pentanol/hexadecane/water	79
MgTPP	MgTPP	SCS/2-pentanol/hexadecane/water	82
Ru(bpy)$_3^{2+}$	Alkyl viologen (C_n—V^{2+})	DAP/toluene/C_n—V^{2+}/water	54

photodegradation and pheophytination of chlorophyll and related molecules. The quenching of chlorophyll triplet states by donors such as ascorbate occurs at the interface region. Kinetic studies on the rate of sensitized photoreduction of dyes such as methyl red and crystal violet have established that the actual mechanisms in these media are different from those found in homogeneous solvents. In place of an oxidative redox cycle observed in homogeneous solvents, the proposed mechanisms involve formation of a triplet exciplex between chlorophyll and methyl red,

$$^3Chl^* + MRed \longrightarrow Chl^+ + MRed^- \qquad (5.24)$$

$$Chl^+ + Asc \longrightarrow Chl + Asc^+ \qquad (5.25)$$

which in turn is reduced by ascorbate:

$$^3Chl^* + MRed \longrightarrow {}^3[Chl^+ \cdot \cdot MRed^-]^* \xrightarrow{Asc} Chl + Asc^+ + MRed^- \qquad (5.26)$$

The quantum yield for the photoreduction in different ionic microemulsions follows the order: anionic > neutral > cationic. An inverse order was obtained for the initial increase in yield with ascorbate concentration. In both cases the results are ascribed primarily to the effect of surface charge on the local concentration of ionic species (H^+ and ascorbate anions).

In studies of photoredox reactions of duroquinone triplets

$$^3DQ^* + DPA \rightleftharpoons DQ^- + DPA^+ \qquad (5.27)$$

$$^3DQ^* + TTF \rightleftharpoons DQ^- + TTF^+ \qquad (5.28)$$

Grätzel and co-workers have observed formation of dimer cation radicals of the electron donors in the microemulsion media.[79] The formation of dimer cations such as those of diphenyl amine (DPA), dimethylnaphthalene (DMN), and tetrathiafulvalene (TTF) [Eq. (5.29)] is unique in itself, since such processes are normally observed only during the low-temperature radiolysis in glass matrices. For DMN, a dimerisation constant of

$$D^+ + D \rightarrow D_2^+ \qquad (D = DPA, DMN, TTF,...) \qquad (5.29)$$

$K = 476\ M^{-1}$ has also been evaluated. Photoinduced electron transfer from TTF to triplet states of water-soluble porphyrins such as $ZnTPPS^{4-}$ [zinc tetrakis (sulfonatophenyl)porphyrin] does not occur in homogeneous solvents such as methanol, but substantial yields of redox products can be observed in microemulsion media.

5.3.3 Other Photoreactions

Microemulsions preferentially containing anionic surfactants have been found to be useful in differentiating between the two activated forms of oxygen, singlet oxygen ($^1O_2^*$) and superoxide (O_2^-), often produced in the oxygen quenching of molecular excited states[83] [cf. Eq. (5.19) and (5.20)]. The method is based on the known ability of microheterogeneous systems containing charged interfaces to reject/repel ionic species of same charge. In microemulsions, singlet oxygen diffuses through the interface and can be analyzed using a suitable trapping agent in the hydrophobic section whereas O_2^- may be analysed in the bulk aqueous phase.

Self-sensitized photooxidation/photodegration of porphyrins in the presence of molecular oxygen is a reaction of much interest in photobiology. The process often involves singlet oxygen ($^1O_2^*$) and superoxide to a much lesser extent. Krieg and Whitten[84,85] have found microemulsions to be a useful medium in helping understand the photoprocesses involved in natural and model membrane systems. In erythrocite ghosts, for example, photolysis of Protoporphyrin IX does not yield the normal photooxidation products but leads to photobleaching involving oxygen. Photolysis in an O/W microemulsion (DTAB/1-butanol/benzene/water) gave the same products as observed in neat benzene. Addition of aminoacids such as histidine, tryptophan, or methonine to the microemulsion caused enhanced photodegradation of the porphyrin with small amounts of photooxidation products – behavior similar to that observed in the ghosts. It has been proposed that porphyrins do sensitize the singlet oxygen production, but they are rapidly scavenged by substrates such as amino acids present in the biological systems.

Photodimerization of isophorone has been examined in microemulsions.[86] In solvents, three dimers (head–tail syn and anti, head–head anti dimers) can be formed.

isophorone   HT syn   HT anti   HH anti

In microemulsions (as in ionic micelles) the photodimerization occurs more rapidly with higher yields. The amount of HH isomer is formed in substantial yields (50–90%). Interestingly, even when the continuous phase is cyclohexanic, a minimum of 50% HH isomer was formed, although isophorone is soluble in cyclohexane but not in water. (HH isomer formation generally is favored in polar solvents, e.g., methanol 80% as compared to 10% in neat cyclohexane.) The results have been interpreted as due to a preferential reactivity of isophorone at the interface. A phenomenon of continuous extraction of reactants and products occurs during the photochemical process allowing higher conversion rates and yields in microemulsion media.

5.4 Photochemistry in Polymerized Microemulsions[87-90]

By incorporating a polymerizable substrate, e.g., styrene and acrylamide, in either the dispersed or the continuous phase, polymerization can be induced in a microemulsion chemically or with UV light. A few exploratory studies are available reporting syntheses of normal, inverse microlatex particles in microemulsions of oil-in-water[87-92] and water-in-oil[93-96] type. Unlike the micelles or microemulsions, polymerized microemulsions are quite stable to dilution and remain intact over a wide range of concentrations. Photochemical studies are beginning to provide information on the nature of the polymer, reactivity, dynamics of the processes in the polymer core and of the surfactant phase in the outside.

Candau and co-workers prepared stable inverse microlatex particles of water-swollen polyacrylamide particles stabilized by Aerosol OT and dispersed in toluene.[93-95] The dimensions of the micellar particles were found to increase with the water or acrylamide content whereas the size distribution remained unaffected. The final dispersions have been shown to contain two species of particles in equilibrium: polymer particles ($d \simeq 400$ Å) and a narrow size distribution of small Aerosol OT particles ($d \simeq 30$ Å). In a related structural study using proton nuclear magnetic resonance,[96] polymerized microemulsions composed of toluene/water/surfactant graft copolymer of polystyrene–polyethyleneglycol/2-propanol were found to contain large, stable water-swollen micelles dispersed in the organic phase. The waterpools in these inverse micelles are very large (with up to 13% water content).

Thomas and co-workers polymerized styrene in CTAB/hexanol/water microemulsion (O/W).[87-92] These polymerized microemulsions provide two distinct solubilization sites for the guest molecules: in the outer surfactant coating (or mantle) or in the inner polymerized styrene core. Quenching studies with I^- and O_2 indicate that the excited state of probes located in the former are readily quenched while that in the polymer relatively unaffected.

The nature of the styrene polymer core in such polymerized microemulsions (cetylpyridinium chloride/styrene–divinylbenzene/water) has been probed using the fluorescence of styrene itself. Polystyrene in cyclohexane exhibits mainly monomer emission (285 nm) with some excimer (325 nm) while polystyrene films show only excimer emission. In P-ME, the emission resembles that of the film rather than that in solution. In CPyC-microemulsions, the pyridinium head group acts as a quencher of the pyrene excited states that are in the outer surfactant mantle. Hence, solubilized pyrene that is in the inner rigid core alone emit. Thus, in these systems, energy and electron reactions can be studied exclusively in the inner core. Studies have shown that the energy transfer to acceptors such as perylene occur by the Förster mechanism. The electron transfer to species such as nitrobenzene takes place via tunnelling.

References

1. H. F. Eicke, *Top. Curr. Chem.* **87**, 86 (1980).
2. A. Kitahara, *Adv. Colloid Interface Sci.* **12**, 109 (1980).
3. K. L. Mittal and E. J. Fendler, eds., "Solution Behaviour of Surfactants," Vol. 2. Plenum, New York, 1981.
4. P. Luisi, ed., "Reversed Micelles: Biological and Technological Relevance of Amphiphilic Structures in Apolar Media." Plenum, New York, 1984.
5. J. H. Fendler, *Acc. Chem. Res.* **9**, 153 (1976).
6. A. H. Kertes and H. Gutman, *Surf. Colloid Sci.* **8**, 194 (1976).
7. K. Shinoda and S. Friberg, *Adv. Colloid Interface Sci.* **4**, 281 (1975).
8. R. A. Mackay, *Adv. Colloid Interface Sci.* **15**, 131 (1981).
9. K. L. Mittal, ed., "Micellisation, Solubilisation and Microemulsions." Plenum, New York, 1977.
10. (a) K. L. Mittal, ed., "Solution Chemistry of Surfactants," Vol. 1,2. Plenum, New York, 1979. (b) K. L. Mittal and B. Lindman, eds., "Solution Chemistry of Surfactants," Vol. 3. Plenum, New York, 1984.
11. L. M. Prince, ed., "Microemulsions." Academic Press, New York, 1977.
12. I. M. Robb, ed., "Microemulsions." Plenum, New York, 1982.
13. M. Zulauf and H. F. Eicke, *J. Phys. Chem.* **83**, 480 (1979).
14. E. Keh and B. Valeur, *J. Colloid Interface Sci.* **79**, 465 (1981).
15. B. H. Robinson, D. C. Steytler, and R. D. Tack, *J.C.S. Faraday I* **75**, 481 (1979).
16. R. A. Day, B. H. Robinson, J. H. R. Clark, and J. V. Doherty, *J.C.S. Faraday I* **75**, 132 (1979).
17. (a) P. D. I. Fletcher and B. H. Robinson, *Ber. Bunsenges. Phys. Chem.* **85**, 863 (1981). (b) P. D. I. Fletcher, M. F. Galal, and B. H. Robinson, *J.C.S. Faraday I* **80**, 3307 (1984). (c) P. D. I. Fletcher, A. M. Howe, N. M. Perrins, B. H. Robinson, C. Toprakcioglu, and J. C. Dore, in ref. 10, p. 1745; in ref. 4, p. 68. (d) A. M. Cazabet and D. Langevin, *J. Chem. Phys.* **74**, 3148 (1981); *J. Colloid Interface Sci.* **73**, 1 (1980). (e) D. Roux, A. M. Bellocq, and P. Botherel, in ref. 10B, p. 1843. (f) D. J. Cebula, R. H. Ottewill, J. Ralston, and P. N. Puey, *J.C.S. Faraday I* **77**, 2585 (1981).
18. B. H. Robinson, C. Toprakcioglu, J. C. Dore, and P. Chieux, *J.C.S. Faraday I* **80**, 13 (1984).
19. C. Toprakcioglu, J. C. Dore, B. H. Robinson, A. Howe, and P. Chieux, *J.C.S. Faraday I* **80**, 413 (1984).
20. F. Menger, J. Donohue, and R. Williams, *J. Am. Chem. Soc.* **95**, 286 (1973).
21. H. F. Eicke, J. C. Shepherd, and A. Steinemann, *J. Colloid Interface Sci.* **56**, 168 (1976).
22. H. F. Eicke and P. Zinsli, *J. Colloid Interface Sci.* **65**, 131 (1978).
23. M. Wong, M. Grätzel, and J. K. Thomas, *Chem. Phys. Lett.* **30**, 329 (1975).
24. M. Wong, M. Grätzel, and J. K. Thomas, *J. Am. Chem. Soc.* **98**, 2391 (1976).
25. M. Wong and J. K. Thomas, ref. 3, p. 647.
26. M. Wong, J. K. Thomas, and T. Nowak, *J. Am. Chem. Soc.* **99**, 4730 (1977).
27. U. Hermann and Z. A. Schelly, *J. Am. Chem. Soc.* **101**, 2665 (1979).
28. U. K. A. Klein, D. J. Miller, and M. Hauser, *Spectrochim. Acta, Part A* **32A**, 379 (1978).
29. T. F. Hunter and A. I. Younis, *J.C.S. Faraday I* **75**, 550 (1979).
30. G. D. Correll, R. N. Cheser, F. Nome, and J. H. Fendler, *J. Am. Chem. Soc.* **100**, 1254 (1978).
31. B. Valeur and E. Keh, *J. Phys. Chem.* **83**, 3305 (1979).
32. P. E. Zinsli, *J. Phys. Chem.* **83**, 3223 (1979).
33. U. Klein and M. Hauser, *Z. Phys. Chem. (Wiesbaden)* **90**, 215 (1974).
34. H. Kondo, I. Miwa, and J. Sunamoto, *J. Phys. Chem.* **86**, 4826 (1982).
35. E. Bardez, B.-T. Goguillon, E. Keh, and B. Valeur, *J. Phys. Chem.* **88**, 1909 (1984).
36. J. R. Escabi-Perez and J. H. Fendler, *J. Am. Chem. Soc.* **100**, 2234 (1978).

37. D. J. Miller, U. K. A. Klein, and M. Hauser, *J.C.S. Faraday I* **73**, 1654 (1977).
38. S. S. Atik and J. K. Thomas, *Chem. Phys. Lett.* **79**, 351 (1981).
39. S. S. Atik and J. K. Thomas, *J. Am. Chem. Soc.* **103**, 3543 (1981).
40. E. Geladé and F. C. DeSchryver, *J. Photochem.* **18**, 223 (1982).
41. M. A. J. Rodgers and J. C. Becker, *J. Phys. Chem.* **84**, 2762 (1980).
42. N. J. Bridge and P. D. I. Fletcher, *J.C.S. Faraday I* **79**, 2161 (1983).
43. K. Tsuiji, J. Sunamoto, F. Nome, and J. H. Fendler, *J. Phys. Chem.* **82**, 423 (1978).
44. E. Geladé, N. Boens, and F. C. DeSchryver, *J. Am. Chem. Soc.* **104**, 6288 (1982).
45. E. Geladé and F. C. DeSchryver, *J. Am. Chem. Soc.* **106**, 5871 (1984).
46. M. A. J. Rodgers and H. D. Burrows, *Chem. Phys. Lett.* **66**, 238 (1979).
47. J. K. Thomas, F. Grieser, and M. Wong, *Ber. Bunsenges. Phys. Chem.* **82**, 937 (1978).
48. M.-P. Pileni, B. Hickel, C. Ferradini, and J. Puchaeult, *Chem. Phys. Lett.* **92**, 308 (1979).
49. G. Bakale, G. Beck, and J. K. Thomas, *J. Phys. Chem.* **85**, 1062 (1981).
50. V. Calvo-Perez, G. S. Beddard, and J. H. Fendler, *J. Phys. Chem.* **85**, 2316 (1981).
51. M.-P. Pileni, P. Brochette, B. Hickel, and B. Lerabours, *J. Colloid Interface Sci.* **98**, 549 (1984).
52. I. Willner, W. E. Ford, J. W. Otvos, and M. Calvin, *Nature (London)* **280**, 823 (1979).
53. I. Willner, J. W. Otvos, and M. Calvin, ref. p. 1237.
54. D. Mandler, Y. Degani, and I. Willner, *J. Phys. Chem.* **88**, 4366 (1984).
55. J. M. Furois, P. Brochette, and M.-P. Pileni, *J. Colloid Interface Sci.* **97**, 552 (1984).
56. I. B. C. Matheson and M. A. J. Rodgers, *J. Phys. Chem.* **86**, 884 (1982).
57. P. C. Lee and M. A. J. Rodgers, *J. Phys. Chem.* **87**, 4894 (1983).
58. M. A. J. Rodgers and P. C. Lee, *J. Phys. Chem.* **88**, 3480 (1984).
59. N. Miyoshi and G. Tomita, *Z. Naturforsch., B* **34B**, 339, 1552 (1979).
60. N. Miyoshi and G. Tomita, *Z. Naturforsch., B* **35B**, 107, 731, 741, 1444 (1980).
61. N. Miyoshi and G. Tomita, *Photochem. Photobiol.* **34**, 417 (1981).
62. S. J. Gregoritch and J. K. Thomas, *J. Phys. Chem.* **84**, 1491 (1980).
63. M. Almgren, F. Grieser, and J. K. Thomas, *J. Am. Chem. Soc.* **102**, 3188 (1980).
64. S. S. Atik and J. K. Thomas, *J. Am. Chem. Soc.* **103**, 4367 (1981).
65. S. S. Atik and J. K. Thomas, *J. Phys. Chem.* **85**, 3921 (1981); *J. Am. Chem. Soc.* **103**, 7404 (1981).
66. P. Lianos, J. Lang, C. Strazielle, and R. Zana, *J. Phys. Chem.* **86**, 1019 (1982).
67. P. Lianos, J. Lang, and R. Zana, *J. Phys. Chem.* **86**, 4809 (1982).
68. P. Lianos, J. Lang, J. Strum, and R. Zana, *J. Phys. Chem.* **88**, 819 (1984).
69. M. Almgren and J. E. Lofroth, *J. Colloid Interface Sci.* **81**, 486 (1981).
70. Y. Tricot, J. Kiwi, W. Niederberger, and M. Grätzel, *J. Phys. Chem.* **85**, 862 (1981).
71. N. S. Dixit and R. A. Mackay, *J. Am. Chem. Soc.* **105**, 2928 (1983).
72. K. Kano, T. Yamaguchi, and T. Ogawa, *J. Phys. Chem.* **88**, 793 (1984).
73. C. E. Jones and R. A. Mackay, *J. Phys. Chem.* **82**, 63 (1978).
74. C. E. Jones, C. A. Jones, and R. A. Mackay, *J. Phys. Chem.* **83**, 805 (1979).
75. C. A. Jones, L. E. Weaner, and R. A. Mackay, *J. Phys. Chem.* **84**, 1495 (1980).
76. J. Kiwi and M. Grätzel, *J. Am. Chem. Soc.* **100**, 6314 (1978).
77. J. Kiwi and M. Grätzel, *J. Phys. Chem.* **84**, 1503 (1980).
78. C. K. Grätzel and M. Grätzel, *J. Phys. Chem.* **86**, 2710 (1982).
79. C. K. Grätzel, A. Kira, M. Jirousek, and M. Grätzel, *J. Phys. Chem.* **87**, 3983 (1983).
80. S. S. Atik and J. K. Thomas, *J. Am. Chem. Soc.* **103**, 3550 (1981).
81. M.-P. Pileni, *Chem. Phys. Lett.* **75**, 540 (1980).
82. M.-P. Pileni and S. Chevalier, *J. Colloid Interface Sci.* **92**, 326 (1983).
83. M.-T. Maurette, E. Oliveros, P. P. Infelta, and A. M. Braun, *Helv. Chim. Acta* **66**, 722 (1983).
84. M. Krieg and D. G. Whitten, *J. Am. Chem. Soc.* **106**, 2477 (1984).
85. M. Krieg and D. G. Whitten, *J. Photochem.* **25**, 235 (1984).

86. R. S. Farques, M.-T. Maurette, E. Oliveros, M. Riveriere, and A. Lattes, *J. Photochem.* **18,** 101 (1982).
87. S. S. Atik and J. K. Thomas, *J. Am. Chem. Soc.* **103,** 4279 (1981).
88. S. S. Atik and J. K. Thomas, *J. Am. Chem. Soc.* **104,** 5868 (1982).
89. S. S. Atik and J. K. Thomas, *J. Am. Chem. Soc.* **105,** 4515 (1983).
90. P. Lianos, *J. Phys. Chem.* **86,** 1935 (1982).
91. J. O. Stoffer and T. Boue, *J. Disp. Sci. Tech.* **1,** 37 (1980); *J. Polym. Sci. Part A-1* **18,** 2641 (1980).
92. L. M. Gau, C. H. Chew, and S. Friberg, *J. Macromol. Sci. Chem. Part A-1* **A19,** 739 (1983).
93. F. Ballet and F. Candau, *J. Polym. Sci. Part A-1* **21,** 155 (1983).
94. F. Candau, Y. S. Leong, G. Pouyet and S. Candau, *J. Colloid Interface Sci.* **101,** 167 (1984).
95. Y. S. Leong and F. Candau, *J. Phys. Chem.* **86,** 2269 (1982).
96. Y. S. Leong, G. Reiss, and F. Candau, *J. Chim. Phys.* **78,** 279 (1981).

Chapter 6

Photoprocesses in Lipids, Surfactant Vesicles, and Liposomes

Apart from the normal micelles, vesicles and liposomes formed by phospholipid molecules are the other molecular aggregated systems that have been widely studied. These are quasi-spherical aggregates much larger in size than those formed by simple surfactants. Though there is not any clear-cut distinction between vesicles and liposomes, for our discussions we will use the term *vesicles* to describe the single bilayer assemblies (unilamellar vesicles, ULV) and *liposomes* to refer to multibilayer systems (multilamellar vesicles, MLV). Most of our discussions deal with vesicles and liposomes formed by synthetic and naturally derived phospholipid molecules of biological interest. Vesicles formed by surfactants with single, double, or triple alkyl chains (surfactant vesicles) and their polymerized versions with cross-linked chains (polymerized vesicles) have also been made in recent years. Toward the end of this chapter, we briefly discuss these systems.

6.1 General Features of Lipid Vesicles and Liposomes[1-12]

Vesicles are quasi-spherical multimolecular aggregates formed by phospholipids and related molecules in which a lipid bilayer separates an inner aqueous compartment from the bulk aqueous phase. As illustrated in Fig. 6.1 the structure of a vesicle is such that the polar head groups of the lipids are exposed to the aqueous phase on both sides of the bilayer. The hydrocarbon chains align themselves in the inner core forming a bilayer. In liposomes we

173

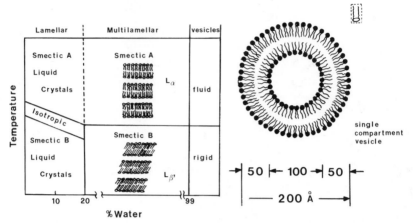

FIG. 6.1. Schematic diagram showing various lipid phases in a phospholipid–water phase diagram. Also shown schematically is the arrangement of lipid chains and headgroups in single compartment vesicles. (From Fendler.[12] Reprinted by permission of John Wiley & Sons, Inc.)

have several such bilayers sandwiched, one on top of the other, with an aqueous phase separating each bilayer. An attractive feature of these systems is that they lend themselves naturally to molecular compartmentalization of reactants. Location of the reactants can be restricted to one of several places: in the inner hydrophobic lipid region, in the head group region, or in the inner or outer aqueous phase. Consequently, these systems are ideal candidates for segregation of high-energy products in photochemical redox reactions. From a biochemical point of view, there is also a large interest in spectroscopic studies of these systems. We can utilize the very same molecules (natural lipids) that go to constitute the biological membranes. Structural and dynamical properties are determined in simpler model systems with gradual addition of other membrane components such as proteins and bile acids.

Various membrane lipids (phospholipids, glycosyldiacetylglycerol, plasmalogen, sphingomyelin, glycosphingolipids, and gangliosides) all form vesicles and liposomes. The term *lecithin* is used to describe the lipid mixtures of different chain length and/or head group structure. In general, surfactants and lipids carrying two long alkyl chains form vesicles while those with single chains form globular micelles. As reflected in their natural abundance in membranes (over 90%), a majority of the physical studies have been with phospholipids (sometimes also called phosphalipids). Biosynthetically, the two fatty acid residues of a phospholipid are esterified to a molecule of glycerophosphate to form phosphotidic acid (PA), which in turn condenses with bases to form either phosphotidylcholine (PC), phosphotidylserine (PS), phosphotidylethanolamine (PE), or phosphotidylinositol (PI):

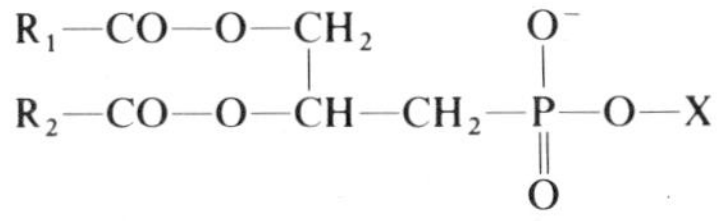

R_1—CO—O—CH$_2$

R_2—CO—O—CH—CH$_2$—P—O—X

R_1, R_2: hydrocarbon tails

$R_1 = R_2$ = mrystoyl (C$_{14}$) DMPX
= palmitoyl (C$_{16}$) DPPX
= stearoyl (C$_{18}$) DSPX
= oleolyl (C$_{18}$(1,9-*cis*)) DOPX
= elaidoyl (C$_{18}$(1,9-*trans*)) DEPX

phosphotidic acid (PA),
    X = —H
phosphotidylcholine (PC),
    X = —CH$_2$CH$_2$N(CH$_3$)$_3^+$
phosphotidylserine (PS),
    X = —CH$_2$CH$_2$NH$_3^+$COO$^-$
phosphotidylethanolamine (PE),
    X = —CH$_2$CH$_2$NH$_3^+$
phosphotidylinositol (PI),
    X = —C$_6$H$_5$(OH)$_5$

The head groups of these phospholipids are very polar containing the negative charge of the phosphohydroxyl ($pK = 1$–2) together with the charges and polar head groups on bases. While PC and PE are neutral at pH 7.0, PS and PI carry a net negative charge.

Depending on the amount of water added, phospholipids assume different structural forms in the aggregates. At low water content, we have liquid crystalline phases of different types:

Dry PL $\xrightarrow{\text{H}_2\text{O}}$ Liquid crystaline phases $\xrightarrow{\text{H}_2\text{O}}$ liposomes $\xrightarrow{\text{sonication}}$ vesicles

Upon further addition of water and gentle shaking, they disintegrate forming large liposomes—multibilayer structures with a milky syrup appearance. The first liposome structures were reported by Bangham. Ultrasonic dispersal of these liposomes leads to smaller vesicles in different sizes ranging from 250 to 2500 Å in diameter. Huang has described preparation of relatively homogeneous preparations from these by gel exclusion chromatography. Uniform vesicles (molecular weight $\simeq 2 \times 10^6$, external diameter about 200 Å, internal diameter = 100 Å, giving bilayers with thickness of about 50 Å) are commonly prepared in this manner. Several other methods not involving sonication have been reported for the preparation of unilamellar vesicles. These include slow injection of an organic solvent solution into an aqueous buffer, cholate dialysis, and ultracentrifugation.

Various physical parameters can be manipulated in vesicles and liposomes such as head-group structure (which includes electrical charges), the hydrocarbon chain length, and the degree of unsaturation. The latter two parameters control the relative fluidity of the bilayer. We can distinguish three main modes in the mobility of the individual phospholipid molecules within the bilayer:

(a) *Fast lateral diffusion within the plane of the bilayer.* This relatively rapid diffusion is constrained to the two dimensions of the horizontal plane of the bilayer and occurs mainly when the lipids are in the liquid crystalline L_∞ state. Diffusion coefficients in the range of 10^{-8} cm^2 sec^{-1} have been measured by NMR, ESR, and fluorescence techniques for a wide variety of model and biological membranes. Mechanistically, the diffusion can be visualized as a hopping or jumping event that occurs with a frequency of about 10^{-7} sec.

(b) *Phospholipid flip-flop.* This is the transveral movement of lipid molecules from one-half of the bilayer to the other on a one-for-one exchange basis. Such a process is extremely slow and can take up to several days. Methods involving ESR, chemical modification, and biochemical modification involving phospholipid exchange proteins have been used to study these slower processes.

(c) *Segmental motion.* In biological membranes at physiological temperature, the bilayer is in a fluid state and fatty acyl chains are continually being distorted by rotation about the C—C bonds yielding kinks in the chains. These are extremely short-lived states lasting for about 10^{-9} sec. Currently there is extensive research activity attempting to derive a detailed quantitative description of such motions by NMR relaxation and ESR spin label techniques. A dynamic picture involving a mobility gradient within the bilayer—increasing mobility as one moves away from the glycerol backbone—has emerged. An order parameter S, which describes the time averaged orientations of groups within the bilayer, can be determined. The order parameter is a quantitative estimate on the relative probability of gauche and trans conformers occurring in the lipid methylene group.

Another area of investigation in these systems is in the characterization of lipid phase transitions. The lipids in the bilayer can exist either in a gel-like or liquid state depending on the temperature. Controlled by the chain length, degree of unsaturation, and head group structure, lipid bilayers exhibit distinct phase transitions at characteristic temperatures. Table 6.1 presents a collection of phase transition temperatures for various lipid systems. Below the phase transition temperature, in the so-called L_β state, the fatty acyl chains are not fully extended in the all-trans form. (In PCs, the acyl chains are not fully extended but are tilted to the normal plane of the bilayer, a form denoted as $L_{\beta'}$.) The phase transition leading to the L_∞ form is accompanied by the occurrence of rotational isomers. These rotomers contain gauche forms, producing overall kinks with gauche–trans–gauche conformations along the chain. The presence of unsaturated double bonds disturbs extensively the order and lowers drastically the phase transition temperatures.

Numerous aspects of the static and dynamic properties of lipid vesicles and liposomes have been investigated using photophysical and photochemical techniques. In this chapter, we broadly review some of these studies. Topics under discussion include lipid-phase transitions, mobility and permeability of solutes, exchange/transfer of probes and lipid molecules between vesicles, intravesicular lateral diffusion of lipids and of probes, microviscosity of the lipid and head group region, influence of additives such as steroids, anesthetics

TABLE 6.1. Phase Transition (gel $L_\beta \rightarrow$ mesomorphic lamellar $L_{\alpha'}$ phase) Temperatures for Lipid Bilayers Composed of Various Phospholipids in Aqueous Solution

Lipid[a]	T_t (°C)	Lipid[a]	T_t (°C)
DLPC	12	DSPE	82
DMPC	24	DOPE	15
DPPC	42	DMPG	23
DSPC	55	DPPG	41
DOPC	−22	DSPG	54
DMPE	51	DPPA	67
DPPE	63		

[a] Abbreviations for head groups: PC = phosphotidylcholine; PE = phosphotidylethanolamine; PG = phosphotidylglycerate, and PA = phosphotidic acid; for hydrocarbon chains: DL = dilauryl(C_{12}); DM = dimyristoyl(C_{14}); DP = dipalmitoyl(C_{16}); DS = disteroyl(C_{18}), and DO = dioleoyl(C_{18}-1,9-*cis*); DPPC = dipalmitoylphosphotidylcholine.

and others, Ca^{2+}-, Mg^{2+}-induced vesicle fusion, surface charge, and potentials, as model systems for antenna chlorophyll and sensitized electron transport across lipid bilayers.

6.2 Fluorescence Probe Analysis

There are numerous organic molecules (covalent as well as ionic) whose fluorescence properties are extremely sensitive to their local concentration and to the nature of the immediate vicinity of the probe. Figure 6.2 shows structures of some of the common fluorescence probes found to be useful in the study of model membrane and biological membrane systems. In Chapter 2, we have already seen the application of these probes in the study of micellar systems. From the observed changes in the fluorescence properties in lipid vesicles as a function of various parameters (probe:lipid ratio, additives, temperature, pH, etc.), useful information can be derived on the properties of lipid vesicles and liposomes. Several reviews covering applications up to the mid-1970s are available.[12–17]

The relative intensities of vibronic bands in pyrene monomer fluorescence is extremely sensitive to the polarity of the environment (cf. Section 2.1). In lipid vesicles, simple monitoring of the (III/I) peak ratios allows determination of the phase-transition temperatures and also the effect of cholesterol addition.[18,19] Below the phase-transition temperature, pyrene is

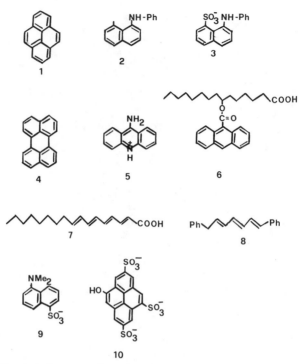

FIG. 6.2. Chemical structures of common fluorescence probes used in the study of lipid vesicles and liposomes: (1) pyrene, (2) N-phenylnaphthylamine (NPN), (3) 1-anilinonaphthalene-8-sulfonate (ANS), (4) perylene, (5) 9-aminoacridine, (6) 9-anthroyl fatty acid, (7) parinaric acid, (8) diphenylhexatriene, (9) dansyl chloride, and (10) pyranine.

solubilized in the lipid bilayer presumably near the head-group (polar) region. Above the phase-transition temperature, pyrene moves deeper into the bilayer as evidenced by the increase in the peak ratios. Cholesterol induces changes which correspond to polarity increases of up to 9 mol.%. Contents over 13 mol.% however cause a decrease in the polarity of the probe environment.

The changes in the (III/I) ratio and the fluorescence lifetimes also serve to monitor the lysis of egg PC vesicles on addition of lysophosphotidylcholine.[20] Up to 40% addition of LPC, the bilayer remains intact, although it exhibits increased fluidity and higher permeability. Lysis leading to formation of mixed micelles occurs at LPC concentrations over 40% (Fig. 6.3). Delayed luminescence from pyrene has also been observed in DPPC and DMPC vesicular dispersions.[21] The luminescence spectrum is similar to the normal fluorescence (delayed fluorescence?). The transient behavior of the luminescence also shows a drastic change at the lipid phase-transition temperature.

Trans-parinaric acid fluorescence can be used to determine the phase transition temperature, as shown with aqueous vesicular dispersions of

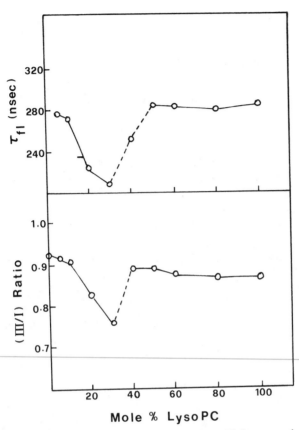

FIG. 6.3. Variation in the pyrene monomer fluorescence lifetime τ_{fl} and vibronic band intensity ratio (III/I) in phosphotidylcholine vesicles during the addition of lyso-phosphotidylcholine. (From Morris et al.[20])

diether, mono-, and diester derivatives of PC.[22] The diether analog of PC has a higher transition temperature T_c while the monoether analog has a lower T_c than their diester counterparts. Similarly, the fluorescence of amphiphatic cyanine dyes has been used to determine various phase-transition temperatures $(T_c, H ...)$ in DMPC and DPPC vesicles.[23] Using the solvent-dependent fluorescence behavior of propanolol, the fluorescing moiety of propanolol has been inferred to be localized in the lipid polar head-group region in the bilayers of PC and PS vesicles.[24] The time-dependent fluorescence enhancement in probes DPH and a series of 9-anthroyl fatty acids allows deduction of the position of the chromophores in the bilayer.[25] The activation energy for the incorporation of these probes is a function of the distance of the chromophore from the polar end and the length of the acyl chain portion of the probes (E_{act} for 16-anthroylpalmitic acid > 2-anthroylpalmitic acid; E_{act} for 12-anthroylstearic acid > 2-anthroylstearic acid).

A major area of focus in membrane biophysics has been the study of interactions between lipids and proteins in the membrane matrix, and there have been several attempts to use polypeptides and lipid vesicles to mimic this process. Dufourcq et al.[26] have used the fluorescence of tryptophan-containing peptides such as Lys–Trp–Lys or Lys–Trp–(OMe) to study the binding of oligopeptides to phospholipid vesicles. The fluorescence of these peptides exhibits a blue shift and increased fluorescence quantum yields on binding to PI or PS vesicles. No binding could be detected with PC vesicles. The binding is strongly dependent on the ionic strength and pH of the medium. The binding of tyrosine-containing peptides to PS and PI vesicles is also characterized by increased fluorescence quantum yields of the peptides without any shift in the emission maxima.

It is a common observation for a variety of fluorescence probes that their fluorescence is quenched at high probe concentrations (concentration or self-quenching). When such molecules are incorporated into the lipid vesicles, their fluorescence shows large changes during any perturbations caused to the bilayer structure by processes such as phase transitions and lipid exchange/transfer during vesicle fusion. Fluorescence of lipophilic probes such as 1-acyl-2-parinaroyl-phosphotidylcholine (AP-PC) has been used to study the lipid exchange/transfer in this manner.[27] Due to extensive self-quenching, the fluorescence of this probe (AP-PC) in PC vesicles is very weak. When phospholipid transfer protein and vesicles containing nonlabelled PC are added to these, the protein catalyzes an exchange of PC between the labelled and nonlabelled acceptor vesicles. The insertion of AP-PC into nonlabelled vesicles is accompanied by an increase in its fluorescence. (Redistribution of the probe occurs over the entire vesicle population and a reduction of self-quenching). Using various amounts of phosphatidic acid PA in the acceptor egg PC vesicles, the rate of probe transfer was found to be stimulated, inhibited, or unaffected by the negative charges at the interface, the nature of the protein, and the donor-to-acceptor ratio used.

The self-quenching of 6-carboxyfluorescein has found applications in the study of phase transitions of lipid vesicles.[28,29] Due to its high solubility in water, the probe is confined to the inner aqueous core of the lipid vesicles. Changes in the fluorescence are observed during phase changes, indicating a decrease in the trapped inner aqueous volume. This volume change is fully reversible and is not caused by vesicle–vesicle fusion.

Dilution of the probe local concentration will occur during the fusion of the vesicles and the process of vesicle fusion can be induced by addition of Ca^{2+} or Mg^{2+} ions (Fig. 6.4). Papahadjopoulos et al. have used this technique to study the fusion capacity of different phospholipid vesicles.[30–32] Dequenching of the carboxyfluorescein fluorescence occurs when phosphitidate vesicles fuse in

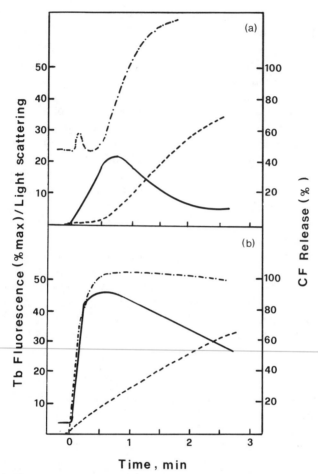

FIG. 6.4. Fluorescence intensity and light scattering changes during Ca^{2+}-induced aggregation and fusion of phosphotidylserine vesicles at 25°C. (a) Large unilamellar vesicles (LUV) and (b) small unilamellar vesicles (SUV). Ca^{2+} was added at $t = 0$ to a final concentration of 3 mM (LUV) and 1.5 mM (SUV). [Lipid] $= 50\ \mu M$. Tb fluorescence (——) (monitored at 545 nm) is given as the percentage of maximal fluorescence attained when all of Tb in the vesicles reacts with excess of dipicolinic acid (DPA); 90° light scattering at 276 nm ($\cdots$) was followed simultaneously with Tb fluorescence. Carboxyfluorescein (CF) release (– – –) (monitored at 530 nm) is given as the percentage of fluorescence obtained when the vesicles are lysed with detergent. (From Wilschutt et al.[33] Copyright 1980 American Chemical Society.)

the presence of calcium or magnesium ions at threshold concentration levels (0.03 mM Ca^{2+} or 0.07–0.15 mM Mg^{2+}). In contrast, PI vesicles do not fuse with either Ca^{2+} or Mg^{2+} even at 50 mM concentration, in spite of aggregation induced by both these cations in the 5–10 mM range. The influence of PE or PC in the fusion of phosphotidate or PI vesicles can be investigated in the same manner.

When solutions of lipid vesicles are mixed in the presence of other additives, the differentiation (or identification) of the membrane fusion process from others such as aggregation and lipid exchange is not a simple matter. Membrane fusion is expected to involve communication between the internal aqueous compartments of the two fusing vesicles. Papahadjopoulos *et al.* have developed a fluorescence assay technique which monitors the interaction of Tb^{3+} ions entrapped in one population of the vesicles and dipicolinic acid entrapped in another.[33,34] Tb^{3+} and dipicolinic acid form a complex that exhibits strong fluorescence, four orders of magnitude more intense than the Tb fluorescence. The formation of this complex is inhibited outside the vesicles by the presence of EDTA and Ca^{2+} ions. Therefore, only the reaction protected from the outside environment by a nonleaky membrane is recorded as fusion. The role of the head-group composition in membrane fusion has been studied by this Tb^{3+} complex fluorescence method in PS, and PC mixed vesicles.

A topic of major concern in the understanding and duplication *in vitro* of the green plant photosynthesis is the functioning of the antenna and reaction centers of chloroplasts. In the thylakoid membrane, energy migration occurs rapidly and very efficiently among the light harvesting chlorophyll molecules present at a concentration of nearly 0.1 *M*. In all of the *in-vitro* systems examined to date, the concentration quenching of chlorophyll fluorescence has been found to occur at relatively low concentrations. This suggests that there may be major structural differences in the molecular organization of chlorophyll molecules between all model systems and that in the chloroplasts. One such possibility is that in thylakoid membranes, the chlorophyll molecules are separated from each other by lipid molecules and one simple way to mimic this is to solubilize chlorophyll molecules in the bilayer lipid vesicles and liposomes. Beddard *et al.*[35] have examined the concentration quenching of Chl fluorescence in various lipid vesicles and liposomes composed of egg lecithins, mono- and digalactosyldiglycerides (Mgdg, Dgdg) and their mixtures. The system having the highest half-quenching concentration (i.e., the concentration at which the fluorescence is reduced to one-half of the unquenched fluorescence) was found to be a 3 : 1 mole-to-mole mixture of Mgdg : Dgdg with a $C_{1/2}$ value of 7×10^{-2} *M*.

Lipid oxidations do occur at a significant rate in multilamellar vesicles of egg PC under normal conditions. Barrow and Lentz[36] have shown that this process can be readily monitored by following the fluorescence intensity of the vesicle-associated probe DPH. Lipid oxidations apparently cause chemical modifications in the fluorophore resulting in a decrease in the fluorescence intensity with time (over a period of 120 h). Consequently caution needs to be exercised in studies of microviscosity by fluorescence measurements using this probe. Fortunately, lifetime-derived anisotropies/microviscosities did not show any change with time.

Studies of Spectral Relaxation

It was mentioned earlier that probe molecules such as ANS or NPN show large spectral shifts with changes in the polarity and viscosity of the medium. Time-resolved studies of the spectral relaxation of fluorescence are very useful in assessing the origin of the environmental effects. Brand and co-workers have investigated the time-resolved emission spectra of 2, 6 TNS (2-toluidinylnaphthalene-6-sulfonate) in lipid vesicles using single photon counting techniques.[37,38] On nanosecond time scales, we can observe the evolution of the red-shifted emission and the observed spectral behavior can be adequately explained in terms of the continuous solvent relaxation models.

The fluorescence of N-phenylnaphthylamine has been examined in egg PC vesicles and in various solvents by Matayoshi and Kleinfeld using the technique of phase modulation spectroscopy.[39] The probe emission decay is a single exponential in low viscous solvents of polar and nonpolar nature. In polar viscous solvents or in nonpolar solvents containing an added polar solute, the decay is heterogeneous and the emission wavelength dependent. In such cases, dielectric relaxation and/or excited-state complexation give rise to emission spectral shifts on nanosecond time scales. Wavelength-dependent emission decay was also observed when NPN was bound to egg PC vesicles. From these results and emission maxima, it has been proposed that NPN probes the ester carbonyl region of the phospholipid acyl chains where it undergoes an excited-state reaction. This inference contradicts the often-made assumption that NPN probes the hydrocarbon layer of the bilayer.

Lakowicz et al. have also made a similar phase modulation fluorometric study of spectral relaxation of 2,6-TNS in the viscous solvent glycerol and at the lipid–water interfaces of DMPC, DPPC, and DOPC vesicles.[40,41] Phase modulation methods allow ready recording of the emission spectra of the relaxed and initially excited states. The relative intensities of these phase sensitive spectra, in combination with the measured phase modulation values on the short and, long-wavelength sides of the emission provide the spectral relaxation times. For saturated and unsaturated PCs, at temperatures ranging from 5 to 50°C, the relaxation times ranged from 5 to 1 nsec. The activation energy for the spectral relaxation were near 4 kcal/mole. The relaxation times decreased smoothly with increasing temperature and did not change abruptly at the phase-transition temperature. These results indicate that small molecular motions of the interface region of the membranes responsible for the observed spectral relaxation are not dramatically influenced by the phase state of the acyl chains of the membranes.

Fluorescence Photobleaching Recovery

A novel fluorescence technique has found increasing use in the study of diffusion coefficients of membrane components.[41a-h] The method can be described as follows. A fluorescently labelled membrane component (a lipid or protein) is initially distributed uniformly in the bilayer membrane. An intense, focused laser pulse of appropriate wavelength bleaches a small circular spot (typically 8-μm radius) on the membrane. Subsequently, the same photolysis beam is attenuated and monitors the reappearance of the fluorescence within the circle due to the arrival (via diffusion and/or flow) of unbleached molecules. The so-called fluorescence recovery curve $F_k(t)$ (fluorescence intensity versus time after bleaching) thereby contains all the information needed to analyse the transport process quantitatively. For isotropic diffusion, the diffusion coefficients are derived using the expression for an uniform circular beam profile:

$$D_T = (0.22 \ w^2/t_{1/2}) \qquad (6.1)$$

where w is the radius of the circular observation area and $t_{1/2}$ is the time required for the fluorescence intensity in the bleached ratio to reach 50% of its intensity after complete redistribution of fluorescence. The method is also abbreviated as FRAP fluorescence recovery after photobleaching. Table 6.2 provides a collection of data on the diffusion coefficients for membrane components determined using this and other photochemical methods.

6.3 Fluorescence Quenching

The quenching of fluorescence of the solubilized probe molecules provides a simple method for the analysis of the factors that control the permeability and interactions of molecules in lipid vesicles and liposomes. Cheng and Thomas found that the fluorescence decay of pyrene solubilized in various phospholipid dispersions is enhanced by ionic and neutral molecules such as I^-, O_2, and CH_3NO_2.[42] The presence of salts and anesthetics like benzyl alcohol or cholesterol affect the quenching rate in a manner that reflects the permeability changes in the lipid aggregates.

The quenching of fluorescence of N-alkylcarbazole derivatives by various chlorinated hydrocarbons has been studied in lipid vesicles.[43] Two fluorescent probes β-11-(9-carbazole) undecanoyl-L-α-phosphatidylcholine (CVA-PC) and β-3-(9-carbazole) propionyl-L-α-phosphatidylcholine (CPA-PC) (which readily incorporate into lipid vesicles at lipid:probe ratios in the 50:1 to 1000:1 range without formation of excimers) have been used:

TABLE 6.2. Lateral Diffusion Coefficients for Lipid Analogs in Lipid Bilayers Determined by Photochemical Techniques

Method	Lipid matrix	Temperature (°C)	Probe	D_{diff} (cm² sec⁻¹)	Reference
Fluorescence photo-bleaching recovery	Egg lecithin	25	N-4-nitrobenz-2-oxa-1,3-diazo-PE	4×10^{-8}	41h
	DMPC	30	N-4-nitrobenz-2-oxa-1,3-diazo-PE	8×10^{-8}	41h
	DMPC/Cholesterol (1:1)	30	N-4-nitrobenz-2-oxa-1,3-diazo-PE	2×10^{-8}	41h
	Egg lecithin	24	2,4,6-trinitrophenylrhodamine–stearoyl dextran	4.5×10^{-8}	41i
	DMPC	40	dansyl-labelled gramicidin	6×10^{-8}	41g
	DMPC	25	dansyl-labelled gramicidin	3×10^{-8}	41g
	DMPC	20	dansyl-labelled gramicidin	5×10^{-11}	41g
	DMPC	15	glycophorin	1.2×10^{-8}	41f
	DMPC	15	glycophorin	5×10^{-11}	41f
Excimer fluorescence	Egg lecithin	25	pyrene lecithin	1.3×10^{-7}	41j
	DPPC	50	pyrene lecithin	1.6×10^{-7}	41j
	DMPC	30	pyrene lecithin	0.65×10^{-7}	41j
Triplet–triplet annihilation	DPPC	50	PE-anthracene	2.3×10^{-7}	41k
Fluorescence correlation spectroscopy	Egg lecithin	24	DPPE–rhodamine	2.4×10^{-7}	41l
	Egg lecithin/cholesterol (1:1)	24	DPPE–rhodamine	1.6×10^{-7}	41l

Lipid–water partition coefficients of chlorinated hydrocarbons were all determined from the fluorescence quenching measurements using the following simple models and expressions.

The lipid–water partition coefficients were determined from

$$(1/k_{app}) = (1/k_{qm}P) + [(1/k_{qm}P) - (1/k_{qm})\alpha_m] \tag{6.2}$$

where k_{qm} is the quenching rate constant in the membrane, P the lipid–water partition coefficient, α_m the volume fraction of the membrane, and k_{app} the apparent quenching rate constant. Here α_m was determined from the lipid concentration and the density of the phospholipid vesicles; k_{app} was calculated from the normal Stern–Volmer relation:

$$(\tau_0/\tau - 1) = k_{app}[Q]\tau_0 \tag{6.3}$$

where τ_0 and τ are the excited-state lifetimes of the probe in the absence and presence, respectively, of quencher [Q]. The diffusion coefficient D_{qm} is derived from k_{qm} using the relationship

$$k_{qm} = \gamma 4\pi\sigma D_{qm}N/1000 \tag{6.4}$$

where γ is the quenching efficiency, σ the sum of the molecular radii of the quencher and fluorophore, and N, the Avagadro's number.

Verkman has used fluorescence quenching of DPH and several anthroyl fatty acids by phloretin and related compounds in PC vesicles to determine the equilibrium binding and permeability properties of the latter.[44] The mechanism of quenching has been determined to be largely diffusional in nature. The trans-membrane movement of phloretin was also observed by stopped-flow techniques in which phloretin is mixed rapidly with a vesicle population containing the fluorescence probe.

Paramagnetic spin labels such as nitroxides are efficient quenchers of fluorescence of solubilized probes in lipid vesicles.[45,46] Bieri and Wallach used the fluorescence quenching of perylene by nitroxides to follow the phase transitions and the influence of additives,[45] on lines similar to that of Cheng and Thomas. The quenching curves have been analyzed in terms of a nonrandom distribution of probes and quenchers (coclustering) below the phase-transition temperature and a statistical distribution above the transition temperature T_c. Due to large interests on the molecular oxygen concentration levels and its diffusion in biological systems, quenching efficiencies of oxygen on the fluorescence of pyrene and pyrenebutyric acid have been studied in lipid vesicles.[47,48]

6.4   Intravesicular pH and Interface Potentials

Fluorescence probe molecules whose emission spectra are pH-dependent are well suited to study intravesicular pH and surface potentials. Lee and

Forte have demonstrated the usefulness of quinine, acridine, and hydroxy-cinnamic acid in such studies.[49] The fluorescence maxima and intensities of probes incorporated into asolectin liposomes reflect the internal pH (i.e., the pH of the medium where the phospholipid vesicles are prepared) rather than the external pH (i.e., the pH of the medium where the vesicles are present). For example, quinine in phospholipid vesicles prepared at pH 2.5 and diluted into a medium of pH 6.5 exhibit fluorescence characteristic of the acid form of the dye (pH 2.5) and not that of the neutral form. Thus, from a knowledge of the pH-dependent fluorescence of the probe and the external pH of the medium, it is possible to estimate the intravesicular pH gradient.

Actually, the use of weak acids and bases to measure pH gradient (Δ pH) across vesicular structures is a well-established method. Amines such as 9-aminoacridine are widely used. With these probes, there is a quenching of the fluorescence intensity as the probe distributes into the intravesicular space according to the existing pH gradient. Concentration-dependent self-quenching has been proposed as the most possible mechanism for the fluorescence quenching observed.[50,51] From the rate of decay, permeability coefficients for H^+ and OH^- across liposomal membranes have also been calculated. (approximately 10^{-4} cm/sec, a value six orders of magnitude greater than that measured for Na^+ under similar conditions.) The use of the fluorescence probe pyranine (pyrenetetrasulfonate) to measure proton per-meability coefficients of lipid membranes has also been described.[52]

Fernandez[53] has examined the use of the fluorescent pH indicator probe 4-heptadecylumbelliferone (4-heptadecyl-7-hydroxycoumarin)

to measure the electrical potential at the surface of negatively charged liposomes (liposomes prepared from mixtures of egg lecithin and dicetylphos-phate at different molar ratios). The interfacial pK_0 characterizing the dissociation of the acid–base indicator located at a neutral surface can be obtained by measuring the dissociation degree (α) and the bulk pH (pH_b) of the solution surrounding the membrane:

$$pK_0 = pH_b - \log(\alpha/1 - \alpha) \qquad (6.5)$$

If the surface carried net charge, the apparent pK of the indicator will be shifted with respect to the value at the neutral interface:

$$pK_{ch} = pK_0 - (\psi F/2.3\ RT) \qquad (6.6)$$

where pK_{ch} is the apparent pK of the indicator at the interface, ψ the surface potential, and F the Faraday constant. Rearranging Eq. (6.6) we have

$$\psi = -(pK_{ch} - pK_0)2.3\, RT/F \tag{6.6a}$$

7-Heptadecylumbelliferone is a good acid–base indicator for the above purpose, for only the basic form fluoresces at 452 nm when excited at 377 nm, and the fluorescence intensity is proportional to the dissociation degree. It was found that the dependences of the experimental surface potential on the proportion of the charged lipid in the vesicle and the salt concentration varied in a manner as predicted by Gouy–Chapman equation.

6.5   Dynamics of Excimers and Exciplexes

Among various fluorescence probe molecules that have found utility in the studies of microheterogeneous systems, pyrene stands unique due to its applications in many different forms. One such application involves its ability to form a transient excited-state complex between an excited and ground-state pyrene molecule, often known as an excimer (Fig. 6.5):

$$^1P^* + P \xrightleftharpoons{k_2C} P{\cdot}P^* \tag{6.7}$$

The excimer formation occurs at a diffusion-controlled rate and is characterized by its distinct fluorescence, structureless and redshifted to the monomer fluorescence. In studies of lipid vesicles and liposomes, this feature of excimer formation has been used to study various processes such as phase transitions, vesicle fusion, lipid probes, lateral diffusion, and exchange/transfer. We briefly review some of these applications here.

The kinetic aspects of excimer formation between isolated molecules in solution has been adequately described by Birks and others (cf. Sections 1.25 and 1.26). The ratio of fluorescence quantum yields of excimer to monomer is related to the association rate k_2C:

$$\Phi_E/\Phi_M = (k_M/k_E)\tau_E k_2 C \tag{6.8}$$

where C is the probe concentration per unit area, k_2 the second-order rate constant for the formation of excimers [Eq. (6.7)], τ_E the lifetime of the excimer, and k_E and k_M represent the rates of radiative decay of excimer and monomer, respectively. (The above equation holds good only if the dissociation of the excimer is negligible!) Provided the fluorescence spectra of monomer and excimers do not change with concentration, temperature etc., the fluorescence quantum yield ratios can also be expressed by their respective fluorescence intensities measured at their respective maxima:

$$I_E/I_M = k(\Phi_E/\Phi_M) \tag{6.9}$$

where k is a proportionality constant characteristic of the probe.

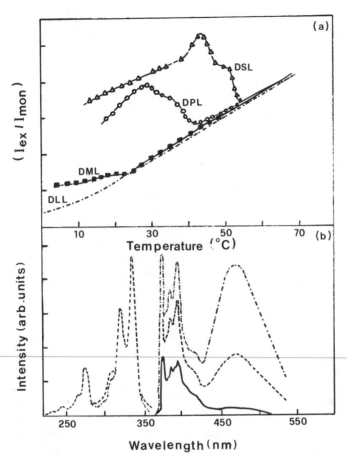

FIG. 6.5. (a) Temperature dependence of fluorescence intensity ratio (I_{ex}/I_{mon}) of excimer to monomer of pyrene embedded in various lecithin vesicles. (From Galla and Sackmann.[54] Copyright 1974 American Chemical Society.) (b) Uncorrected excitation and emission spectra of pyrene in dipalmitoyllecithin vesicles. Pyrene concentration is 0.7 μM (———), 2.7 μM (– – –) and 6.7 μM (–·–·–). (From Vanderkooi and Callis.[57] Copyright 1974 American Chemical Society.)

Lipid Phase Transitions

An empirical application of the diffusion-controlled nature of the excimer formation is in the determination of lipid phase-transition temperatures in vesicles and liposomes and in the study of the influence of additives such as cholesterol and anesthetics. The fluorescence intensity ratio I_E/I_M plotted as a function of temperature shows abrupt changes in the temperature ranges corresponding to well-known phase-transition temperatures for various lipid and lecithin preparations.[54-57] At the phase transition temperature, melting of the acyl chains leads to increased fluidity (at levels that outweigh incremental increases in the rate of excimer formation at higher temperatures) and a sharp drop in the excimer-to-monomer emission intensity ratio. Figure 6.5 illustrates typical intensity ratio plots.

Probe Exchange/Transfer

The concentration dependence of the excimer to monomer ratio has also been used in stopped-flow experiments to study the kinetics of exchange of probes (labelled lipid molecules) between lipid vesicles.[58-65] If two vesicle preparations, one labelled with pyrene and the other empty, are mixed and if an exchange of probe molecules occurs the I_E/I_M ratio will decrease with respect to the original intensity ratio of the doped vesicle. The ratio of exchange can simply be observed by recording the increase in the monomer fluorescence intensity (and/or the decrease in the excimer emission) as a function of time after rapid mixing. Control experiments need to be performed to ascertain as to whether the observed ratio changes are merely due to rapid vesicle fusion or indeed due to exchange of probes at rates much faster than the lipid exchange/transfer. This is readily done by monitoring the temperature dependence of I_E/I_M ratios of two different lipid vesicles, DPPC, DSPC for example, at various times after mixing. It turns out that the latter is indeed true.

Some of the principal conclusions of these studies are as follows: (1) for a variety of pyrene derivatives (alkyl, alcohol, and carboxylic acid derivatives) the migration of the probes occurs mainly via the aqueous phase at a rate determined by the exit of pyrene from the vesicles; (2) the measured overall exchange rate k_{ex} vary in the range 10^2-10^{-2} sec^{-1}, depending on the nature of the probe and lipid vesicles; (3) for carboxylic acid labels, the rate of transfer is greater for the ionized form (pH > pK_a) than that for the neutral form; (4) for pyrene in soybean lecithins containing 10% dicetylphosphate, the exit rate constant k has been estimated to be 4×10^2 sec^{-1} by Almgren using an appropriate kinetic model; (5) the process is first order with activation energies in the range of 5–15 kcal/mole; (6) for pyrene lecithin derivatives, the exchange kinetics are biphasic (a fast component with a half-life of 11 sec and a slow component with a half-life of about 8; the slow process has been attributed to an intrabilayer lipid flip–flop process); (7) with cerobroside-labelled pyrene derivatives hardly any exchange of probes was observed in phospholipid vesicles for periods up to 30 h.

Lateral Phase Separation and Vesicle Fusion

The pyrene excimer technique has also been applied to study phase separation in membranes of lipid mixtures. Consider a lipid monolayer which undergoes a conformational change leading to a heterogeneous (mosaic-like) structure of alternating fluid and rigid domains. Due to the very low solubility in rigid lipid layers, the probe molecules will be squeezed out from the rigid domains into the fluid phases. The effective label concentration and therefore the I_E/I_M ratio will increase and the relative amounts of rigid and fluid domains can be determined quantitatively.[66] The technique has also been used to follow vesicle fusions caused by the addition of Ca^{2+}, Mg^{2+} or bovine serum albumin (BSA).

Intramolecular Excimers

A disturbing factor in the intermolecular excimer formation in aggregated systems such as vesicles and micelles is the difficulty in estimating the actual local concentration of the probe molecules in the interior. The dependence of the intensity ratio I_E/I_M on the lipid/probe ratio (for isolated pyrene molecules) precludes, as in the case of micelles, a simple correlation of the intensity ratios with the microviscosity of the probe environment. We can, however, circumvent this problem to some extent by utilizing covalently linked pyrene molecules. Using standard calibration curves determined in isotropic solvents of different viscosity, the intramolecular excimer-to-monomer ratio can be converted to microviscosity values. Zachariasse et al.[67] have demonstrated such applications using dipyrenylalkanes in various lipid vesicles. It should be pointed out, however, that caution needs to be exercised in the use of such techniques for the following reason: probe concentrations must be sufficiently low in all the studies, otherwise, contributions from the intermolecular excimers can lead to large errors. The calibration curves are dependent on the probe concentration, except at very low concentrations (pure intramolecular excimer case).[68,69]

Lateral Diffusion Coefficients

There is growing evidence that indicates that artificial and biolgical membranes are to be considered as two-dimensional fluids, characterized by a high lateral mobility of the phospholipid molecules. A direct consequence of the lipid self-diffusion is a considerable lateral and rotational mobility of species present (bound?) within the membranes. There are arguments that reduction of dimensionality can, in fact, lead to a more effective directional transport than a normal three-dimensional diffusion. Bimolecular reactions such as fluorescence quenching, intermolecular excimer formation and triplet–triplet annihilation are ideal candidates for the determination of lateral diffusion coefficients.

Bimolecular reactions in homogeneous solutions (isotropic), in general, are analyzed in terms of Smoluchowski's theory. The probe distribution in the solution is uniform. The derivations assume an infinite volume for the solvent phase whereas in aggregated systems such as vesicles, the inner volume is finite. Consequently, new theoretical/kinetic models need to be developed to treat quantitatively such bimolecular reactions in lipid vesicles. In the introductory chapters, we indicated such necessities and the approaches currently available.

For the analysis of excimer formation in lipid vesicles, Sackmann and co-workers have adapted a statistical model proposed earlier by Montroll to treat the diffusion in two dimensions.[54,55,66,70–72] The probe/lipid mobility is measured in terms of jump frequencies rather than as diffusion coefficients. Without going into the details of the derivation, we outline below some of the key results. Taking into acount the dimensionality of the random walk process on a lipid lattice, the following relation between the molecular jump frequency v_j and the excimer-to-monomer ratio has been derived:

$$v_j = \langle n_s \rangle (I_E/I_M k)(k_E/k_M)(1/\tau_E) \tag{6.10}$$

$$= \langle n_s \rangle v_{coll} \tag{6.11}$$

The average step number $\langle n_s \rangle$ is given by:

$$\langle n_s \rangle = (2/\pi X_L) \ln(2/X_L) \tag{6.12}$$

where X_L is the mole fraction of the label molecules with respect to the lipid.

The jump frequency v_j is related to the lateral diffusion coefficient D_{diff} according to the relation

$$D_{diff} = \tfrac{1}{4} v_j \lambda \tag{6.13}$$

where λ is the average jump length given by the average distance of the lattice. (A value of $\lambda = 8$ Å has been used to estimate $D_{diff}, \tfrac{1}{4}\lambda^2 = 1.6 \times 10^{-15}$ cm^2). The method has been applied to pyrene and several of its derivatives (pyrene decanoic acid, pyrene lecithin, and 1-palmitolyl-2-pyrenedecanoylphosphotidylcholine) in various lipid vesicles. Data on the jump frequencies derived are collected in Table 6.3. (A comparison of diffusion coefficients determined by various methods can be found in Table 6.2.)

While the above-mentioned studies were carried out in the steady state, Owen and Vanderkooi et al.[57,73,74] attempted to measure lateral diffusion coefficients by time-resolved fluorescence studies on monomer and excimer fluorescence growth and decay. In isotropic solvents, the diffusion-controlled rate constant k has the familiar form of the Einstein–Smoulouski expression:

$$k = 4\pi NRD \tag{6.14}$$

where R is the sum of the interaction radii, D is twice the diffusion coefficient, and N is Avogadro's number per millimole. It has been pointed out by Noyes and others that generally the rate constant k is time dependent (especially in viscous solvents). The exact form of the rate constant within the assumptions of the diffusion theory is

$$k = 4\pi NRD[1 + R/(\pi Dt)^{1/2}] \tag{6.15}$$

In this expression the rate constant is time dependent but approaches asymptotically the time-independent value with increasing time. Physically, the origin of the time-dependence of k arises from the existence of a finite probability for the two probe molecules being separated by a distance

TABLE 6.3. Jump Frequencies v_j for Various Pyrene Derivatives in the Lipid Bilayer Region of Phospholipid Vesicles Determined using the Excimer Technique[a,b]

Lipid/T_t (°C)	Temperature (°C)	Probe		
		Pyrene	Pyrene decanoic acid	Pyrene lecithin
DOPC (−22)	30	2.7	0.9	n.d
DPPC (42)	45	1.9	1.4	0.8
DPPC (42)	50	2.5	1.6	1.0
DPPC (42)	60	3.9	2.4	1.5
DMPC (24)	25	0.6		
DMPC (24)	35	1.0	0.8	
DMPC (24)	50	2.1	1.7	
DSPC (55)	60	3.5		
DLPE	30		1.4	1.7
DLPE	60		4.1	3.7
DPPE (63)	65		4.2	3.7
Egg lecithin	35		0.9	
Egg lecithin	50		1.4	
DPPA (pH 7.0)	70		3.2	

[a] Values in $\times 10^8$ sec^{-1}. Lateral diffusion coefficients can be obtained from the jump frequencies using the following relation $D_{\text{diff}} = v_j(1.6 \times 10^{15} \text{ cm}^2)$ (see text for details).
[b] From Sengupta et al.[61] and Galla et al.[70,71]

comparable to their interaction radii at $t = 0$. Light excitation in such systems will lead to instant excimer formation from these near neighbors. After these pairs are extinguished, the system will come into a steady state in which the relative distribution of ground-state pyrene around an excited pyrene is independent of time. For low-viscosity solvents, the transient term is significant only for a very short time (a few picoseconds) and may be safely neglected. For viscous fluids such as lipid membranes, the transient term can be of paramount importance. Vanderkooi et al. in fact found that the excimer kinetics in the case of pyrene solubilized in lipid vesicles (DML, DPL, and egg lecithins) are better described using the diffusion model that takes into account the time dependence. For similar systems, the diffusion coefficients derived via time-resolved studies are an order of magnitude smaller than those derived from steady state analysis of Sackmann et al. ($D_{\text{diff}} \sim 1-6 \times 10^{-8}$ cm^2/sec versus $1-5 \times 10^{-7}$ cm^2/sec of Sackmann et al.).

Intermolecular Exciplexes

The formation and emission from heteroexcimers (or exciplexes) during the quenching of arene singlet excited states by donor molecules are extremely sensitive to polarity and fluidity of the environment. The features of the pyrene–dimethylaniline system have been studied in DMPC vesicular membranes by Waka et al.[75] A very weak exciplex emission was observed, and it was suggested that the interior of the lipid bilayer is highly polar:

$$^1\text{Py}^* + \text{DMA} \longrightarrow [\text{Py} \cdots \text{DMA}]^* \tag{6.16}$$

Based on the dependence of the exciplex yield with solvent dielectric constant, the dielectric constant for the exciplex environment in the lipid bilayer has been estimated to be approximately 10. The exciplex formation occurs very rapidly and has a lifetime of about 61 nsec.

Kano *et al.* have extended these studies using pyrene and pyrenedodecanoic acid as probes and DMA, dicetylaniline, and dimethylanilinesulfonate as quenchers in DPPC vesicles.[76] In addition to confirming the results of Waka *et al.* these studies discuss in detail the kinetic features of the amine quenching. Below the phase-transition temperature T_c of the membranes, nonexponential fluorescence decay and nonlinear Stern–Volmer plots were observed. The results were interpreted in terms of a continuum theory for two-dimensional processes. Above T_c, the quenching obeyed the Stern–Volmer equation without the transient term, on lines similar to that of Owen and Vanderkooi *et al.* for the excimers mentioned earlier.

6.6 Depolarization of Fluorescence

An important landmark in the structural studies of biological membranes has been the introduction of the fluid mosaic membrane picture by Singer and Nicholson in 1972. Since then all biophysical studies have been directed toward quantifying various aspects of this model. One such area concerns the fluidity of the membrane bilayer. Hydrodynamic descriptions of fluids suggest viscosity to be an ideal parameter to describe fluidity. For a photochemist, monitoring the depolarization of fluorescence of embedded probes is probably the simplest method to determine this property. In low-viscosity, isotropic solvents at least, any depolarization of fluorescence observed subsequent to excitation by polarized light is caused by the viscous drag/forces to the brownian motion of the probe. Consequently there have been numerous efforts made to determine the microviscosity of the bilayer using different types of molecules. (The term *microviscosity*, or *microfluidity*, is used to distinguish the viscosity of the probe environment from the viscosity of the medium where the vesicles are present.)

Steady-State Analysis

The relevant parameter here is the steady-state fluorescence anisotropy r_s defined as the ratio of the difference to the total polarized intensities:

$$r_s = (I_h - I_v)/(I_h + 2I_v) \tag{6.17}$$

[This relation holds strictly for δ-pulse excitation. In time-resolved studies, the existence of a finite pulse width requires the calculation of $r(t)$ by normalization of $I_D(t)$ and $I_T(t)$ profiles.] Steady-state fluorescence anisotropy values are fairly easy to determine, though difficult to interpret. A simple approach is to use classical hydrodynamic expressions of the Perrin type to determine microviscosity η values:

$$(r_0/r_s) = 1 + (kT\tau/\eta v_0) \qquad (6.18)$$

Here r_s and r_0 are the measured and limiting fluorescence anisotropies, respectively, τ is the excited-state lifetime, and η the viscosity of the medium. In early studies the limiting anisotropy values (r_0) at the excitation wavelengths were determined using highly viscous solvents such as mineral oil.[77-79]

In order to have a proper appreciation for the quantity microviscosity it is useful to digress a little on the hydrodynamic treatments of brownian motion of solutes that lead to viscosity. Focusing on the solute, hydrodynamic treatments consider the solvent medium as being continuous (or at least the size of the solute is considered quite large as compared to the solvent). This approximation known as the stick-boundary condition is quite appropriate for the brownian motion of large colloidal particles (even the aggregated system as a whole) and large macromolecules such as proteins. It may possibly be true down to molecular sizes of the order of the dye molecule dimensions, at least if these solutes are solvated. On the other hand, for unsolvated, flat aromatic hydrocarbons such as perylene, the rotational diffusion may largely be determined by slipping (or partial slip) hydrodynamic boundary conditions. It is questionable whether the viscosity concept is appropriate at all for the slip (or partial slip) conditions. In structurally anisotropic media such as lipid bilayers in vesicles and liposomes, the above considerations become even more important. Nevertheless, following the early work of Weber et al. for the hydrophobic regions of micelles and lipid vesicles, there have been numerous investigations[80-88] of fluorescence anisotropy in model and biological membranes, often extrapolating measured anisotropies into microviscosities.

The extrapolation implies that the nature of the depolarizing rotations of the probe (i.e., the freedom and the isotropic rotation) in membrane and reference solvent is identical. Time-resolved fluorescence depolarization studies have shown that the probe diffusion is hindered, as evidenced by the existence of nonzero values for the fluorescence anisotropy at times long compared to fluorescence lifetimes ($r_\infty \neq 0$ as $t \to 0$).[90] For a uniformly distributed probe tumbling isotropically, the fluorescence anisotropy at long times should decay to zero ($r_\infty = 0$ as $t \to 0$), as has been verified for several fluorescence probes in mineral oils. Presumably in lipid bilayers, the probe molecules do not assume all possible orientations with equal probability due to restrictions imposed by the lipid bilayer architecture. The nonzero values for the limiting anisotropy, r_∞, in fact, can be related to the square of the lipid order parameter S according to the relation

$$(r_\infty/r_0) = S^2 \tag{6.19}$$

where r_0 is the fluorescence anisotropy value in the absence of any rotational motion. The theoretical value for r_0 is 0.395. There have been proposals to derive a modified microviscosity, η_0, even for anisotropic media by taking into account the order parameter S:

$$\eta_0 = V\left[\frac{r_0(r_s - r_\infty)}{(r_0 - r_s)(r_0 - r_\infty)}\right] = \eta - \frac{V}{r_0 - r_\infty}, \quad \text{Shinitzky}[88a] \tag{6.20}$$

$$\frac{r_s}{r_0} = \frac{S}{1 + S - S^2}, \quad \text{Pottel}[88b] \tag{6.21}$$

Thus, there are limitations on the extrapolation of measured steady-state fluorescence anisotropy values into microviscosity values for the lipid bilayer region. The anisotropy values themselves, however, still provide a simple means of monitoring the processes where the fluidity of the bilayer is affected in any way (hence the rotational motion of the probe) by temperature, pH, and additives such as cholesterol. Figure 6.6 presents data on one such application. Measurements of steady-state anisotropy values have other applications as well. Strehlow and Jahnig[89] have used the steady-state fluorescence anisotropy of DPH to study the kinetics of proton- (or Ca^{2+}-) induced phase transitions in charged lipid vesicles of DMPA in a stopped-flow apparatus. Fluorescence depolarization of DPH has also been used to study the binding of protein β-lactalbumin to DMPC vesicles and of peptide antibiotic polymixin to PG and PA vesicles.

Time-Resolved Studies of Fluorescence Anisotropy

Given the restrictions on the possible orientations and rotational diffusion the probe can have in the lipid bilayer matrix, quantitative description of the probe motion is obtained via time-resolved fluorescence polarization measurements. Probes such as 1,6-diphenylhexatriene (DPH), anthroyl fatty acids, parinaric acid, and anilinonaphthalene derivatives have been found to be useful. Several publications dealing with the theoretical aspects[90-101] and experimental results[102-120] are available.

Ikegami et al.[90] were the first to give a theoretical analysis relating the characteristics of the fluorescence anisotropy decay curves $r(t)$ to membrane properties. Picturing the rotation of probes such as DPH as that of a cylinder, they assume that the symmetric axis of the probes tumbles isotropically within a cone of semi-angle θ_c. That is, the rod-like DPH molecule is assumed to exist in a square well potential such that its rotation is unhindered until a certain critical angle θ_c is reached. Rotation beyond this angle is assumed to be

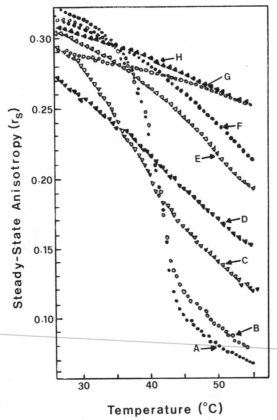

FIG. 6.6. Influence of various steroidal additives on the temperature dependence of steady-state fluorescence anisotropy values (r_s) of diphenylhexatriene (DPH) in sonicated DPPC vesicles. Additives: A = none, B = cholestane, C = 5α-androstan-3β-ol, D = 4-cholesten-3-one, E = 5α-cholestan-3-one, F = epicholesterol, G = cholesterol, and H = 5α-cholestan-3β-ol. (From Vincent and Gallay.[115])

energetically impossible (cf. Figure 6.7). (The restrictions in the form of square well potential can be replaced by other potential forms such as Gaussian.[97]) The model allows a quantitative interpretation of the decay of anisotropy $r(t)$ in terms of a wobbling diffusion constant D_w and a degree of orientational constraint.

The cone angle θ_c was calculated from r_∞ using the relation

$$r_\infty/r_0 = \cos^2 \theta_c (1 + \cos \theta_c)^2/4 \qquad (6.22)$$

The wobbling diffusion constant D_w was estimated from the following

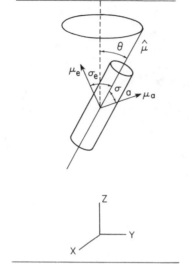

FIG. 6.7. Schematic representation of a cylindrical probe embedded in a lipid bilayer and executing isotropic rotation within a cone of semiangle θ_c. (From Lipari and Szabo.[94] Reproduced with copyright permission of the Biophysical Society.)

equation (where $x = \cos\theta_c$):

$$\frac{D_w \phi (r_0 - r_\infty)}{r_0} = -x^2(1+x)^2 \frac{\{\ln[(1+x)/2] + (1-x)/2\}}{2(1-x)}$$

$$+ (1+x)\frac{(6 + 8x - x^2 - 12x^3 - 7x^4)}{24} \qquad (6.23)$$

In order to determine the required r_0, the nanosecond fluorescence total decay $I_T(t)$ and the anisotropy decay $r(t)$ were analysed by assuming exponential decays of the following forms:

$$I_T(t) = \alpha_1 \exp(-t/\tau_1) + \alpha_2 \exp(-t/\tau_2) \qquad (6.24)$$

$$r(t) = (r_0 - r_\infty)\exp(-t/\phi) + r_\infty \qquad (6.25)$$

where ϕ is the rotational relaxation time. The parameters α_i, τ_i, r_∞, and ϕ were so determined that the convoluted products $g(t) * I_T(t)$ and $g(t) * I_T(t)r(t)$ best fitted the observed total intensity decay curve $I_T(t)$ and difference decay curve $I_D(t)$. The rotational relaxation time ϕ refers to the rotational diffusion of the probe in an anisotropic environment where the probe prefers to be oriented along the normal to the membrane plane. As mentioned earlier, the degree of orientational constraint r_∞/r_0 is proportional to the square of the second-rank orientational order parameter S:

$$\frac{r_\infty}{r_0} = S^2 = P_2^2 = \left\langle \frac{3\cos^2\theta_c}{2} \right\rangle^2 \qquad (6.26)$$

Thus, the time dependence of $r(t)$ which reflects the dynamics of the probe can be described as rotational diffusion in an anisotropic environment. Figure 6.8 presents typical time-resolved anisotropy decay curves. Table 6.4 presents data on the derived fluorescence anisotropy parameters in various systems.

The technique of differential polarized phase fluorimetry has been employed by Lakowicz and co-workers to study the hindered rotational diffusion of probe molecules in lipid bilayers of model membranes and vesicle systems.[116-121] The technique is complementary to the time-resolved measurements and utilizes incident light that is modulated instead of being pulsed. It

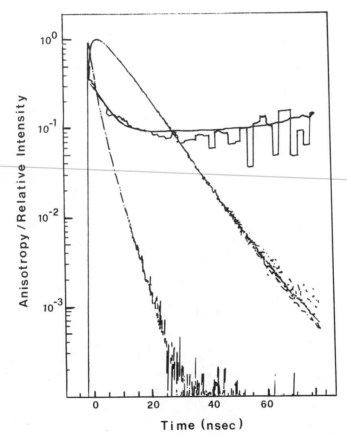

FIG. 6.8. Time-resolved fluorescence anisotropy/relative intensity decay curves for diphenyl-hexatriene in 1-palmitoyl,2-linoleoylphosphotidylcholine (PLPC) liposomes at 10°C: $g(t)$ = instrument response function, $I_T(t)$ = total fluorescence intensity, and $r(t)$ = fluorescence anisotropy. The broken and solid lines superimposed on $I_T(t)$ are calculated best fit curves for single and double exponential approximations, respectively. (From Stubbs et al.[107] Copyright 1981 American Chemical Society.)

TABLE 6.4. Fluorescence Anisotropy Parameters for Rotational Diffusion of
Diphenylhexatriene in Lipid Bilayers Derived using Cone Model[a]

Lipid[b]	Temperature (°C)	r_s	r_∞	ϕ (nsec)	D_w (nsec)	θ_c (deg)
DPPC ($T_t = 42°C$)	10	0.363	0.358	1.37	0.014	14.6
	25	0.357	0.353	0.86	0.029	15.6
	37	0.331	0.326	0.78	0.049	20.4
	50	0.091	0.054	0.75	0.290	60.9
DOPC ($T_t = -22°C$)	10	0.149	0.031	3.51	0.069	66.4
	25	0.089	0.013	1.84	0.136	73.5
	37	0.063	0.007	1.19	0.215	77.6
	50	0.045	0.001	0.76	0.340	84.8
POPC ($T_t = -5°C$)	10	0.171	0.096	2.93	0.062	52.3
	25	0.104	0.040	1.80	0.127	63.8
	37	0.071	0.021	1.17	0.209	69.8
	50	0.050	0.013	0.72	0.348	73.7

[a] Data from Stubbs et al.[107]
[b] Abbreviations: DPPC = dipalmitoylphosphotidylcholine; DOPC = dioleoylphosphotidyl-choline; and POPC = 1-palmitoyl-2-oleoylphosphosphotidylcholine r_s = steady-state fluores-cence anisotropy; r_∞ = residual equilibrium anisotropy; ϕ = apparent relaxation time; D_w = wobbling diffusion constant; and θ_c = the cone angle (see text for details).

allows measurement of r_0 starting from lifetime measurements and the steady-state anisotropy r_s. The study of complete transient decay $r(t)$ is replaced by the study of influence of quenchers (oxygen, for example) introduced into the medium at various concentrations. This introduces changes in r_s that are coupled to a decrease in lifetime. So plots of $(r_0 - r_s)/\tau$ versus quencher yield r_∞ and ϕ.

6.7 Excitation Energy Transfer and Light Harvesting Phenomena

Apart from the fluorescence quenching and excimer techniques, excitation energy transfer provides another useful method for the study of structural changes in lipid vesicles and for the study of vesicle fusion. Rehorek et al.[121] have used a novel fluorescence energy transfer technique involving DPH (donor) and retinal (acceptor) to test the current theories of fluorescence energy transfer in two dimensions and to obtain information on the effect of intrinsic membrane protein bacteriorhodopsin on the order and dynamics of a lipid bilayer in DMPC vesicles. Increasing the surface concentration of acceptors by raising the protein:lipid ratio leads to a decrease in the mean fluorescence lifetime by a factor of four. When the acceptor concentration is reduced at a fixed protein:lipid ratio by photochemical bleaching of retinal, the lifetime increases and reaches approximately the value in protein-free vesicles when the bleaching is complete. The shape of the decay curve and the dependence of the mean lifetime on the surface concentration are in good agreement with theoretical predictions for a two-dimensional random distribution of donors and acceptors. From this analysis, a distance of closest approach between donors and acceptors to 18 Å was derived (a value close to the effective radius of bacteriorhodopsin), which is consistent with the current theories about the location of retinal in the interior of the protein.

The internal organization of bilayers in the vesicles composed of DMPC and a sugar-containing membrane protein, glycophorin, has been investigated by energy transfer involving labelled fluorescein (donor) and Eosin (acceptor) dyes.[122] The concentration dependence of the energy transfer efficiency has been interpreted in terms of a two-conformation model: at low concentrations (≤ 0.08 mol.%) the carbohydrate-carrying head groups of the protein spread at the lipid–water interface forming a two-dimensional pancake-like structure; at higher concentrations, the head groups start to assume a three-dimensional conformation protruding into the aqueous phase.

Fusion of two lipid vesicles and the exchange of lipid molecules between vesicles are being investigated by various physical methods. Kano et al.[123] have used the time-dependent energy transfer from alloxazines to isoalloxazines to measure the exchange rates after mixing of two separately prepared DPPC vesicles. The intervesicle exchange rates of fluorescent probes decreased with increasing alkyl chain length. Considerably fast exchange occurs when the probes have short or intermediate alkyl side chain length (e.g., butyl or octyl) while the transfer rate of the probes with a long alkyl chain (CBIA) is markedly reduced. The probes having two alkyl chains (DDA) are completely fixed at lipid bilayers.

A fluorescence stopped-flow technique has also been used by Almgren[124] in which excitation energy transfer of the Förster type between pyrene and perylene has been used to monitor the migration of these aromatic molecules between the lipid vesicles in aqueous solution. It is shown that the migration occurs via the aqueous phase, the rate being controlled by the exit rate of the solubilized molecules from the vesicles. The exit rate and hence the migration rate is diffusion controlled and is simply related to the size of the vesicles and the partition coefficient of the molecules between the lipid bilayer and the aqueous phase. The exit rate constants at 18°C were obtained as 3.0 ± 0.5 sec^{-1} for perylene and 77 ± 6 sec^{-1} for pyrene in soybean lecithin vesicles containing 10% dicetylphosphate in 2 mM tris buffer and 8 vol% of ethanol. Similar results have been obtained with egg lecithin and DMPC vesicles, whereas no relaxation process was detected in pure dicetylphosphate vesicles.

As mentioned earlier, resonance energy transfer is one of the photophysical methods that has found utility in the study of membrane fusion. The methodology can be described as follows.[125,126] Vesicles are prepared using lipid molecules carrying two different chromophore groups. When both fluorescent lipids are in the same vesicle at appropriate surface densities efficient energy transfer is observed. This is observed, for example, in the pair N-7-nitro-2,3-benzoxadiazolyl-4-yl (donor) and rhodamine B (acceptor) labelled at the free amino group of PE present at about 0.01 ratio of fluorescent lipid to the total lipid.

When such labelled vesicles are mixed with a population of pure lipid vesicles and the process of vesicle fusion induced by the addition of Ca^{2+} or Mg^{2+}, the two probes mix with other lipids present to form new membranes. This mixing reduces the surface density of the acceptor resulting in a decreased efficiency of the resonance energy transfer process being measured experimentally. These changes in the transfer efficiency allow kinetic and quantitative measurements on the fusion process. By comparison of derived surface densities with estimates based on various models, Struck et al.[125] have concluded that the vesicle fusion process involves a complete randomization of all input lipid molecules and that the final fusion product involves coalescence of at least ten or more of the input, small unilamellar vesicles.

Closely related to the vesicle fusion process is the phase separation of lipids which can also lead to self-quenching of the donor fluorescence due to an increase in the local concentration. Addition of Ca^{2+} to a mixture of two PS vesicle populations one containing a small amount of donor (NBD) and the other acceptor (N-rhodamine-PE), for example, leads to vesicle fusion and quenching of the NBD fluorescence via energy transfer quenching. Addition of Ca^{2+} to PS/N-NBD-PE vesicles alone leads to NBD self-quenching via the phase separation process. With Ca^{2+}, the fusion proceeds much more rapidly ($t_{1/2} \sim 5$ sec) than the process of phase separation ($t_{1/2} \sim 1$ min). Mg^{2+} also induced fusion, albeit at higher concentrations than Ca^{2+}. However, Mg^{2+}-induced phase separation was not detected. Fusion/phase separation studies in the presence of dehydrating agent polyethylene glycol (PEG) suggested that macroscopic phase separation may facilitate but does not induce the fusion process and, hence, is not directly involved in the actual fusion mechanisms.[126]

Vanderwerf and Ullman[127] have used similar methods to study Ca^{2+}-induced fusion of vesicles composed of PS, PC, and PE. More than 85% quenching results when PS vesicles labelled with dansyl-PE (donor) are fused with vesicles labelled with rhodamine-PE (acceptor) in the presence of 5 mM CaCl$_2$ or 10 nM MgCl$_2$. Higher concentrations of divalent cations are required to obtain maximal quenching when PS is partiallly replaced by PE or PC. The rate of vesicle fusion is dependent on the concentration of both the cation and the vesicles.

6.8 Photoredox Reactions and Charge Separation Phenomena

Photochemical properties of pigments incorporated into phospholipid bilayer membranes have been extensively studied in recent years. The goals of these studies are on one hand, to duplicate the photoredox chemistry of photosynthetic systems *in vitro* and, in doing so, adapt such systems for the photochemical conversion of solar energy. The reasons for picking lipid vesicles as reaction media for studies *in vitro* are the striking similarity of these systems with the thylakoid membranes and the unresolved questions as to how the lipid architecture efficiently brings about the two primary events of photosynthesis, namely, the collection followed by extraordinarily rapid transport of the excitation energy in the antenna and quantitative charge separation in the reaction center.

Very early studies in lipid bilayers were concerned with photoelectric effects in pigmented membranes.[128] Chloroplast, for example, extracts were incorporated into the phospholipid bilayer (in the form of thin films) separating two aqueous solutions of different redox potentials and photocurrents through the bilayer were monitored. Such thin membrane films appeared black (hence the name black lipid membranes for these films), and we will return to some of these studies in the next chapter. Nichols *et al.* demonstrated photoredox reactivity of chlorophyll in lipid liposomes (probably for the first time!).[129] They observed that illumination of chlorophyll incorporated into lipid liposomes caused oxidation of added ferrocytochrome *c*. Later, Tomkiewicz and Corker[130] detected by ESR light-driven chlorophyll radical formation in chlorophyll-containing liposomes in the presence of electron acceptors such as $Fe(CN)_6^{4-}$ or Eu^{3+} at 77 K. Even photooxidation of water in vesicles containing chlorophyll and $Fe(CN)_6^{4-}$ have been claimed[13] though attempts to duplicate this experiment have been unsuccessful.[132] Soon after more detailed room-temperature ESR studies of photoinduced electron transfer in chlorophyll-containing liposomes were reported by Oettmeier *et al.*[133] Acceptors having access to the lipid membrane such as $Fe^{3+}-$ pyrophosphate and methylviologen (MV^{2+}) gave rise to chlorophyll radical cation formation and with quinone formation of semiquinone radicals under illumination. Mangel[134] reported light-induced charge transport in liposomes solutions on introduction of a gradient of redox potential across the liposome membrane. Sudo and Todo[135] similarly observed photoreduction of fast red by ascorbate (in/on liposome?) using stearyl-anthraquinone-2-sulfonate as a sensitizer. These early studies, though largely qualitative and lacking precision on the structural organization of the photoredox partners, clearly established that redox reactions do take place at the membrane–solution interface under the influence of light absorption by incorporated dye molecules.

The presence of aqueous phases on either side of the lipid bilayer in vesicles and liposomes provides chemists with an opportunity to organize reactants on a microscopic (molecular) level. By appropriate choice of the electron donors, acceptors, and the preparation techniques, it is possible to construct vesicular systems that carry either the electron donor or the acceptor in the inner aqueous phase and the other component in the outer phase. Hydrophobic, water-insoluble photosensitizers are bound to the lipid phase. Such systems should allow a more direct assay on the possibility of achieving vectorial electron transport from the outer to the inner aqueous phase through the lipid bilayer.

Starting with the early experiments of Ford et al.[136] on these lines, there have been numerous studies of this kind[136-142] including a few studies using millipore membranes.[143-146] Figure 6.9 presents schematically various steps involved in the preparation of such asymmetrically organized systems. Egg lecithin vesicles carrying a surfactant derivative of tris(bipyridyl)-ruthenium(II), RuL_3^{2+}, were prepared initially in an aqueous buffer solution containing the electron donor EDTA. Untrapped EDTA in the outer aqueous phase was rapidly removed by gel filtration using Sephadex G-25. The eluted vesicle fractions were subsequently transferred to an aqueous solution containing the electron acceptor (MV^{2+}) and zinc acetate. (The latter was added to complex any residual EDTA present in the outer aqueous phase.) Visible light irradiation of such organized vesicular dispersions leads to a net

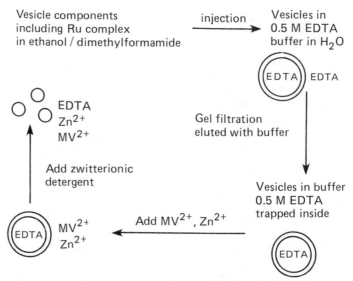

FIG. 6.9. Schematic diagram of the method used for the preparation of asymmetric lipid vesicles and liposomes. (From Ford et al.[136] Copyright © 1978 Macmillan Journals Limited.)

photoreduction of MV^{2+} sensitized by the Ru complex:

$$[MV^{2+}]_0 + [EDTA]_i \xrightarrow[\text{egg PC}]{hv, RuL_3^{2+}} [MV^+]_0 + \text{oxidation products of } [EDTA]_i \quad (6.27)$$

The quantum yield for the production of MV^+ radical was found to depend on the phospholipid : Ru complex ratio. The rate constant for the transmembrane electron transport was estimated to be of the order of 10^4–10^6 sec^{-1}, several orders of magnitude faster than the transmembrane diffusion of lipids. Because of the large differences in the bilayer thickness compared to the diameter of the Ru complex, the electron transport is pictured to occur via electron exchange (from the oxidized Ru complex on one side of the bilayer to a Ru^{2+} complex on the other side transversally). The exchanged electron may have to tunnel through part of the hydrocarbon-like core of the bilayer membrane. Using procedures analogous to Calvin et al., Kurihara et al.[139,140] found photoreduction of Cu^{2+} in the solution-side outside liposomes containing chlorophyll was enhanced by a reductant such as ascorbate localized in the solution inside the liposome:

The need for any electron mediator or others (if any) to promote such electron transport across the lipid bilayer has not been clearly established. In the in-vivo systems, β-carotene is believed to play an essential role. In early studies, Calvin et al.[136] used vitamin K_1, quinone, and decachlorocarborane to assist the electron transport. Later studies by these authors[137] found these additional components not necessary. Matsuo et al.[141,142] found the photoreduction of anthraquinone sulfonates in Zn–porphyrin containing vesicles to be stimulated by the addition of reagents such as 1,3-dibutyl (or 1,3-didodecyl)alloxazine. In transmembrane electron transport, protons or other cations must also be transported for charge neutralization. In chlorophyll-containing liposomes, Kurihara et al. found the photoreduction of $Fe(CN)_6^{3-}$ to be enhanced by the addition of proton carriers.[140] Laane et al.[138] found the photosensitized reduction of heptylviologen ($(C_7)_2V^{2+}$) to be enhanced by a factor of 6.5 on addition of the ionophore valinomycin in the presence of K^+. Addition of gramicidin lead to a three-fold increase in the rate, results that indicate that the rate of photoinduced electron transport across vesicle walls in the absence of ion carriers is limited by cotransport of cations. As an alternative to the electron exchange mechanism, Matsuo et al.[142] have proposed a two-photon mechanism:

Electron exchange mechanism   Two photon mechanism

Yablonskaya and Shafirovich[147] have described yet another version of photosensitized electron transport in lipid vesicles. An electron acceptor such as octadecylviologen is incorporated in the lipid vesicles. Photosensitizer $Ru(bpy)_3^{2+}$ or a $ZnTMPyP^{4+}$ (Zn tetrakismethylpyridyl porphyrin) was encapsulated in the inner aqueous phase and an oxidant (methylene blue) in the outer phase. Illumination of such a system leads first to the sensitized reduction of methylene blue (in the outer phase via reduced viologen) and when this process is complete, then there is a net photoreduction of viologen.

Tollin and co-workers have made extensive flash photolysis studies[148-152] on the kinetic and mechanistic aspects of these electron transport processes across lipid walls, especially those involving chlorophyll a. In the photoredox quenching of chlorophyll triplets by MV^{2+} or benzoquinone (BQ), the reverse electron transfer reaction:

$$^3Chl^* + A \rightleftharpoons Chl^+ + A^- \tag{6.28}$$

was biphasic containing fast and slow components with decay half-lifes that differ by about two orders of magnitude. It has been proposed that the kinetics reflect two different environments: the fast component is due to the recombination within the membrane bilayer and the slow component is due to a recombination across the water–bilayer interface. The radical recombination was also independent of light intensity, suggesting first-order pathways (probably an intra-aggregate process as observed in micellar solutions). Interestingly, for a given electron acceptor, the kinetics of quenching and reverse electron transfer were different, depending on whether the acceptor was present in the inner or outer aqueous phase.

A lower rate of 10^4 sec^{-1} has been set for the rate constant for electron transport through the lipid bilayer. This is based on simple arguments that EDTA could effectively compete for Chl^+ in the reverse reaction. The magnitude of this rate constant strengthens arguments for electron transport mechanisms that involve electronic rather than molecular charge carriers. Based on the kinetic behavior of the redox products in the quenching of $^3Chl^*$

by a water-soluble naphthoquinone (S-(2-methyl-1,4-naphthoquinonyl)-3-glutothione) the rate constant for electron exchange between Chl^+ and Chl present within the lipid monolayer has been estimated to be 3.2 × $10^6 \ M^{-1} \ sec^{-1}$.

The overall quantum efficiency of charge separation across the bilayer thus depends on the relative rates of electron transfer reactions across the lipid–water interface and across the bilayer itself. Attempts have been made to manipulate the rates of these individual reactions in order to optimize the efficiency of energy conversion by incorporating either neutral spacer molecules such as cholesterol or ionic surfactants such as DHP (dihexylphosphate) or DODAB (didodecyldimethylammonium bromide). Incorporation of cholesterol at $\geqslant 15$ mol% produced two main effects: alteration in the distribution of chlorophyll molecules within the vesicles and an increase in the accessibility of these chlorophyll molecules to water and other quencher molecules.

6.9 Photoprocesses in Surfactant Vesicles

6.9.1 General Aspects of Surfactant Vesicles

Apart from the synthetic and naturally derived phospholipids, there exists a wide variety of surfactant molecules with double or triple hydrocarbon chains attached to a polar head group that also form closed bilayer structures found in vesicles and liposomes.[153–156] Gebicki and Hicks observed the formation of bilayer structures in aggregates of simple surfactants.[156] With the elegant synthesis of hundreds of surfactant molecules of assorted types and a systematic study of their aggregated forms, Kunitake and co-workers have greatly expanded the scope and utility of surfactant vesicles. Basing their strategy on the known ability of diphenylazomethine dyes to form liquid crystalline structures, these authors have demonstrated that certain single-chain surfactants are also capable of forming closed bilayer structures.[157,158] (Theories of self-assemblies by Tanford, Israelachvili, and others suggest the presence of two alkyl chains as requirements for the formation of bilayer structures.)

Based on the aggregation behavior of numerous surfactants (single, double, and triple chain hydrocarbon or fluorocarbon chains attached to polar head groups), Kunitake et al. have identified three essential structural elements in surfactants for the formation of bilayer structures:[157] (1) a flexible tail, which consists of linear methylene chains (C_7 or longer) or related structures; (2) a rigid segment, which consists usually of two benzene rings as in biphenyl and azobenzene, and (3) a hydrophilic head group, which consists of groups such as quarternary ammonium, phosphate, or sulfonate:

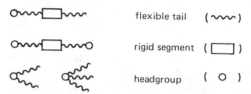

The presence of additional spacer groups (methylene groups C_{10} or more inserted between the rigid segment and the head group) and other interacting groups such as esters promotes vesicle formation. Table 6.5 illustrates some of the surfactant structures that have been studied.

Surfactant vesicles in general exhibit most of the properties characteristic of the lipid bilayers of vesicles and liposomes, e.g., thermotropic phase transitions. Synthetic surfactant vesicles, however, exhibit stability over a relatively narrow range of solution conditions. In general, high salt concentrations or the presence of oxyanions or polyanions leads to destabilization of the vesicle structures. Most of the photochemical studies have been principally

TABLE 6.5. Morphologies of Membrane Structures formed by Various Single, Double, and Triple Chain Amphiphilic Molecules in Aqueous Solution[a]

Surfactant type	Molecular structure	Type of membrane formed	Tail
Single chain		Bilayer	Hydrocarbon
Single chain		Bilayer	Fluorocarbon
Single chain		Monolayer	Hydrocarbon
Double chain		Bilayer	Hydrocarbon
Double chain		Bilayer	Fluorocarbon
Triple chain		Bilayer	Hydrocarbon
Triple chain		Bilayer	Fluorocarbon
Mixed chain		Monolayer	Hydrocarbon
Mixed chain		Monolayer	Hydrocarbon

[a] From Kunitake et al.[158]

with surfactant vesicles formed by dialkyldimethylammonium halides such as DODAC or dialkylphosphates such as dihexadecylphosphate (DHP).

$$C_{18}H_{37} \diagdown \overset{+}{N} \diagup CH_3 \quad (Cl^-) \qquad C_{16}H_{33}\!-\!O \diagdown \underset{O}{\overset{O^-}{P}} $$

DODAC (cationic)   DHP (anionic)

Studies using DHP in particular have revealed that the size, and stability of vesicles, their adsorption and entrapment capabilities, and the permeability of the vesicular membrane are strongly affected by the pH and the ionic strength. Hence, preparation, handling, and usage of surfactant vesicles warrrants careful scrutiny.

6.9.2 Photochemical Studies in Surfactant Vesicles

Fluorescence Probe Analysis

McNeil and Thomas made a comparative study of cationic CTAB micelles and DODAB (C_{12}) vesicles using pyrene and pyrenecarboxaldehyde (PyCHO) as probes.[159] The vibronic band intensity ratio (III/I) of pyrene and the emission maximum of PyCHO indicate a low dielectric constant ($\varepsilon \sim 10$) for the probe environment. The positive charges on the surfaces of the vesicles cause an enhancement in the I^- quenching and a decrease in Tl^+ quenching of pyrene fluorescence as one would anticipate. The quenching by neutral-molecule oxygen is about 5.7 times more efficient as compared to that in CTAB micelles. The formation of the amine cation in the quenching of pyrene fluorescence by dimethylanilene:

$$^1Py^* + DMA \rightleftharpoons \underset{\text{exciplex}}{[Py\cdots DMA]^*} \longrightarrow Py^- + DMA^+ \qquad (6.29)$$

was found to be quite low and consist with the low polarity of the probe environment.

Fluorescence Depolarization

The polarization of solubilized probes 2-methyl anthracene[160] or DPH[158,161] in surfactant vesicles is quite high (0.3–0.4) at temperatures below the phase transition temperature T_c, implying a high microviscosity for the bilayer region. As with lipid vesicles, the steady-state emission anisotropy value can be used to detect phase-transition temperatures.[158] Table 6.6 presents a comparison of T_c values for a few surfactant vesicle systems measured using the fluorescence polarisation and differential calorimetry methods.

TABLE 6.6. Comparison of Phase-Transition Temperatures
for Bilayers Formed by Triple Chain Surfactants,
Determined by Fluorescence Polarization and by
Calorimetric Methods[a]

Surfactant	T_t (fluorescence polarization)	T_t (calorimetry)
$3C_{16}$-N^+	43, 60	48.6, 60
$3C_{12}$-te-N^+	38	37, 41.5
$3C_{16}$-te-N^+	63	60.5
$3C_{12}$-tris-C_2N^+	31	28.7
$3C_{16}$-tris-C_2N^+	62	61.5
$3C_{12}$-tris-$C_{10}N^+$	41	41.5

[a] Values in °C. From Kunitake *et al.*[158] Abbreviations: $3C_n$-$N^+ = [CH_3(CH_2)_{n-1}]_3\overset{+}{N}(CH_3)$, $3C_n$-te-$N^+ = [CH_3(CH_2)_{n-2}COOCH_2CH_2]_3\overset{+}{N}(CH_3)$, and $3C_n$-tris-$C_mN^+ = [CH_3(CH_2)_{n-2}COOCH_2]_3CNHCO(CH_2)_{m-1}\overset{+}{N}(CH_3)_3$.

Excimers

In surfactant vesicles composed of double- or triple-chain amphiphiles carrying a benzene group in the spacer region, we observe the formation of excimers (emission maxima around 357 nm and monomer emission at 327 nm).[158,161] Phase transitions cause a significant change in the bilayer fluidity to affect the excimer/monomer emission intensity ratios (I_{357}/I_{327}) that they can be used, as in lipid vesicles, to monitor phase transitions. Illustrative cases can be found in the following surfactant-based systems:

$$CH_3—(CH_2)_{11}—O—CO—CH_2$$
$$CH_3—(CH_2)_{11}—O—CO—CH—NH—CO—\langle\bigcirc\rangle—O—(CH_2)_4\overset{+}{N}(CH_3)_3$$

$$(2C_{12}—L—Glu—ph—C_4\overset{+}{N})$$

$$[CH_3—(CH_2)_{10}—O—CO—CH_2]_3C—NH—CO—\langle\bigcirc\rangle—O—(CH_2)_m—\overset{+}{N}(CH_3)_3$$

$$(3C_{12}—tris—ph—C_m\overset{+}{N})$$

Determination of aggregation number for DODAB vesicles using the excimer formation process with unbound molecules (cf. Section 2.6) indicates $N > 1000$.[159] Detailed studies have been made on the intramolecular excimer formation[162,162a] in surfactant vesicle systems composed of DHP and DODAB and in lipid vesicles composed of DPPC and DSPC, using 1,1′-dipyrenylmethylether. Below the phase-transition temperature, the behavior was similar in both types. For surfactant vesicles above T_c however, there was no significant change in the bilayer structure as evidenced by the absence of any changes in the (excimer/monomer) emission intensity ratio. For DHP vesicles, the method of preparation drastically affects the excimer behavior. Vesicles prepared in pure water show a blue-shifted emission compared to that prepared in buffer solutions (480 nm in water instead of 505 nm found in buffer and in DODAB vesicles) presumably reflecting different solubilization sites.

Energy Transfer

Nomura et al.[163] have demonstrated intermolecular Förster-type energy transfer between a pyrene chromophore (donor) labelled onto a lipid and pyranine attached to the surface of DODAC vesicles. Depending on the concentration of the acceptor, energy transfer efficiencies up to 43% were observed:

$$^1\text{Pyrene*} + \text{Pyranine} \longrightarrow \text{Pyrene} + {}^1\text{Pyranine*} \qquad (6.30)$$

In the absence of vesicles the efficiency of the energy-transfer process is $\leqslant 3\%$.

Excitation singlet–singlet energy transfer between pyrene and proflavine in DODAC vesicles has been subjected to a detailed kinetic investigation.[164] Compared to that in DPPC (lipid) vesicles, the energy transfer process is very efficient in surfactant vesicles ($>90\%$). The energy transfer process occurring in vesicles has been treated in terms of two-dimensional energy transfer. According to Förster theory for three-dimensional energy transfer in a rigid matrix, the decay of excited donor molecule is given by:

$$P(t) = \exp[-(t/\tau_D) - \sqrt{\pi} N_A R_0^3 R_g^{-3}(t/\tau_D)^{1/2}] \qquad (6.31)$$

where τ_D is the fluorescence lifetime of the donor in the absence of quenching N_A the number of acceptors, and $R_g = \frac{4}{3}\pi V^3$, where V is the volume of the reaction vessel. Several theoretical equations have been proposed for the case of two-dimensional energy transfer.[165–168] In the above case, the pyrene fluorescence decay was found to fit the following equation:

$$P(t) = \exp[-(t/\tau_D) - \varepsilon C(t/\tau_D)^{1/3}] \qquad (6.32)$$

where

$$\varepsilon = \pi\Gamma(2/3) \simeq 4.25409 \quad \text{and} \quad C = R_0^2 C \qquad (6.33)$$

$\Gamma(2/3)$ is the gamma function and C is the two-dimensional concentration of the acceptor (i.e., the number of molecules per unit area).

Photoredox Reactions and Charge Separation Phenomena

Principles and applications of charge separation in photoredox reactions have been elaborated for the case of ionic micelles (Section 3.5) and lipid vesicles Section 6.9). There have been several studies in surfactant vesicles with similar goals and strategies.[169-181] Interests in vesicles in particular has been in the demonstration of transmembrane electron transport:

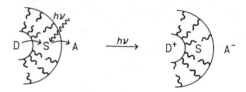

The architecture of the vesicles with the presence of charged interfaces can be exploited to advantage in photoionization and photoredox processes. The photoionization of pyrene in DHP vesicles,[169] for example, has shown that the efficiency of the process is enhanced by the stabilization of the pyrene cation

$$Py^* \xrightarrow{h\nu} Py^+ + e_{aq}^- \qquad (6.34)$$

at the anionic interface and ejection of electrons to the aqueous phase. The Ag^0 species generated via photoreduction of Ag^+ complexes using cyanine dyes is stabilized in the lipid environment of the surfactant vesicles if the substrate is in the form of Ag^+–crown ether as in crown ether surfactants[170]

$$LAg^+ + D^* \longrightarrow LAg^0 + D^+ \qquad (6.35)$$

The electron-transfer quenching of $Ru(bpy)_3^{2+*}$ by viologens shows striking differences to that observed in homogeneous solvents if we employ a functional surfactant vesicle formed by a alkylviologen surfactant $(R, Me)V^{2+}$ as[171]

$$R = (CH_2=C(CH_3)-CO-O-(CH_2)_{11}-)$$

The rate of the forward reaction is enhanced as compared to the $Ru(bpy)_3^{2+}-MV^{2+}$ pair in water, but the reverse reaction is slowed down. Actually the reverse electron transfer shows a biphasic character with a slow component reflecting the ejection of RuL_3^{3+} by the positive charges on the cationic vesicle. In a related study, ESR monitoring of the UV photolysis of MV^{2+} in cationic DODAB and anionic DHP vesicles showed that viologens bind strongly to DHP vesicles.[172]

Three different pathways have been recognized in the reverse electron transfer reaction that follows the reductive quenching of surfactant derivatives of $Ru(bpy)_3^{2+*}$ by N-methylphenothiazine and its surfactant derivatives in DODAC vesicles[173]

$$R-PTH + RuL_3^{2+} \underset{DODAC}{\overset{h\nu}{\rightleftharpoons}} R-PTH^+ + RuL_3^+ \qquad (6.36)$$

(1) a rapid recombination of the initially formed redox pairs (2) escape of R—PTH$^+$ into the inner aqueous phase where it can react with RuL_3^+, and (3) escape into the bulk aqueous phase where RPTH$^+$ survives for extended periods (several milliseconds).

Transmembrane photoredox processes are studied in asymmetrically organized vesicle systems—systems where either the donor or acceptor is located in the inner aqueous phase, the other is in the outer aqueous phase and a hydrophobic sensitizer bound to the vesicles (cf. scheme, p. 212). Tunuli and Fendler[174] observed the formation of MV$^+$ as a permanent product during the irradiation of the organized vesicle system $MV_i^{2+}/DHP-RuL_3^{2+}/EDTA_o$ (The subscripts i and o indicate location of redox reagents in the inner and outer aqueous phases.) The process was pictured as transmembrane photoredox reaction. Later investigations using the zinc porphyrin ZnTMPyP^{4+} and surfactant derivatives of RuL_3^{2+} revealed that photoinduced transmembrane diffusion of the acceptor viologen occurs first followed by photoredox reactions at the outersurface[175-178]

The experiments clearly point out some of the potential permeability/leakage problems associated with these surfactant vesicles that complicate unambiguous demonstration of transmembrane photoredox processes.

Pronounced magnetic field effects have been observed in the photochemistry of benzophenone in surfactant vesicles incorporated with colloidal particles of magnetite.[181] The colloidal particles play the role of large external magnets *in situ* and the field effect observed is comparable to those in DTAC micelles in the presence of a 2000 G external magnetic field.

6.10 Photoprocesses in Polymerized Vesicles

Given the problems of long-term stability and lack of controllable permeability of reagents in normal vesicles, attention is being focused on polymerized versions of these vesicle systems. A wide variety of surfactant and lipid molecules with polymerizable double bonds (vinyl, methacrylate, diacetylene, isocyano, stryl, etc.) are being synthesized.[182–184] Subsequent to vesicle formation, the constituents are polymerized by irradiation with UV light or gamma rays in the presence of chain initiators such as persulfate or azobis(isobutyro)nitrile. Depending on the position of the double bonds, vesicles could be polymerized across either bilayers or their head groups. It has been reported that polymerized vesicles do have lifetimes of several months, remain stable up to 20% alcohol content, and have controllable size and permeability properties.

Fendler and co-workers have made some early photochemical studies in these systems.[185–189] The kinetics of photopolymerization has been studied by the technique of laser photolysis and details of intravesicular surface polymerization deduced.[185] Fluorescence polarization studies using negatively charged probes erythrosine and pyranine[186,187] in polymerized and nonpolymerized vesicles have indicated formation of clefts in the polymerized vesicles:

Presumably cleft formation is caused by pulling together of some of the aromatic rings. Gradual penetration of pyranine with increasing incubation time observed in normal micelles is suppressed in polymerized vesicles.

In a novel line of investigations, polymerized vesicles are being examined as potential media for the deposition of finely divided redox catalysts and colloidal semiconductors of interest in studies of photochemical solar energy conversion.[188–192] Stable uniformly small Pt colloids have been prepared *in situ* in the interiors of vesicles prepared from mixtures of DPPC and a polymerizable surfactant. Finely divided Rh-coated CdS particles generated *in situ* inside DHP vesicles have been shown to evolve H_2 efficiently on photolysis in the presence of electron donor molecules.

References

1. A. D. Bangham, *Prog. Biophys. Mol. Biol.* **18**, 29 (1968).
2. A. D. Bangham, M. W. Hill, and N. G. A. Miller, *in* "Methods in Membrane Biology" (E. D. Korn, ed.), pp. 1–68. Plenum, New York, 1974.
3. G. Gregoriadis and A. C. Allison, eds., "Liposomes in Biological Systems." Wiley (Interscience), New York, 1980.
4. D. Papahadjopoulos, ed., "Liposomes and Their Use in Biology and Medicine, Annals of the New York Academy of Sciences Vol. 308. N.Y. Acad. Sci., New York, 1976.

5. D. Chapman, ed., "Biomembrane Structure and Function," Molecular and Structural Biology Series, Vol. 4. Macmillan, London 1983.

6. M. K. Jain and C. Wagner, "Introduction to Biological Membranes." Wiley, New York, 1980.

7. E. Grell, ed., "Membrane Spectroscopy," Molecular Biology, Biochemistry and Biophysics, Vol. 31. Springer-Verlag, Berlin and New York, 1983.

8. D. Chapman, in "Membrane Structure and Function" (E. E. Bittar, ed.), pp. 103–152. Wiley, New York, 1980.

9. D. Chapman and D. F. H. Wallach, eds., "Biological Membranes." Academic Press, New York, 1976.

10. R. B. Cundall and R. E. Dale, eds., "Time Resolved Fluorescence Spectroscopy in Biochemistry and Biology," NATO Advanced Institute Series A69. Plenum, New York, 1983.

11. J. Seelig and A. Seelig, $Q. Rev. Biophys.$ **13,** 19 (1980).

12. J. H. Fendler, "Membrane Mimetic Chemistry." Wiley, New York, 1982. (a) J. H. Fuhrhop and J. Mathieu, $Angew. Chem. Int. Ed. Engl.$ **23,** 100 (1984).

13. G. K. Radda and J. Vanderkooi, $Biochim. Biophys. Acta$ **265,** 509 (1972).

14. A. Azzi, $Q. Rev. Biophys.$ **8,** 237 (1975).

15. E. Sackmann, $Z. Phys. Chem.$ $(Wiesbaden)$ **101,** 391 (1976).

16. E. Sackmann, $Ber. Bunsenges. Phys. Chem.$ **82,** 891 (1978).

17. L. Brand and J. R. Gohlke, $Annu. Rev. Biochem.$ **41,** 843 (1972).

18. P. Lianos, A. K. Mukhopadhyay, and S. Georghiou, $Photochem. Photobiol.$ **32,** 415 (1980).

19. S. Georghiou and A. K. Mukhopadhyay, $Biochim. Biophys. Acta$ **645,** 365 (1981).

20. D. A. N. Morris, R. McNeil, F. J. Castellino, and J. K. Thomas, $Biochim. Biophys. Acta$ **599,** 380 (1980).

21. N. Wakayama and Y. Kondo, $Biochim. Biophys. Acta$ **647,** 155 (1981).

22. T.-C. Lee and V. Fitzgerald, $Biochim. Biophys. Acta$ **598,** 189 (1980).

23. K. Onuki, K. Kurihara, Y. Toyashi, and M. Sukihara, $Bull. Chem. Soc. Jpn.$ **53,** 1914 (1980).

24. W. K. Surewicz and W. Lekyo, $Biochim. Biophys. Acta$ **643,** 384 (1981).

25. M. G. Rockley and D. S. Najjar, $Biochim. Biophys. Acta$ **644,** 96 (1981).

26. J. Dufourcq, J. F. Faucon, R. Maget-Dana, M.-P. Pileni, and C. Helene, $Biochim. Biophys. Acta$ **649,** 75 (1981).

27. P. Somerharju, H. Brockelhoff, and K. W. A. Witz, $Biochim. Biophys. Acta$ **649,** 521 (1981).

28. D. Lichtenberg, P. L. Felgner, and T. E. Thompson, $Biochim. Biophys. Acta$ **684,** 277 (1982).

29. J. L. Slater, D. Lichtenberg, and T. E. Thompson, $Biochim. Biophys. Acta$ **734,** 125 (1982).

30. R. Sundler and D. Papahadjopoulos, $Biochim. Biophys. Acta$ **649,** 743 (1981).

31. R. Sundler, N. Duzgunes, and D. Papahadjopoulos, $Biochim. Biophys. Acta$ **649,** 751 (1981).

32. D. Hoekstra, $Biochim. Biophys. Acta$ **692,** 171 (1982).

33. J. Wilschutt, N. Duzgunes, and D. Papahadjopoulos, $Biochemistry$ **19,** 6011 (1980).

34. N. Duzgunes, J. Wilschut, R. Fraley, and D. Papahadjopoulos, $Biochim. Biophys. Acta$ **642,** 182 (1981).

35. G. S. Beddard, S. E. Carlin, and G. Porter, $Chem. Phys. Lett.$ **43,** 27 (1976).

36. D. A. Barrow and B. R. Lentz, $Biochim. Biophys. Acta$ **645,** 17 (1981).

37. R. E. DeToma, J. H. Easter, and L. Brand, $J. Am. Chem. Soc.$ **98,** 5001 (1976).

38. J. H. Easter, R. P. DeToma, and L. Brand, $Biophys. J.$ **16,** 571 (1976).

39. E. D. Matayoshi and A. M. Kleinfeld, $Biochim. Biophys. Acta$ **644,** 233 (1981).

40. J. R. Lakowicz, R. B. Thompson, and H. Cherek, $Biochim. Biophys. Acta$ **734,** 295 (1981).

41. J. R. Lakowicz and D. Hogan, $Biochemistry$ **20,** 1366 (1981). (a) D. Axelrod, D. E. Koppel, J. Schlessinger, E. Elson, and W. W. Webb, $Biophys. J.$ **16,** 1055 (1976). (b) D. E. Koppel, D. Axelrod, J. Schlessinger, E. Elson, and W. W. Webb, $Biophys. J.$ **16,** 1315 (1976). (c) D. E. Koppel, $Biophys. J.$ **28,** 281 (1979). (d) B. A. Smith and H. M. McConnel, $Proc. Natl. Acad. Sci. U.S.A.$ **75,** 2759 (1977). (e) K. Derzko and K. Jacbson, $Biochemistry$ **19,** 6050 (1980). (f) W. Vax, H. Kapitza, J. Stumpel, E. Sackmann, and T. Jounin, $Biochemistry$ **20,** 1392 (1981).

(g) D. W. Tank, E. S. Wu, P. R. Meers, and W. W. Webb, *Biophys. J.* **40,** 129 (1981). (h) E. S. Wu, K. Jacobson, and D. Papahadjopoulos, *Biochemistry* **16,** 3936 (1977). (i) D. E. Wolf, J. Schlessinger, E. L. Elson, W. W. Webb, R. Blumenthal, and P. Reinhart, *Biochemistry* **12,** 3476 (1972). (j) H.-J. Galla, W. Hartmann, U. Thielen, and E. Sackmann, *J. Membr. Biol.* **48,** 215 (1979). (k) K. Razi Naqvi, J.-P. Behr, and D. Chapman, *Chem. Phys. Lett.* **26,** 440 (1976). (l) P. F. Fahay, D. E. Koppel, L. S. Banak, D. E. Wolf, E. L. Elson, and W. W. Webb, *Science* **195,** 305 (1976).

42. S. Cheng and J. K. Thomas, *Radiat. Res.* **60,** 268 (1974).
43. G. M. Omann and J. R. Lakowicz, *Biochim. Biophys. Acta* **684,** 83 (1982).
44. A. S. Verkmann, *Biochim. Biophys. Acta* **599,** 370 (1980).
45. V. G. Bieri and D. F. H. Wallach, *Biochim. Biophys. Acta* **382,** 175 (1975).
46. E. London and G. W. Ferguson, *Biochim. Biophys. Acta* **649,** 89 (1981).
47. M. W. Geiger and N. J. Turro, *Photochem. Photobiol.* **26,** 221 (1975).
48. S. Fischoff and J. M. Vanderkooi, *J. Gen. Physiol.* **65,** 663 (1975).
49. H. G. Lee and J. G. Forte, *Biochim. Biophys. Acta* **601,** 152 (1980).
50. J. W. Nichols, M. W. Hill, A. D. Bangham, and D. W. Deamer, *Biochim. Biophys. Acta* **596,** 393 (1980).
51. J. W. Nichols, M. W. Hill, A. D. Bangham, and D. W. Deamer, *Proc. Natl. Acad. Sci. U.S.A.* **77,** 2038 (1980).
52. M. Rossignol, P. Thomas, and C. Grignon, *Biochim. Biophys. Acta* **684,** 195 (1982).
53. M. S. Fernandez, *Biochim. Biophys. Acta* **601,** 152 (1980).
54. H.-J. Galla and E. Sackmann, *Ber. Bunsenges. Phys. Chem.* **78,** 949 (1974).
55. H.-J. Galla and E. Sackmann, *Biochim. Biophys. Acta* **339,** 103 (1974).
56. A. K. Soutar, H. J. Pownall, and L. C. Smith, *Biochemistry* **13,** 2828 (1974).
57. J. M. Vanderkooi and J. B. Callis, *Biochemistry* **13,** 4000 (1974).
58. M. Almgren, *Chem. Phys. Lett.* **71,** 539 (1980).
59. S. C. Charlton, J. S. Olson, K. Y. Hong, H. J. Pownall, D. Louie, and L. C. Smith, *J. Biol. Chem.* **251,** 7952 (1976).
60. S. C. Charlton, K. Y. Hong, and L. C. Smith, *Biochemistry* **17,** 3304 (1978).
61. P. Sengupta, E. Sackmann, W. Kühnle, and H. P. Scholz, *Biochim. Biophys. Acta* **436,** 869 (1976).
62. M. A. Roseman and T. E. Thompson, *Biochemistry* **19,** 439 (1980).
63. M. C. Dooty, H. J. Pownall, Y. J. Kao, and L. C. Smith, *Biochemistry* **19,** 108 (1980).
64. M. C. Correa-Freire, Y. Barenholz, and T. E. Thompson, *Biochemistry* **21,** 1244 (1982).
65. S. Schenkman, P. S. Araujo, R. Dijkman, F. H. Quina, and H. Chiamovich, *Biochim. Biophys. Acta* **649,** 633 (1981).
66. H.-J. Galla and E. Sackmann, *J. Am. Chem. Soc.* **97,** 4115 (1975).
67. K. A. Zachariasse, W. Kuhnle, and A. Weller, *Chem. Phys. Lett.* **73,** 6 (1980).
68. M. L. Viriot, M. Bouchy, M. Donner, and J.-C. Andre, *J. Chim. Phys.* **79,** 525 (1982).
69. M. L. Viriot, R. G. Willard, I. Kaufman, J.-C. Andre, and G. Siest, *Biochim. Biophys. Acta* **733,** 34 (1983).
70. H.-J. Galla, U. Thielen, and W. Hartmann, *Chem. Phys. Lipids* **23,** 239 (1979).
71. H.-J. Galla, W. Hartmann, U. Thielen, and E. Sackmann, *J. Membr. Biol.* **48,** 216 (1979).
72. H.-J. Galla and J. Luiśetti, *Biochim. Biophys. Acta* **596,** 108 (1980).
73. J. M. Vanderkooi , S. Fischkoff, M. Andrich, F. Pod, and C. S. Owen, *J. Chem. Phys.* **63,** 3661 (1975).
74. C. S. Owen, *J. Chem. Phys.* **62,** 3204 (1975).
75. Y. Waka, F. Tanaka, and N. Mataga, *Photochem. Photobiol.* **32,** 335 (1980).
76. K. Kano, H. Kawazumi, T. Ogawa, and J. Sunamoto, *J. Phys. Chem.* **85,** 2204 (1981); *Chem. Phys. Lett.* **74,** 511 (1980).
77. U. Cogan, M. Shinitzky, G. Weber, and T. Nishida, *Biochemistry* **12,** 521 (1973).

78. M. Shinitzky, A. C. Dianoux, C. Gitler, and G. Weber, *Biochemistry* **10,** 2106 (1971).
79. M. Shinitzky and Y. Barenholz, *J. Biol. Chem.* **249,** 2652 (1974).
80. J. Suurkuusk, B. R. Lentz, Y. Barenholz, R. L. Brittonnen, and T. E. Thompson, *Biochemistry* **15,** 1393 (1976).
81. B. R. Lentz, Y. Barenholz, and T. E. Thompson, *Biochemistry* **15,** 4521 (1976).
82. M. P. Andrich and J. M. Vanderkooi, *Biochemistry* **15,** 1257 (1976).
83. K. Jacobson and D. Papahadjopoulos, *Biochemistry* **16,** 152 (1977).
84. B. Mely-Goubert and M. H. Freeman, *Biochim. Biophys. Acta* **601,** 315 (1980).
85. J. Sunamoto, K. Iwamoto, K. Inoue, T. Endo, and S. Nojima, *Biochim. Biophys. Acta* **685,** 283 (1982).
86. F. Sixl and H.-J. Galla, *Biochim. Biophys. Acta* **643,** 626 (1981).
87. N. Duzgunes, J. Wilschutt, K. Hong, R. Fraley, C. Perry, D. S. Friend, J. L. James, and D. Papahadjopoulos, *Biochim. Biophys. Acta* **732,** 289 (1983).
88. F. V. Cauwelaert, I. Hanssens, W. Herrmann, J.-C. Van Ceunebroeck, J. Baert, and H. Berghmans, *Biochim. Biophys. Acta* **727,** 273 (1983). (a) M. Shinitzky and I. Yuli, *Chem. Phys. Lipids* **30,** 261 (1982). (b) H. Pottel, W. Van der Meer, and W. Herrman, *Biochem. Biophys. Acta* **730,** 181 (1983).
89. U. Strehlow and F. Jahnig, *Biochim. Biophys. Acta* **641,** 301 (1981).
90. A. Ikegami, K. Kinoshita, T. Kouyama, and S. Kawato, in "Structure, Dynamics and Biogenesis of Membranes" (R. Sato and S. Ohnishi, eds.), p. 1.
91. K. Kinoshita, S. Kawato, and A. Ikegami, *Biophys. J.* **20,** 289 (1977).
92. M. P. Heyn, *FEBS Lett.* **108,** 359 (1979).
93. F. Jahnig, *Proc. Natl. Acad. Sci. U.S.A.* **76,** 6361 (1979).
94. G. Lipari and A. Szabo, *Biophys. J.* **30,** 489 (1980).
95. G. Lipari and A. Szabo, *J. Chem. Phys.* **75,** 2971 (1981).
96. P. Wahl, *Biophys. J.* **42,** 205 (1979).
97. C. Zannoni, *Mol. Phys.* **39,** 1813 (1979); **42,** 1303 (1981).
98. K. Kinoshita, A. Ikegami, and S. Kawato, *Biophys. J.* **37,** 461 (1982).
99. Van Blitternijk, R. P. Van Hoeven, and B. W. Van der Meer, *Biochim. Biophys. Acta* **644,** 323 (1981).
100. S. Kawato, K. Kinoshita, and A. Ikegami, *Biochemistry* **16,** 2319 (1977).
101. L. A. Chen, R. E. Dale, S. Roth, and L. Brand, *J. Biol. Chem.* **252,** 2163 (1977).
102. R. E. Dale, L. A. Chen, and L. Brand, *J. Biol. Chem.* **252,** 7500 (1977).
103. S. Kawato, A. Ikegami, S. Yoshida, and Y. Orii, *Biochemistry* **19,** 1598 (1980).
104. M. G. Badea, R. P. DeToma, and L. Brand, *Biophys. J.* **24,** 197 (1978).
105. W. R. Veatch and L. Stryer, *J. Mol. Biol.* **117,** 1109 (1977).
106. P. L. Wolber and B. S. Hudson, *Biochemistry* **20,** 2800 (1981).
107. C. D. Stubbs, T. Kouyama, K. Kinosita, and A. Ikegami, *Biochemistry* **20,** 4257 (1981).
108. S. Kawato, S. Yoshida, Y. Orii, A. Ikegami, and K. Kinosita, *Biochim. Biophys. Acta* **634,** 85 (1981).
109. K. Kinosita, S. Kawato, A. Ikegami, S. Yoshida, and Y. Orii, *Biochim. Biophys. Acta* **647,** 7 (1981).
110. K. Kinosita, S. Mitaku, A. Ikegami, N. Ohbo, and T. Kunii, *Jpn. J. Appl. Phys.* **15,** 2433 (1976).
111. K. Kinosita, R. Kataoka, Y. Kimura. O. Gotoh, and A. Ikegami, *Biochemistry* **20,** 4270 (1981).
112. K. Hildebrand and C. Nicholau, *Biochim. Biophys. Acta* **553,** 365 (1979).
113. M. Vincent, B. deForesta, J. Gallay, and A. Alfsen, *Biochem. Biophys. Res. Commun.* **107,** 914 (1982).
114. M. Vincent, B. deForesta, J. Gallay, and A. Alfsen, *Biochemistry* **21,** 708 (1982).
115. M. Vincent and J. Gallay, *Biochem. Biophys. Res. Commun.* **113,** 799 (1983).

116. J. R. Lakowicz and F. G. Prendergast, *Biophys. J.* **24**, 213 (1978).
117. J. R. Lakowicz and F. G. Prendergast, *Science* **200**, 1399 (1978).
118. J. R. Lakowicz, F. G. Prendergast, and D. Hogan, *Biochemistry* **18**, 508, 520 (1979).
119. J. R. Lakowicz and R. B. Thompson, *Biochim. Biophys. Acta* **732**, 359 (1983).
120. J. R. Lakowicz and J. R. Knutson, *Biochemistry* **19**, 905 (1980).
121. M. Rehorek, N. A. Dencher, and M. P. Heyn, *Biophys. J.* **43**, 39 (1983).
122. D. Ruppel, H. G. Kapitza, H.-J. Galla, F. Sixl, and E. Sackmann, *Biochim. Biophys. Acta* **692**, 1 (1982).
123. K. Kano, T. Yamaguchi, and T. Matsuo, *J. Phys. Chem.* **84**, 72 (1980).
124. M. Almgren, *J. Am. Chem. Soc.* **102**, 788 (1980).
125. D. K. Struck, D. Hoekstra and R. E. Pagano, *Biochemistry* **20**, 4093 (1981).
126. D. Hoekstra, *Biochemistry* **21**, 2883 (1982).
127. P. Vanderwerf and E. F. Ullman, *Biochim. Biophys. Acta* **596**, 302 (1980).
128. For review see H. Ti Tien, "Bilayer Lipid Membranes." Dekker, New York, 1974.
129. P. Nichols, J. West, and A. D. Bangham, *Biochim. Biophys. Acta* **363**, 190 (1974).
130. M. Tomkiewicz and G. A. Corker, *Photochem. Photobiol.* **22**, 249 (1976).
131. Y. Toyoshima, M. Morino, H. Motoki, and M. Sukigara, *Nature (London)* **265**, 188 (1977).
132. W. Stillwell and H. Ti Tien, *Biochem. Biophys. Res. Commun.* **81**, 212 (1978).
133. W. Oettmeier, J. R. Norris, and J. J. Katz, *Z. Naturforsch.*, **31C**, 163 (1976).
134. M. Mangel, *Biochim. Biophys. Acta* **430**, 459 (1976).
135. Y. Sudo and F. Toda, *Chem. Lett.* p. 1011 (1978); *Nature (London)* **279**, 807 (1979).
136. W. E. Ford, J. W. Otvos, and M. Calvin, *Nature (London)* **274**, 507 (1978).
137. W. E. Ford, J. W. Otvos, and M. Calvin, *Proc. Natl. Acad. Sci. U.S.A.* **76**, 3590 (1979).
138. C. Laane, W. E. Ford, J. W. Otvos, and M. Calvin, *Proc. Natl. Acad. Sci. U.S.A.* **78**, 2017 (1981).
139. K. Kurihara, M. Sukigara, and Y. Toyoshima, *Biochem. Biophys. Res. Commun.* **88**, 320 (1979).
140. K. Kurihara, Y. Toyoshima, and M. Sukigara, *Biochim. Biophys. Acta* **547**, 117 (1979).
141. T. Matsuo, K. Itoh, K. Takuma, K. Hashimoto, and T. Nagamura, *Chem. Lett.* p. 1009 (1980).
142. T. Matsuo, K. Takuma, Y. Tsutsui, and T. Nishijima, *J. Coord. Chem.* **10**, 187 (1980).
143. S. S. Anderson, I. G. Lyle, and R. Paterson, *Nature (London)* **259**, 147 (1976).
144. J. J. Grimaldi, S. Boileau, and J.-M. Lehn, *Nature (London)* **265**, 229 (1977).
145. T. Sugimoto, J. Miyazaki, T. Kokubo, S. Tanimoto, M. Okano, and M. Matsumoto, *J.C.S. Chem. Commun.* p. 210 (1981).
146. T. Sugimoto, J. Miyazaki, T. Kokubo, S. Tanimoto, M. Okano, and M. Matsumoto, *J.C.S. Chem. Commun.* p. 186 (1982).
147. E. E. Yablonskaya and V. Y. Shafirovich, *Nouv. J. Chim.* **8**, 117 (1984).
148. J. K. Hurley, F. Castelli, and G. Tollin, *Photochem. Photobiol.* **32**, 79 (1980).
149. J. K. Hurley and G. Tollin, *Photochem. Photobiol.* **34**, 623 (1981).
150. W. E. Ford and G. Tollin, *Photochem. Photobiol.* **35**, 809 (1982); **36**, 647 (1982).
151. W. E. Ford and G. Tollin, *Photochem. Photobiol.* **38**, 441 (1983); **40**, 249 (1984).
152. Y. Fang and G. Tollin, *Photochem. Photobiol.* **38**, 429 (1983); **39**, 685 (1984).
153. J. H. Fendler, *Acc. Chem. Res.* **13**, 7 (1980).
154. T. Kunitake, *J. Macromol. Sci., Chem.* **A13**, 587 (1979).
155. T. Kunitake and S. Shinkai, *Adv. Phys. Org. Chem.* **17**, 435 (1980).
156. J. M. Gebicki and M. Hicks, *Chem. Phys. Lipids* **16**, 142 (1976); see also W. R. Hargreaves and D. W. Deamer, *Biochemistry* **17**, 3759 (1978).
157. T. Kunitake, Y. Okahata, Y. Shiomomura, S. Yasumuni, and K. Takarabe, *J. Am. Chem. Soc.* **103**, 5401 (1981), and references cited therein.

158. T. Kunitake, N. Kimizuka, N. Higashi, and N. Nakshima, *J. Am. Chem. Soc.* **106**, 1978 (1984), and references cited therein.

159. R. McNeil and J. K. Thomas, *J. Colloid Interface Sci.* **73**, 522 (1980).

160. K. Kano, A. Romero, B. Djemuri, H. J. Ache, and J. H. Fendler, *J. Am. Chem. Soc.* **101**, 4030 (1979).

161. T. Kunitake, S. Tawaki, and N. Nakashima, *Bull. Chem. Soc. Jpn.* **56**, 3235 (1983).

162. D. Goregescauld, J. P. Desmasez, P. Lapouyade, A. Badeau, H. Richard, and M. A. Winnik, *Photochem. Photobiol.* **31**, 539 (1980). (a) S. Lukac, *Photochem. Photobiol.* **36**, 13 (1982).

163. T. Nomura, J. R. Escabi-Perez, J. Sunamoto, and J. H. Fendler, *J. Am. Chem. Soc.* **102**, 1484 (1980).

164. K. Kano, H. Kawazumi, and T. Ogawa, *J. Phys. Chem.* **85**, 2998 (1981).

165. M. Hauser, U. K. A. Klein, and U. Gösele, *Z. Phys. Chem. (Wiesbaden)* **101**, 255 (1976).

166. T. N. Estep and T. E. Thompson, *Biophys. J.* **26**, 195 (1979).

167. P. K. Wolber and B. S. Hudson, *Biophys. J.* **28**, 197 (1979).

168. T. G. Dewey and G. G. Hammes, *Biophys. J.* **32**, 1023 (1980).

169. J. R. Escabi-Perez, A. Romero, S. Lukac, and J. H. Fendler, *J. Am. Chem. Soc.* **101**, 2231 (1979). (a) D. J. W. Barbar, D. A. Morris, and J. K. Thomas, *Chem. Phys. Lett.* **37**, 481 (1976).

170. K. Monserrat, M. Grätzel and P. Tundo, *J. Am. Chem. Soc.* **102**, 5527 (1980).

171. K. Kurihara, P. Tundo, and J. H. Fendler, *J. Phys. Chem.* **87**, 3777 (1983).

172. S. Lukac and J. R. Harbour, *J. Am. Chem. Soc.* **105**, 4249 (1983).

173. P. P. Infelta, M. Grätzel, and J. H. Fendler, *J. Am. Chem. Soc.* **102**, 1479 (1980).

174. M. S. Tunuli and J. H. Fendler, *J. Am. Chem. Soc.* **103**, 2507 (1981).

175. L. K. C. Lee, J. K. Hurst, M. P. Politi, K. Kurihara, and J. H. Fendler, *J. Am. Chem. Soc.* **105**, 370 (1983).

176. M. S. Tunuli and J. H. Fendler, *ACS Symp. Ser.* No. 177, 53 (1982).

177. J. K. Hurst, L. Y. C. Lee, and M. Grätzel, *J. Am. Chem. Soc.* **105**, 7048 (1983).

178. M. P. Pileni, *Chem. Phys. Lett.* **71**, 317 (1980).

179. K. Monserrat and M. Grätzel, *J.C.S. Chem. Commun.*, p. 182 (1981).

180. Y. M. Tricot, D. N. Furlong, W. H. F. Sasse, and I. Snook, *Aust. J. Chem.* **36**, 609 (1983).

181. Y. M. Tricot, D. N. Furlong, W. H. F. Sasse, P. Davis, I. Snook, and W. van Megen, *J. Colloid Interface Sci.* **97**, 380 (1984). (b) P. Herve, F. Nome, and J. H. Fendler, *J. Am. Chem. Soc.* **106**, 8291 (1984).

182. (a) J. H. Fendler, *Science* **223**, 888 (1984). (b) J. H. Fendler and P. Tundo, *Acc. Chem. Res.* **17**, 3 (1984). (c) L. Gros, H. Ringsdorf, and H. Schuff, *Angew. Chem. Int. Ed. Engl.* **20**, 305 (1981).

183. (a) H. Bader and H. Ringsdorf, *J. Polym. Sci. Part A-1* **20**, 1623 (1982). (b) B. Hupfer, H. Ringsdorf, and H. Schupp, *Macromol. Chem.* **182**, 247 (1981). (c) H. Gaub, E. Sackmann, R. Büschl, and H. Ringsdorf, *Biophys. J.* **45**, 725 (1984). (d) B. Hupfer, H. Ringsdorf, and H. Schupp, *Chem. Phys. Lipids* **33**, 263, 355 (1983). (e) K. Dorn, R. T. Klingsbiel, D. P. Specht, P. N. Tyminski, H. Ringsdorf, and D. F. O'Brien, *J. Am. Chem. Soc.* **106**, 1627 (1984). (f) S. L. Regen, B. Czech, and A. Singh, *J. Am. Chem. Soc.* **102**, 6638 (1980). (g) S. L. Regen, A. Singh, G. Oehme, and M. Singh, *Biochem. Biophys. Res. Commun.* **101**, 131 (1981); *J. Am. Chem. Soc.* **104**, 791 (1982). (h) A. Kusumi, M. Singh, D. A. Tirrell, G. Oehme, A. Singh, N. K. P. Samuel, J. S. Hyde, and S. L. Regen, *J. Am. Chem. Soc.* **105**, 2975 (1983). (i) P. Tundo, D. J. Kippenberger, P. L. Klahn, N. E. Prieto, T. C. Jao, and J. H. Fendler, *J. Am. Chem. Soc.* **104**, 456 (1982). (j) E. Lopez, D. F. O'Brien, and T. H. Whitesides, *J. Am. Chem. Soc.* **104**, 305 (1983). (k) T. Kunitake, N. Nakashima, K. Takarabe, M. Nagai, A. Tsuge, and H. Yanagi, *J. Am. Chem. Soc.* **103**, 5945 (1981).

184. J. H. Fendler *in* "Surfactants in Solution" (K. L. Mittal and B. Lindman, eds.), p. 1947. Plenum, New York, 1984.

185. W. Reed, L. Guterman, P. Tundo, and J. H. Fendler, *J. Am. Chem. Soc.* **106**, 1897 (1984).

186. M. Politi and J. H. Fendler, *J. Am. Chem. Soc.* **106**, 265 (1984).
187. F. Nome, W. Reed, M. Politi, and J. H. Fendler, *ibid.*, **106**, 8086 (1984).
188. K. Kurihara and J. H. Fendler, *ibid.*, **105**, 6152 (1983).
189. Y-M Tricot and J. H. Fendler, *ibid.*, **106**, 2475 (1984).
190. R. Rafaeloff, Y-M. Tricot, F. Nome, and J. H. Fendler, *J. Phys. Chem.* **89**, 533 (1985).
191. R. Rafaeloff, Y-M. Tricot, F. Nome, P. Tundo, and J. H. Fendler, *J. Phys. Chem.* **89**, 1236 (1985).
192. Y. -M. Tricot, A. Emeren, and J. H. Fendler, *J. Phys. Chem.* **89**, 4721 (1985).

Chapter 7

Photoprocesses in Monolayers, Black Lipid Membranes, and Liquid Crystalline Solvents

7.1 Photoprocesses in Monolayer Assemblies

7.1.1 General and Structural Features of Monolayer Assemblies

The formation of monolayer films at the air–water interface on introduction of surface active compounds onto the surface of water has been known and studied for a long time.[1-3] The discovery by Langmuir in 1917 that the films spread on water could be transferred to solid supports has provided a new technique in which we can manipulate these monolayers to probe several phenomena. Early pioneering studies by Langmuir and Blodgett laid the foundations for this Langmuir–Blodgett technique.[4-7] Kuhn and co-workers have greatly extended the scope and utility of this monolayer deposition technique to allow deposition of successive monolayers (composed of the same or different molecules).[8-13] By combining monolayers in a stepwise procedure, assemblies can be built with a structure designed according to desired properties.

The technique of monolayer deposition is a very delicate procedure involving a Langmuir-type surface balance (consisting of a trough and a moveable float). The method consists of initially forming a monolayer array of fatty acids by dropping a solution of the material in a volatile solvent such as benzene or chloroform and allowing the solvent to evaporate. The solutes at the surface are packed by applying appropriate surface pressure. The process is monitored by surface pressure–area isotherms. The monolayer is then transferred by dipping a well-cleaned glass slide into the trough and raising it slowly. The entire operation of dipping and raising is carried out very slowly and mechanically, lasting as long as 6–10 min for transfer of a single monolayer. For the deposition of several monolayers, the procedure is repeated. By deposition of one or more inert (nonlabelled) fatty acid monolayers in between, the distance between a given donor monolayer and an acceptor monolayer can be varied easily. [One monolayer of arachidic acid (C_{20}) corresponds to a spacing of about 27 Å.] Figure 7.1 illustrates sche-

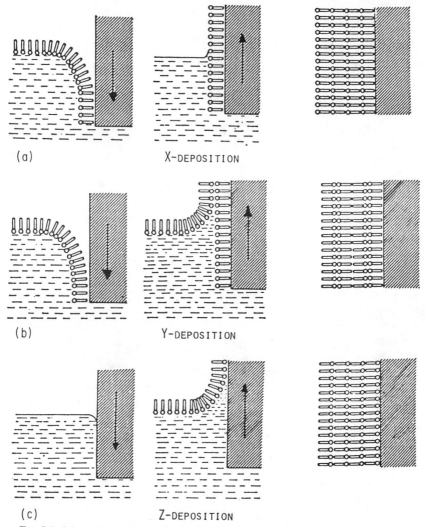

(a)   X-DEPOSITION

(b)   Y-DEPOSITION

(c)   Z-DEPOSITION

FIG. 7.1. Schematic representation of different modes of multilayer deposition: (a) X-deposition: plate-tail-head-tail-head..., (b) Y-deposition: plate-tail-head-head-tail..., and (c) Z-deposition: plate-head-tail-head-tail,....(From Kuhn et al.[3] Reprinted by permission of John Wiley & Sons, Inc.)

matically the formation of different multilayers by the appropriate choice of monolayer forming materials and mode of deposition. The reader is referred to original papers and review articles for detailed descriptions of the technique and procedures.[1-3,8-16]

The immobilization of reactant molecules at known (chosen) distances and the wide range of distances over which this can be achieved (25–200 Å) make the monolayer assembly technique unique and ideally suited for the examination of various theories of distance dependence in bimolecular electron and energy-transfer processes. The close resemblance of monolayer films and assemblies to biological membranes further suggests that phenomena occurring in monolayers might in some cases provide reasonable models for the processes occurring within the membranes or at the membrane–solution interface. Another feature of the monolayer arrangement in a Langmuir trough is that we can influence the intermolecular interactions by variation of the applied surface pressure. The molecules forming the spread monolayer can be confined to decreasing areas in a controlled manner, thus forcing accommodation in the available space by appropriate molecular rearrangement. Polarized absorption spectra of the dye molecules as a function of the surface pressure is a good indicator of the molecular arrangements.

Sagiv and co-workers have developed a new technique for the construction of artificial mono- and multilayer assemblies.[16a-c] It is based on chemisorption of a surfactant trichlorosilane derivative, $X—(CH_2)_n—SiCl_3$ on surfaces rich in hydroxyl groups and subsequent intralayer cross linking to give monolayers of unusual mechanical and chemical stability. The monolayer method can be modified to yield multilayers by using a two-step sequence: monolayer adsorption of a bifunctional surfactant, e.g., $CH_2=CH—(CH_2)_{14}—SiCl_3$ [Hexadecenyltrichlorosilane (HTS)] followed by chemical activation of the exposed surface in order to provide polar adsorption sites for the anchoring of host monolayers. The scheme involved is outlined on p. 224.

Ringsdorf and co-workers have described the synthesis, characterization, and behavior on monolayers of several saturated and polymerizable amphiphiles with fluorocarbon chains,[16d] e.g.,

$$[C_8F_{17}—CH_2—CO—O—(CH_2)_2]_2N(CH_3)$$

$$CH_2=C(COOH)—CH_2—CO—O—(CH_2)_2—C_{10}F_{21}$$

$$[C_8H_{17}—CH_2—CO—O—(CH_2)_2]_2\overset{+}{N}(CH_3)(CH_2—CH=CH_2)(Br^-)$$

Monolayers formed by lipids containing fluorocarbon chains were found to be more stable than those formed from their hydrocarbon analogs. The monolayers of polymerizable fluorocarbon lipids display an increased stability with retention of configuration after UV-light photopolymerization.

Before we start our discussions on the photochemistry in monolayer assemblies, a few cautionary remarks for the novice are in order. The monolayer assemblies are complex (delicate?) systems to handle and require utmost care and cleanliness. Recent studies indicate that under certain conditions movement of donors and/or acceptors between the layers occurs and attention must be paid to this disturbing possibility.

7.1.2   Energy-Transfer Processes

The occurrence of energy transfer in monolayer assemblies and their control by manipulation of the monolayer architecture was demonstrated elegantly in the 1960s by Kuhn and Möbius[8] using the following setup. Layer systems are built up by deposition of mixed layers containing dyes and fatty acid layers on suitable supports. A few layers were first deposited on a glass plate as a base and diluted into three zones as illustrated in Figure 7.2. A layer of the donor dye (D) was deposited on two-thirds of the surface (zones 1 and 2), and a pure arachidic acid layer was deposited over the last zone:

donor $R = C_{18}H_{37}$ acceptor

Zone 1 was now covered with a layer of arachidic acid and zones 2 and 3 with five layers of arachidic acid. Finally, a mixed layer of long-chain substituted cyanine dye was deposited over zones 2 and 3. The system is irradiated by UV light absorbed only by the donor. In zone 1, the distance between D* and A is about 50 Å and energy transfer from D* to A takes place leading to yellow A*

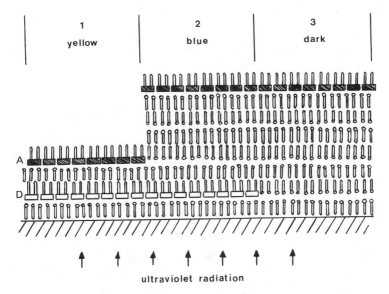

FIG. 7.2. Schematic representation of multilayer assemblies for the study of energy transfer processes: ▬▬, donor D; ▬▬, acceptor A; and ══ arachidic acid spacer. (From Kuhn *et al.*[3] Reprinted by permission of John Wiley & Sons, Inc.)

fluorescence:

$$\underset{\text{(blue fluorescence)}}{D^*} + A \longrightarrow D + \underset{\text{(yellow fluorescence)}}{A^*} \tag{7.1}$$

In zone 2, the distance is quite large (150 Å) for energy transfer to occur and only blue D^* fluorescence is observed. No fluorescence occurs in zone 3 since D is absent and A does not absorb the UV light.

The variation in the distance between D and A can, in fact, be controlled in a systematic manner and energy transfer efficiency followed as a function of the distance. Perrin–Förster theory as expressed by Eq. (7.2) has been verified in this manner:

$$(I_d/I)_s = 1 + (d_0/d)^4 \tag{7.2}$$

where I_d is the fluorescence intensity of donor at distance d, and d_0 is about 73 Å. The binary system D/A can also be extended to the ternary case as well.

$$h\nu \longrightarrow \quad D \xleftarrow{\;\;} d_1 \xrightarrow{\;\;} A_1 \xleftarrow{\;\;} d_2 \xrightarrow{\;\;} A_2$$

Efficient energy transfer observed between a dioctadecyloxacyanine dye and a surfactant derivative of $Ru(bpy)_3^{2+}$ in mixed monolayers of arachidicic acid and methyl arachidate has been used to illustrate a monolayer-based method for the indirect determination of luminescence quantum yields.[17]

J-Aggregates of Cyanine Dyes

Depending on the mode of incorporation, organic dye molecules can be either uniformly dispersed (diluted) in the monolayer matrix or form aggregates of dimer or higher order. Amphiphilic cyanine dyes in particular have been found to form the so-called J-aggregates (also known as Scheibe aggregates). These are composed of 500–1000 molecules stacked together like bricks interacting in a collective manner. The J-aggregates are characterized by their sharp absorption band, red shifted with respect to that of the monomers (depending on the dye, the shift can be as much as 45 nm), dimers, and a near-resonant emission (see Fig. 7.3). Excited aggregates of this type behave as a large array of coupled resonant oscillators. A different molecule with resonance at a smaller frequency would act as an energy trap provided it is incorporated into this array of cooperating molecules.

The above possibility has been demonstrated in the monolayer assemblies by adding a guest thiocyanine dye to a solution of oxacyanine host dye and

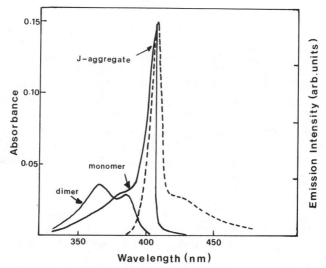

FIG. 7.3. Absorption (solid line) and fluorescence (broken line) spectra of a transferred monolayer of oxacyanine and hexadecane at a molar ratio of dye:hydrocarbon = 1:1. The absorption spectrum indicating the position of the monomer and dimer bands is obtained with a pure dye layer spread without hexadecane. (From Möbius.[9])

hexadecane in chloroform and forming the monolayer assembly:[9,11]

thiocyanine dye $R = C_{18}H_{37}$ oxacyanine dye

In transferred monolayers of the *J*-aggregates, the host fluorescence was observed even up to a host : guest ratio of 5×10^4! Since the guest absorptions occur at a region where the *J*-aggregates emit, the quenching occurs via Förster-type energy transfer. (One trap per 10,000 molecules reduces the fluorescence to one-half!) This efficient energy trapping by a very low concentration of acceptors in a large array of host molecules is certainly reminescent of the energy harvesting phenomenon in photosynthetic units (one trap per 300–400 chlorophyll molecules).

Chlorophyll Monolayers

The lipid environments and the possibility of controlled distribution of reactants in them have led many authors to choose monolayers as the principal model for reproducing chloroplast conditions.[18-35] Early classical work of Bellamy et al.[18,19] established that energy transfer between like and unlike molecules such as chlorophyll *a* and chlorophyll *b* (Chl *a*, Chl *b*) in diluted monolayers can be described by a Förster mechanism. A marked decrease in fluorescence quantum yield of chlorophyll observed with increasing concentrations of chlorophyll. The concentration quenching (or self-quenching) occurs at low fluorophore concentrations where there is no spectral evidence for the formation of associated pigment (Chl* + Chl) molecules. Absorption spectral features of Chl *a* multilayers deposited on a solid hydrophobic surface have been described by Sperling and Ke.[20] Trosper et al.[21] studied monolayers of Chl *a* diluted with various chloroplast lipids. It was found that fluorescence depolarization was brought about by a single Förster-type energy transfer, and there is also concentration quenching.

De Costa et al.[22] prepared monolayers containing Chl *a* and/or Chl *b* and transferred them from a water surface to glass slides. Self-quenching of fluorescence of both chlorophylls was observed and derived half-quenching concentrations compared with similar data for monolayers and solution systems. Energy transfer between Chl *a* and Chl *b* within monolayers was found to involve essentially irreversible energy transfer from *b* to *a*. In studies of chlorophyll monolayers diluted with hexadecane it was found that hexadecane instead of conventional fatty alcohols caused a substantial increase in Φ_{fluor} of Chl *a*. The emission intensity increases monotonically with respect to the intermolecular distance between the chromophores. A model for the orientation of chlorophyll and its role in influencing Förster-type energy transfer has been proposed. For a random distribution of Chl molecules, an average value of 43° has been calculated for the angle between the molecular planes and the water surface.[35]

In a series of investigations, Heithier et al.[23-26] have studied interactions of Chl a/pheophytin a with other molecules (chlorophyll a, lipids, and proteins) in monolayer assemblies. The existence of a phase separation between pheophytin a and lipids has been inferred with a pressure-induced reorganization (aggregation?) of pheophytin at the water interface. Chlorophyll a, on the other hand, forms crystalline Chl arrangements on increase of pressure. Presumably, the presence of Mg in the porphyrin core prevents the phase separation processes.

Bardwell and Bolton[27] examined the behavior of a porphyrin [5-(4-COO-C_6H_4-)10,15,20-tri(tolyl)porphyrin] both at the air–water interface and in films transferred to solid substrates. When the subphase is pure unbuffered water, transferred films were unstable and quickly rearranged to give a film which has been assigned as being amorphous. This amorphous film contains a wide range of aggregated species including monomers. When the monolayer was prepared in buffered subphases, the behavior was different. An equilibrium of the type $A + B \rightleftharpoons C$, has been inferred to exist, the equilibrium shifting to the right at higher pH values. (Species A has been assigned to monomers and B and C to different, distinct forms of aggregates.)

7.1.3 Electron-Transfer Processes

As with the studies of energy transfer, evaluation of the dependence of the electron transfer efficiencies in photoredox reactions as a function of donor–acceptor separation is a challenging task. In solution, the electron transfer occurs via molecular collisions brought about by diffusion of molecules toward one another. Immobilization of the reactants in monolayers at controlled distances is certainly an elegant way of studying distance dependences. Möbius and co-workers have examined such dependences during the quenching of fluorescence of cyanine dyes (J-aggregates) by a surfactant viologen $(C_{18})_2$—V^{2+}.[9,13] With this pair, energy-transfer quenching is ruled out, as the acceptor has no absorption bands in the spectral region of the donor emission. The distance between the electron donor and acceptor was varied by using fatty acid spacer layers. The nondiffusional electron transfer such as the present one usually occurs via electron tunneling mechanisms. The rate constant in such cases should decrease exponentially with increasing distance between the donor chromophore and the acceptor. Plots of $\log[(I_0/I) - 1]$ versus distance, as shown in Fig. 7.4 demonstrate such an occurrence.[12] The tunnelling mechanism has also been confirmed by the observation that the quenching rate constant is independent of temperature in the range of 77–298 K.

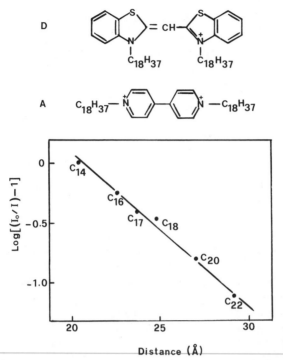

FIG. 7.4. Dependence of $(\log I_0/I - 1)$ with the donor–acceptor (indicated by D and A, respectively, shown above) separation distance during the quenching of a thiacyanine dye by dialkylviologen in monolayer assemblies. (From Kuhn.[12])

Penner and Möbius[28] have shown that in these systems, the plot of $[(I_0/I) - 1]$ versus [quencher] is also linear (as with the conventional Stern–Volmer kinetics) though exciton migration within the dye layer is mainly responsible for the quenching and not the molecular diffusion $(k \simeq 3 \times 10^{-3}$ cm^2 molecule^{-1} sec$^{-1})$. Electron transfer quenching by viologen leads to the buildup of its radical cation (V^+) in quantitatively measurable amounts. The slow formation of this persistent radical cation has been explained by a scheme involving a rapid photoredox reaction, an efficient back reaction, and a reaction of low probability that leads to optically detectable radical cations. The yield of radical cation is enhanced by the incorporation of a sacrificial third component in the monolayer, a species such as daphnetin (7,8-hydroxycoumarin):

$$S^* + C_n-V^{2+} \rightleftharpoons S^+ + C_n-V^+ \tag{7.3}$$

$$S^+ + D \longrightarrow S + D^+ \tag{7.4}$$

Depending on the nature of the particular donor and the conditions, yield enhancements as large as 3.5 have been obtained. This supersensitization incidentally occurs only when the cyanine dye is aggregated in the J-form. Earlier Möbius has examined an analogous three component system and obtained similar results!

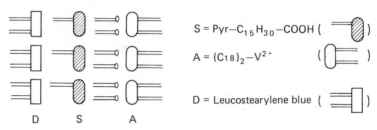

$$S = Pyr-C_{15}H_{30}-COOH \quad (\quad)$$

$$A = (C_{18})_2-V^{2+} \quad (\quad)$$

$$D = \text{Leucostearylene blue} \quad (\quad)$$

Studies of photochemical reactions of chlorophyll in monolayers are sparse due to the extreme sensitivity of chlorophyll to chemical attack and its low absorptivity per layer. The early studies of photochemical reactions of chlorophyll a in monolayers situated at an air–water interface and the interactions between the layer and various subphase components (electron donors and acceptors) have been that of Brody et al.[29,32] Later Porter and co-workers studied electron transfer quenching of chlorophyll a fluorescence by quinones and also chlorophyll a-sensitized photoreduction of tetrazolium blue and methyl red using ascorbate as donor.[33] Tocopherylquinone deposited on top of a layer doubled the rate of photoreduction of tetrazolium blue in the aqueous phase but when the quinone was mixed with chlorophyll in the monolayer, the reaction is quenched. ESR studies have established that quenching of chlorophyll a fluorescence in monolayers by acceptor species such as chloranil (tetracyanobenzoquinone) occurs via electron transfer:

$$^1Chl^* + Q \rightleftharpoons Chl^+ + Q^- \tag{7.5}$$

Chlorophyll cations and semiquinone radical anions are formed in low yields, due to concomitant self-quenching of chlorophyll a singlet states.[34]

Photoprocesses occurring in monolayer and multilayer assemblies of water-insoluble surfactant tin(IV)porphyrins supported on glass, quartz, SnO_2, PtO_2, and Pt in contact with aqueous solutions containing reductants have also been explored.[36] On glass/quartz supports, irradiation of Sn(IV)P layers in contact with amines, hydroquinones, or dimethoxybenzene leads to net photoreduction of the porphyrin and the formation of chlorin and isobacteriochlorin-like products:

$$^3Sn(IV)P^* + D \rightleftharpoons Sn(IV)P^- + D^+ \tag{7.6}$$

$$Sn(IV)P^- + H^+ \longrightarrow Sn(IV)PH\cdot \longrightarrow products \tag{7.7}$$

On supports such as SnO_2 or PtO_2, the support did not quench the porphyrin excited states but can intercept effectively the reduced species formed in the photoreactions. Thus, porphyrin-deposited supports can serve as photo-anodes in photoelectrochemical cells.

Studies have been made on the electron transfer quenching of monolayer-bound surfactant derivatives of $Ru(bpy)_3^{2+}*$ to appropriate electron acceptors located on the same interface as the donor:[17,37,38]

With viologen acceptors (R_2-V^{2+}) the Ru-complex luminescence is quenched at average distances of about 10 Å. This distance is 30, 60, and 75 Å for different cyanine dyes used instead of the Ru complex. A correlation has been shown between the distance and the ionization energy in the excited state of the donor.

It was reported that visible light photolysis of glass slides of the above Ru complex immersed in water lead to an efficient photochemical cleavage of water:[37]

$$2H_2O \xrightarrow[\text{[Ru(bpy)}_2\text{L]}^{2+}/\text{monolayers}]{h\nu} 2H_2 + O_2 \qquad (7.8)$$

(Estimates on the redox properties of the excited state and of the oxidized Ru complex do suggest that these two species are indeed capable of reducing and oxidizing water, respectively.) Due to its potential application in systems for the photochemical conversion of solar energy, the report aroused intense interest among the scientific community at large and in chemists on the photoredox chemistry of polypyridyl ruthenium complexes and on the application of the monolayer techniques. Unfortunately, more detailed experiments[17,38-40] could not confirm these early results, due to complications arising from the in-situ hydrolysis of the Ru complex, etc. Nevertheless, they illustrate the extreme sensitivity and delicate nature of the monolayer-bound species.

7.1.4 Lateral Diffusion of Probes and Aspects of Excimer Formation

In Chapter 6, mention was made of the usage of intermolecular formation of excimers with probe molecules such as pyrene to evaluate the lateral diffusion coefficients of molecules in the bilayers of lipid vesicles and liposomes. Loughran *et al.*[41] have made a combined experimental and theoretical study of the monomer–excimer dynamics of 12-(1-pyrenyl)-dodecanoic acid (PDA) in spread monolayers with oleic acid. The excimer-to-monomer fluorescence intensity ratio in the steady state was found to depend linearly on the mole fraction of PDA in oleic acid. The data has been interpreted theoretically using a reaction dynamic model to describe the lateral diffusion and subsequent interaction of pyrene species.

From the theory of random walk on lattices with traps, an estimate of 1.7×10^{-6} cm^2 sec^{-1} has been derived for the two-dimensional (lateral) diffusion coefficient of pyrene in monolayers. [The diffusion coefficient D_2 is given by:

$$D_2 = \tfrac{1}{3}n\langle l \rangle \qquad (7.9)$$

where l is the length of each step and n the number of displacements per unit time; $\langle l \rangle^2$ has been taken as equal to $\simeq 47$ Å^2.] In a related experiment, Teissie *et al.*[42] have studied the migration of anthroyl stearate molecules in a spread monolayer using a photobleaching technique (in which the fluorescence recovery of the probe is monitored after its bleaching with an intense, short light pulse). The data have been analysed using a continuum approach based on Fick's second law of diffusion.

7.1.5   Organic and Inorganic Photoreactions

Environmental effects on the excited-state relaxation, interactions, and mobility are often reflected in the overall reaction course and product distribution in numerous organic and inorganic photochemical reactions. In earlier chapters, we already encountered such effects for reactions in micellar media. Herein we consider some of these reactions in monolayers, mostly examined by Whitten and co-workers.[14,15,43–53]

Photoisomerization of Olefins[43–45]

Cis–trans isomerization is a widely studied photoreaction with a large medium sensitivity. In general, the isomerization occurs via a rapid relaxation of the cis or trans excited state to a twisted intermediate which can decay to the ground state of both isomers. With olefins such as stilbenes, an increase in the solvent viscosity retards the trans–cis process selectively while allowing the reverse process to proceed. The viscosity effect has been rationalized in terms of an increase in the molecular volume along the reaction path trans–cis process. If such is the case, then in highly ordered monolayer arrays, we can anticipate effects similar to that of highly viscous medium.

Photoisomerization of olefins such as (1)–(3) (spiropyran, stibazole, and styrene derivatives) have confirmed such predictions:

(1)

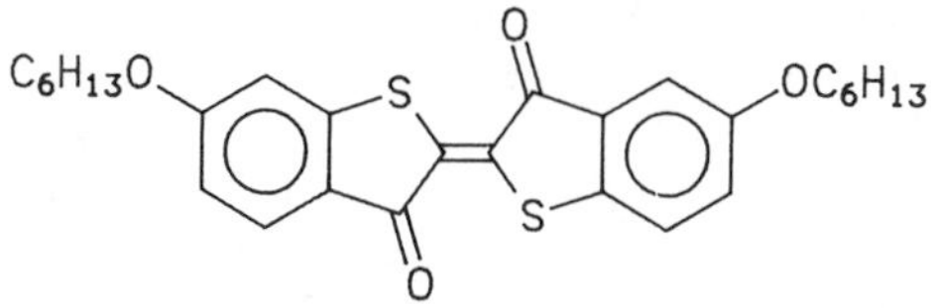

(2)

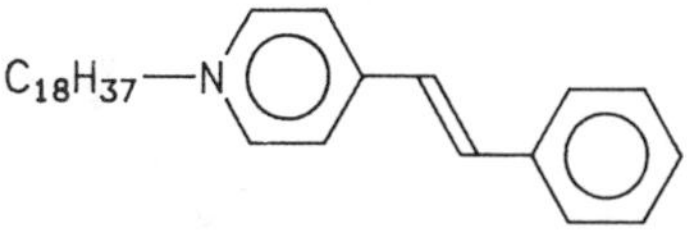

(3)

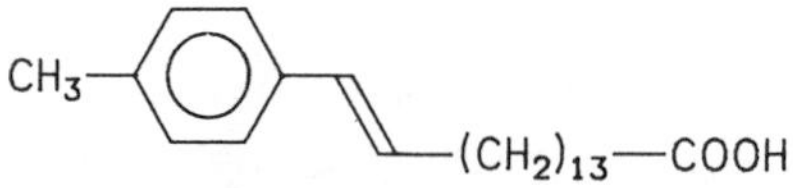

Irradiation of cis olefins in monolayer films or assemblies results in a rapid and irreversible conversion to the trans isomers. In contrast, films or assemblies containing trans were found to be photostable, with no conversion to the cis isomer or other photoproducts. Studies of the surface pressure–area isotherms indicated that a greater cross-sectional area per molecule was required for the cis isomer and that a significant reduction of pressure occurs when films of cis isomer are irradiated and undergo conversion to the trans form.

The photoisomerization of spiropyran (SP) in monolayers has been adapted as a photochromic device.[46] Upon irradiation at 366 nm light, spiropyran isomerizes readily in octadecanol monolayers to merocyanine (MC) and the reverse reaction can be achieved with visible light (545 nm):

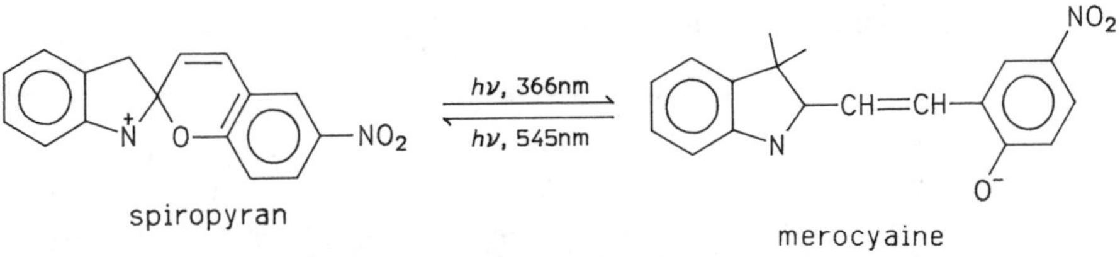

spiropyran

merocyaine

In a monolayer arrangement where the spiropyran and a thiocyanine dye (D) are separated by $\simeq 50$ Å (two monolayers of fatty acids) the arrangement functions as a photochromic device. In the absence of merocyanine, which acts as a quencher of the excited state of cyanine dye D^*, excitation of D with blue light (405 nm) leads to the green fluorescence of D^*.

$$D \xrightarrow[405\ mn]{hv} D^* \longrightarrow D + hv \text{ (green)}$$

$$SP \underset{545\ nm}{\overset{366\ nm}{\rightleftharpoons}} MC$$

The whole assembly functions as a flip-flop element, turned on with 366 nm radiation and turned off with 545 nm light. The fluorescence of D* is quenched as long as the flip-flop element is turned on.

Photoreactions of Ketones (Norrish Type II Reactions)[47]

Photolysis of aliphatic and aromatic ketones carrying γ-hydrogen atoms often leads to the so-called Norrish type II photoelimination:

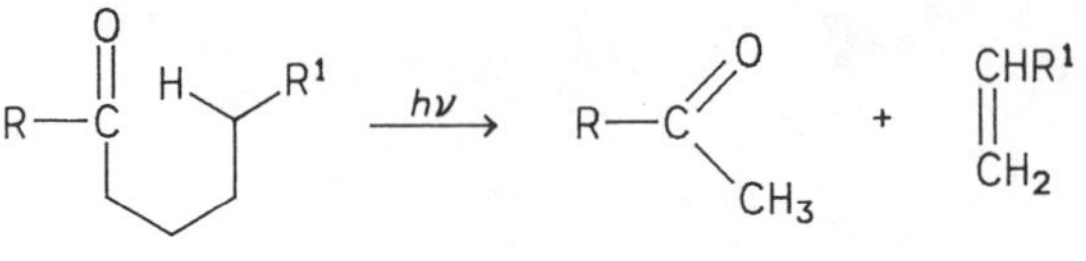

For example, irradiation of surfactant ketone CH_3—C_6H_4—CO—$(CH_2)_{14}$—$COOH$ in benzene yields p-methylacetophenone as the main product with a quantum yield of 0.2. The reaction proceeds via γ-hydrogen abstraction in a cyclic intermediate followed by the cleavage of the biradical to form an enol of the methylketone and an olefin. The packed linear arrangement in a monolayer imposes restrictions on the ketone to form such a cyclic intermediate, and a reduction in quantum yield is anticipated. In monolayers, the disappearance of quantum yield for the ketone via type II processes is 0.06. An alternate pathway for the excited-state ketone is type I (α-cleavage) fragmentation:

$$CH_3—C_6H_4—CO—(CH_2)_{14}—COOH \xrightarrow{h\nu}$$

$$[CH_3—C_6H_4—CO\cdot\cdot CH—(CH_2)_{13}—COOH] \longrightarrow$$

$$CH_3—C_6H_5 \quad \text{or} \quad CH_3—C_6H_4—CHO \quad (7.10)$$

Interestingly, analysis of photoproducts has failed to indicate the occurrence of type I process.

Photodimerization[38,45]

Earlier it was mentioned that trans olefins such as t-stilbazoleoctadecanoic acid do not undergo trans–cis photoisomerization in monolayers. However, they undergo rapid photobleaching which has been shown to be due to formation of dimers:

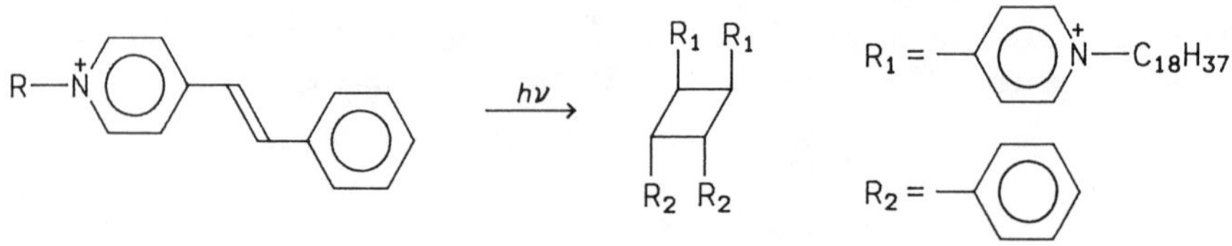

Although the absorption spectrum of stilbazole in monolayers is nearly identical to that observed in dilute solvents such as acetonitrile, the fluorescence in the assemblies is red-shifted by approximately 50 nm, as compared to that in CH_3CN. Based on an analogy to the similar behavior observed in crystals, it has been proposed that the red-shifted fluorescence is due to an excimer which is probably an intermediate in the formation of the photodimer. The absence of any excimer fluorescence in concentrated homogeneous solutions suggests that the tight packing in the monolayer assemblies plays a major role in this process. Photolysis of the styrene derivative $[CH_2{=}CH{-}C_6H_4{-}CO{-}(CH_2)_{14}{-}COOH]$ was found to exhibit similar behavior (formation of photodimers of cyclobutane).

Photoreactions of Metalloporphyrins

Photolysis of metalloporphyrins containing carbonyl groups as ligands often leads to photoelimination of the bound CO and its replacement by a new ligand:[48]

$$L(Porph)M(CO) \underset{-CO}{\overset{h\nu}{\rightleftharpoons}} [L(Porph)M] \xrightarrow{\ L'\ } L(Porph)ML' \qquad (7.11)$$

The CO-free metalloporphyrin intermediate that is formed in the photoejection cannot be isolated in solution, though it can be readily detected by flash photolysis techniques. Irradiation in dry monolayer assemblies containing complexes such as L(Porph)Ru(CO) (where L = pyridine or water) *in vacuo* has lead to the isolation of the extremely reactive CO-free complex and a study of its substitutional chemistry!

In studies of photoaddition, reduction reactions of Zn^{II}, Sn^{IV}, and Pd^{II} porphyrins (anchored on cadmium arachidate multilayers supported on glass slides) using reagents such as dialkylanilines, $SnCl_2$, triethylamine, or ascorbate, it was found that reagents present in the aqueous solution penetrate the assemblies readily on contact.[49] Surfactant amines such as N,N'-dioctadecylaniline (DOA) also cause reductive photoaddition of Pd porphyrins when both were incorporated in multilayers. The results indicate that considerable migration of supposedly anchored reagents such as DOA does occur in the assemblies. Small molecules such as oxygen also readily diffuse throughout the monolayers as evidenced by the efficient photooxidation (singlet-oxygen mediated) of Protoporphyrin IX.[50]

The photochemistry of a number of surface active diazo and azide compounds has been studied in monolayers at the air–water boundary.[52] Irradiation of long-chain α-diazoketones with UV light leads to rapid loss of nitrogen. The resulting ketenes react with the subphase to generate carboxylic acids and dimerise to give β-lactones as side products. The following scheme summarizes the photochemistry of long-chain diazoketones in monolayers:

$$R-CO-CH=\overset{+}{N}=N^{-} \xrightarrow[(-N_2)]{h\nu} R-CO-\overset{\cdot}{CH} \xrightarrow{H_2O} R-CO-CH_2OH$$

with a branch $\xrightarrow{h\nu}$ leading downward, and from $R-CO-\overset{\cdot}{CH}$ a vertical arrow down to:

$$R-CH=C=O \xrightarrow{H_2O} R-CH_2COOH$$

then a vertical arrow down to:

$$R-CH-C=O \xrightarrow{H_2O} R-CH-COOH$$
$$R-CH=C-O \qquad\qquad RCH_2-C=O$$

These reactions cause also pronounced changes in the spreading behavior of the monolayers, e.g., collapse of monolayers to give oily films or disappearance of monolayers by dissolution in the subphase.

7.1.6   Photoelectric and Photoelectrochemical Effects

Photoelectrical and photoelectrochemical properties of chlorophyll a in Langmuir–Blodgett films have been investigated in relation to the primary processes of photosynthesis.[54–60] In these studies, one or more monolayers of chlorophyll a have been deposited along with quinones, lecithins, or carotenoids on electrodes, and the photocurrents were examined. In their study of chlorophyll a and chlorophyll b monolayers on SnO_2 optically transparent electrodes (OTE), Watanabe et al.[54] found photocurrents and photovoltages to be always anodic and negative, respectively. The process corresponds to an injection of electrons from excited chlorophyll molecules to the conduction band of SnO_2:

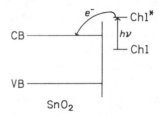

The quantum efficiency for the photocurrent generation reaches values of 0.12–0.16 in chlorophyll a–stearic acid mixed monolayers at a (chlorophyll/stearic acid) mole ratio of approximately 1.0.[54]

In studies of photovoltaic effects of chlorophyll on semitransparent Al electrodes, the quantum efficiency of the processes has been found to be very low.[55–60] (Most studies use Al electrodes due to their low work function, mechanical characteristics, and the very hydrophilic behavior of fresh vacuum evaporated Al films). Jansen and Bolton[57a] have observed that the presence of substituted quinones with unsaturated linkages in the side chain, such as plastoquinone or ubiquinone, increases drastically the quantum efficiencies (Φ increases from 2×10^{-4} to about 2×10^{-3}).

7.2 Photoprocesses in Black Lipid Membranes

7.2.1 General Structural Features of Black Lipid Membranes[61-65]

The preparation of black lipid membranes (BLM) uses the spontaneous ability of the lipids to form membraneous films over narrow void areas in the aqueous phase such as those found in a pinhole or between two narrow plates. Thus, brushing of an organic solution of a lipid across a pinhole separating two aqueous phases or dipping a small teflon loop and pulling it out through a lipid solution results in the formation of thin lipid films that are black in color. Figure 7.5 presents schematically various methods of preparation of BLM and an experimental setup for the measurements on photosensitive membranes. Optical (refraction) and electrical (capacitance) measurements of these membranes indicate the thickness of the lipid films to be in the range of lipid bilayers, namely, about 70 Å. Hence, the common description of these thin films as black (or bilayer) lipid membranes. Negative or positive surface charges can be conferred to these BLM by incorporating in the membrane-forming solution an anionic compound or cationic compound, respectively. Structurally, the planar BLM formed in the above manner is analogous to the *spherical* bilayer lipid membrane encountered earlier (in Chapter 6) in lipid vesicles and liposomes.

A wide variety of natural and synthetic lipids and surfactant molecules have been found to form BLMs. Although a BLM separating two aqueous solutions can be considered as two monomolecular layers with their hydrocarbon chains joined together, the physical properties of BLMs are quite different and not additive. A BLM is unique in possessing two identical interfaces (bifaces). The stability of a BLM is related to the very low bifacial tension (< 6 dynes cm^{-1}). The intrinsic resistivity of an unmodified BLM is quite high ($> 10^8$ Ω cm^2), and the dc conductivity is linear up to a potential differences of $50-100$ mV. For a BLM of thickness ~ 70 Å, a 70-mV potential difference would correspond to a field strength of nearly 100 kV/cm. This high field intensity can be withstood by most BLMs for a long period without any apparent effect.

The mechanical stability of most BLMs is usually enhanced by the addition of cholesterol and derivatives. The intrinsic lecithin BLM or those with oxidized cholesterol show little ion selectivity, which is consistent with the expected properties of ultrathin hydrocarbon layers. The poor ion-selective properties, however, are improved on incorporation of ionophores such as valinomycin.

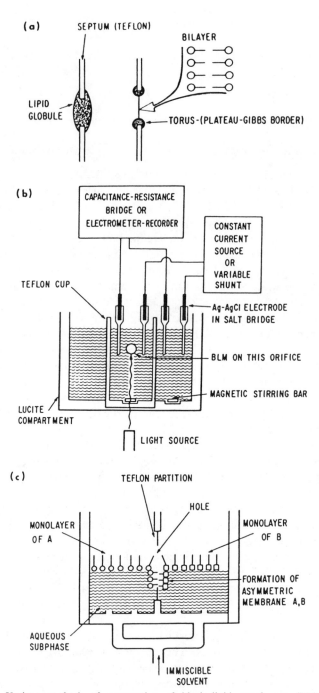

FIG. 7.5. Various methods of preparation of black lipid membranes (BLM) and an experimental arrangement for measurements on photosensitive BLM. (a) Schematic representation of lipid globule placed on septum and spontaneously thinning to a bilayer with torus. (b) Typical experimental arrangement for the measurement of photoelectric effects on pigmented BLM. (c) Schematic representation of experimental process for forming an asymmetric bilayer from two monolayers. (From Berns.[65] Copyright 1976, Pergamon Press.)

7.2.2 Studies of Photoelectric Phenomena

A property of the BLM that is being widely studied is the photoelectric effects observed upon incorporation of chloroplast extract or other light-sensitive pigments (chlorophylls, retinals, or bacteriorhodapsin).[64-77] The occurrence of the photoelectric effect in BLMs demonstrates the creation and efficient mobility of charge carriers in the lipid bilayer.

In the early studies, BLMs were formed using the chloroplast extracts (Chl–BLM). The magnitude of the observed photo-emf was generally small, of the order of few millivolts. It was found later that the open-circuit photo-emf can be increased to $\sim 250\,mV$, by placing the BLM under asymmetric conditions. This can be achieved by the addition of a modifier (a redox reagent) on one side or different modifiers on the opposite sides of the bathing solutions, e.g., Fe^{3+} ions on one side and ascorbic acid on the other. Table 7.1 provides some data on the range of photovoltages observed using various modifiers in chloroplast BLM.

Certain BLMs such as those composed of oxidized cholesterol which are otherwise photoinactive can be sensitized using water-soluble dyes (such as chlorophyllin) or inorganic salts present in the bathing solution. Apparently, a chlorophyllin–BLM shows an enhanced photoresponse compared to Chlorophyll–BLM. Detailed discussions on various aspects of photoelectric effects in pigmented BLMs are available.[64,65,73,74]

Alternative to the chemical bias outlined earlier, photoeffects can also be enhanced by applying electric fields. Electrically, a pigmented BLM can be

TABLE 7.1. Influence of Redox Reagents on the Photo-emf of Chloroplast Bilayer Lipid Membranes[a]

Modifier (in inner solution)	Photo-emf (mV)
None	20–25
NaI	74–127
$FeCl_2$	80–86
$Na_2S_2O_3$	74–93
$K_4Fe(CN)_6$	150
Ascorbic acid	125–188
Flavin mononucleotide	138–167
Hydroquinone	79–96
Cysteine/cystine	108
Methylviologen(MV^{2+})	73

[a] In 0.1 M acetate buffer at pH 5.0. System: salt bridge/0.1 M acetate containing 1 mM $FeCl_3$ (outer solution)/Chl-BLM (prepared using spinach chloroplast) inner solution/salt bridge.[34]

considered as a capacitor (C_m) shunted by a resistor (R_m). When a voltage is applied to the system, either through an external voltage or via the action of light, voltage transients appear across the effective membrane (R_m-C_m) combination. Photoresponses of pigmented BLM have been examined both under continuous illumination and using pulsed light sources. Analysis of current or voltage transients induced across the BLM by short light pulses is currently a major area of research in BLM.

Various mechanisms have been proposed for the origin of photoelectric effects. One or more of the following processes is believed to occur: charge transport across the low dielectric membrane barrier, interfacial potentials, physical displacement of the dyes across the bilayer following photoisomerization, proton displacement, or transport. Early mechanistic studies in this context are those of Ullrich and Kuhn,[70] who studied the photoeffects in egg lecithin–BLM in the presence of an oxacyanine dye added to one side of the bathing solution. On illumination, photovoltages are seen, reaching a maximum and a constant value after about a minute. The action spectrum of the photocurrents corresponded to that of the dye. It was proposed that the dye slowly inserted into the BLM, caused a photoreduction with a donor E (unidentified) present in the electrolyte [Eq. (7.12)] and resulting polarization of the membrane:

$$A^* + E \longrightarrow A^- + E^+ \tag{7.12}$$

Trissel and Läuger performed analogous experiments to reach similar conclusions.[71] Huebner[72] has proposed that photoionic effects (caused by light-induced charge rearrangements within the dye molecules) resulting in a physical movement of the dye within the interface can also be responsible. Hong[73,74] has discussed various possible effects in detail.

7.2.3   Electron and Energy-Transfer Processes

There have been a few studies of electron and energy-transfer processes in BLM, and these are devoted to an examination of the influence of the lipid membrane in the local distribution of the dye/reactant molecules. As indicated earlier in the studies in monolayer assemblies, quenching of fluorescence of chlorophyll at high concentrations (self-quenching) is a process widely studied to understand the antenna chlorophylls and Förster-type energy transfers. Alamuti and Läuger[78] found that the fluorescence intensity of chlorophyll a increases with the mole ratio of chlorophyll to lipid (lecithin) in a BLM forming solution, reaching a maximum at about 0.3. The average distance between the porphyrin planes has been estimated to be about 70 Å. Steinemann et al.[79] performed similar experiments and found the loss of fluorescence at high chlorophyll mole ratios is accompanied by a decrease in the fluorescence polarization.

Pohl[80] used the fluorescence to study energy transfer from chlorophyll b to chlorophyll a molecules and found the process efficient whenever the mean distance between the reactants was less than 100 Å. Strauss and Tien also studied the energy transfer between chlorophyll b and chlorophyll a and with carotenoids. The fluorescence of chlorophyll in BLM was found to be ca. 50 times stronger than that in corresponding homogeneous solutions and the energy transfer-efficiency between chlorophyll b and chlorophyll a in the range of 40–53%. The efficiency in a chloroplast in which the D–A distance was the same as in BLM was only about 23%, results which nicely illustrate the influence of the BLM in organizing the reactants in the lipid matrix.

Fromherz and Arden have made an educative study on the use of dye-coated BLMs as a photosensitive antenna.[81] A BLM of docosylamine was deposited onto an indium–tin oxide (ITO) semiconductor substrate. The bilayer was doped on one side with a cyanine dye (acceptor) in direct contact with the semiconductor and on the other side with a coumarin dye (donor) contact with the electrolyte:

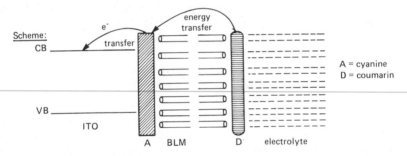

The effect of the antenna is revealed by the action spectrum of sensitized photocurrent which indicates a successful transfer of energy and electrons in sequence:

$$\text{Coumarin* + Cyanine} \longrightarrow \text{Coumarin + Cyanine*} \qquad (7.13)$$

$$\text{Cyanine* + CB/ITO} \longrightarrow \text{Cyanine}^+ + e_{CB}^-(\text{ITO}) \qquad (7.14)$$

The efficiency of the entire process can be modulated by varying the pH. (The absorption spectrum of coumarin dye changes drastically upon protonation.)

Photoinitiated transport of H_3O^+ and/or OH^- across lipid bilayers doped with magnesium octaethylporphyrin (MgOEP) has been studied by Feldberg and co-workers using 1 μs laser flashes and monitoring the electrical response.[82] It was found that MgOEP$^+$ (photogenerated via oxidative quenching with methylviologen) mediated the electron transport:

$$\text{MgOEP* + MV}^{2+} \longrightarrow \text{MgOEP}^+ + \text{MV}^+ \qquad (7.15)$$

Magnesium etiochlorin was found to be equally effective in the above process.[83] Becquerel effects have also been examined using oleic acid-derived BLM containing methylene blue, coated onto Pt electrodes. Studies with chlorophyll *b* cholylhydrazone, a positively charged derivative has indicated significant differences in the interfacial photoreactions as compared to neutral chlorophyll *a* or *b* molecules.[83a]

7.3 Photoprocesses in Liquid Crystalline Solvents

7.3.1 General and Structural Features of Liquid Crystalline Solvents[84–88]

It has been known for quite a long time that certain long-chain fatty acid molecules associate themselves in a unique structural fashion called liquid crystals. Their unusual structural assembly confers certain properties unique to these solvents. In contrast to the isotropic solvents, these systems possess both long-range and short-range order. Although their existence has been known for over a hundred years, only recently they have received some attention as a medium for photoreactions. Weiss and co-workers have made some early pioneering studies (Table 7.2) in this area which we shall review here.[89–100]

Figure 7.6 illustrates schematically some of the common structural forms observed in different liquid crystalline solvents. Depending on the mode of destruction of the order in the parent solid state, liquid crystals are classified into two main categories: thermotropic and lyotropic. The former are prepared by mere heating of a substance or substances while the latter are formed when certain compounds are treated with controlled amounts of water or other polar solvents. Liquid crystals exhibit unique properties that are not found in either solids or liquids, that they are often referred to as the fourth state of matter: optical activity, birefringence, dichroism, electric and magnetic field effects, etc. Mention was made earlier in Chapter 6 of lyotropic liquid crystals formed by lecithins and double alkyl chain surfactants. Depending on the water content and the temperature, amphiphilic surfactants exhibit one of the following forms:

solid $\underset{}{\overset{H_2O}{\rightleftharpoons}}$ lamellar liquid crystal (neat or G phase) $\underset{}{\overset{H_2O}{\rightleftharpoons}}$ cubic liquid crystal (viscous iso-tropic V_1 phase) $\underset{}{\overset{H_2O}{\rightleftharpoons}}$

hexagonal liquid crystal (middle or M_1 phase) $\underset{}{\overset{H_2O}{\rightleftharpoons}}$ micellar phase $\underset{}{\overset{H_2O}{\rightleftharpoons}}$ homogeneous solution

TABLE 7.2. Liquid Crystalline Solvents and Photoreactions Studied in Them

Photoreaction	Reactants	Liquid crystalline solvent (phase)[a]	T_t (°C)	Reference
Fluorescence quenching (intermolecular exciplex)	Pyrene/C_n—DMA	CM1 (C,I)	C – I 58°	92
Fluorescence quenching (intramolecular exciplex)	Pyrene—$(CH_2)_3$—DMA	CM1 (C,I)	C – I 58°	95
Fluorescence self-quenching (intermolecular excimers)	Pyrene/pyrene	CM1 (C,I)	C – I 58°	94
Fluorescence self-quenching (intramolecular excimers)	Pyr—$(CH_2)_n$—pyr, n = 3,7,10,13,22	CM1 (C,I) BCCN (S,N)	C – I 58° S – N 54°	97,100
Photodimerization	Acenaphthylene	BS (S,I) CM3 (C,I) CN (S,C)	S – I 25° C – I 58° S – C 77°, C – I 92°	91,92
Photofragmentation (Norrish type I)	Dibenzylketones	BCCN (S,N) CM2 (S,C,I) BS (S,I)	S – C 45°, C – I 79°	98
Photoinduced H-abstraction (Norrish type II)	Alkyl, aryl diones	BS CM4, CM2		99
Isomerization	Azobenzene, merocyanine	BS CM2		89,93,96

[a] Abbreviations: C = cholesteric, S = smectic, N = nematic, and I = isotropic; BS = n-butyl stearate, CN = 5-cholestanyl nonanoate, BCCN = trans,trans-4-n-butylbicyclohexane-4-carbonitrile, CO = 5-cholestanyl oleate; CN = 5-cholestanyl nonanoate; CA = cholestanyl acetate, and CCl = cholestanyl chloride. CM1 = 59.1/15.6/24.9 (w/w/w) mixture of CO/CN/C—Cl and CM2 = 35/65 (w/w) mixture of C—Cl/CN; CM3 = 1:1 (w/w) mixture of CA/CN; CM4 = 60:29:14 (w/w/w) mixture of CO/CN/CCl.

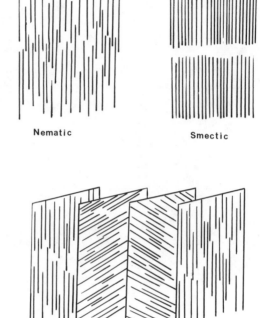

Fig. 7.6. Schematic representation of molecular arrangements in nematic, smectic, and cholesteric liquid crystals. (From Brown.[88] Copyright 1983 American Chemical Society.)

Thermotropic liquid crystals are the one we will consider here. These are formed in general by molecules with one of the following geometry: cylindrical (e.g., MBBA, PAA), cholesteric (e.g., cholestanyl nonanoate) or disk-like (e.g., hexabutoxytriphenylene). Different structural arrangements lead to a classification into nematic and smectic types. Among the various nematic types, cholesteric and ordinary nematic phases are the common ones:

```
                           ┌──→ smectic          ┌──→ cholesteric
    liquid  crystal  ──────┤                      │
                           └──→ nematic  ─────────┴──→ ordinary  nematic
```

In a smectic mesophase, the molecules are arranged in layers with their long molecular axis parallel to one another and perpendicular to the plane of the layers. A nematic phase, although not layered, is still characterized by a parallel arrangement of molecules. In a cholesteric mesophase, the individual molecules are arranged in layers, which individually have nematic-like order.

The long axes of molecules within a layer are parallel and the layers are tilted with respect to each other, forming a twisted helical microstructure. Of the three types, the smectic phase is considered to be the most ordered. Solutes in liquid crystals align themselves in the best packing arrangement based on steric considerations, e.g., planar molecules lie in the plane of the cholesteric layers with long molecular axes oriented parallel to the long axis of the liquid crystalline solvent molecules.

Liquid crystals in the nematic, smectic, and cholesteric phases can be transformed to isotropic phases by heating and the processes are defined by characteristic phase transition temperatures. Table 7.2 presents some data on the various liquid crystalline mesophases and their transition temperatures. The bulk viscosity of liquid crystalline mesophases is rather high. Cholesteric phases are extremely syrupy with macroscopic viscosities in the range of $10-10^3$ P. Smectic phases have viscosities usually between $100-1000$ cP, while those of nematic phases are below 100 cP. Liquid crystalline solvents are non-Newtonian fluids. Hence, it is not straightforward to use macroscopic viscosities to derive rates of solute diffusion (even incorrect!). As the solute motions are highly anisotropic, we need to consider diffusion coefficients along each of the specific solvent axes.

Cholesteric phases which can be considered as twisted nematic phases, are characterized by a pitch, which measures the distance between the layers of solvent molecules whose long axes are parallel. When the helical axes of the solvent are aligned parallel to a beam of incident radiation, the pitch p can be determined experimentally using the relation:

$$p = \lambda/2n \tag{7.16}$$

where $n (\simeq 1.5)$ is the refractive index of the medium and λ the maximum of the reflectance band. Pitches of compensable cholesteric mixtures vary strongly with temperature, diverging and changing sign at the compensation temperature T_n. At T_n, the host loses its helicity and possess properties of a nematic phase. For example, a $65:35$ mol% mixture of cholestanyl chloride and cholestanyl nonanoate compensates to a nematic liquid crystal at about 46°C.

7.3.2 Dynamics of Intermolecular and Intramolecular Excimers and Exciplexes

An area of major concern in photochemical studies of excited-state bimolecular reactions has been the orientational requirements of the reactants and the potential role of the medium (solvent structure) in influencing them. Liquid crystalline solvents, especially in the cholesteric and nematic phases provide medium. Excited-state processes of interest in this context are the formation of excited-state dimers in a transient manner (as in excimers or exciplexes) or as permanent products (as in photodimers). Intramolecular modes of formation of dimers particularly are of special interest, for they also provide information on the mobility of the chromophores and the linking alkyl chains. Weiss and co-workers have obtained some interesting results in this area.

Quenching of Pyrene Fluorescence by Amines[92,95]

Quenching of the pyrene monomer fluorescence by tertiary amine cholestanyl–dimethylaniline (C-DMA) in a cholesteric mixture (CM) [35:65 w/w of cholestanyl chloride (C-Cl) and cholestanyl nonanoate (C-N)] has revealed the effects of cholesteric and isotropic phases. Steady-state emission measurements indicate that the intermolecular exciplex (Py$\cdots$C-DMA)* is nonemissive. Monitoring of the fluorescence intensity of pyrene as a function of temperature (either with or without the amine) allows a ready detection of the phase transitions at temperatures very close to optically detected transition temperatures (e.g., cholesteric to isotropic at 58°C). Kinetic analysis indicate that the activation parameters for the quenching process are phase dependent: E_{act}, A (cholesteric) < E_{act}, A (isotropic). With increasing amine concentration, the activation energies in the cholesteric phase approach the values found in isotropic phases, presumably due to the rupture of the liquid crystalline order of the CM by the solute. The probe vibronic band intensity ratio (III/I) indicates that pyrene resides in a predominantly low polarity hydrocarbon-like environment.

In analogous experiments with covalently linked molecules, Py—$(CH_2)_3$—CDMA (abbreviated as P3D), the broad, structureless emission of the exciplex is readily observed in the CM (emission maximum at 480 nm). The emission intensity of the exciplex relative to that of the monomer fluorescence is greatly reduced largely due to the high viscosity of the cholesteric mixture. The lifetime of the exciplex in CM however is very close to that observed in *n*-hexane solutions, about 88 nsec. The kinetics of the disappearance of ^{1}Py* correlate well with that of the formation of the fluorescent exciplex. In cholesteric mixtures, ^{1}Py* fluorescence exhibits a mono-exponential behavior, in distinct contrast to biexponentials observed in isotropic solvents. Presumably in CM, different ground-state conformations yield the exciplex with a common rate. The activation parameters are slightly larger than those of corresponding isotropic phase of CM, data that are reconcilable with the phase viscosity differences.

Self-Quenching of Pyrene Fluorescence Leading to Excimers[94,97,100]

With increasing concentration, pyrene forms excimers efficiently in the cholesteric phases of cholesteric mixtures CM (emission maxima $\simeq$470 nm). The monomer fluorescence vibronic band intensity ratio (III/I) indicates probe local environments to be nonpolar. The monomer fluorescence decay as a single exponential for pyrene concentrations of 0.012–0.22 M:

$$^1Py^* + Py \rightleftharpoons [Py\cdot\cdot Py]^* \qquad (7.17)$$

The kinetic data have been analyzed in terms of conventional models for intermolecular excimers formed by isolated molecules. In distinct contrast to the results obtained in the amine quenching case, data on the activation parameters indicate that the mesophase order of the cholesteric mixture plays a negligible role in the dynamics of the intermolecular quenching process. Apparently, the ^1Py–Py collisional orientations necessary for the quenching are those which favored in the mesophases (pyrene molecules oriented with their long molecular axis parallel to the cholesteryl derivatives).

The dynamics of intramolecular excimers have also been examined in various cholesteric solvents (cholesteric and isotropic phases of a cholesteric mixture and in nematic and smectic phases of BCCN) using a series of bispyrenylalkanes [Py—(CH$_2$)$_n$—Py, n = 3, 7, 10, 13, and 22]. Table 7.3 provides a summary of the activation parameters that have been determined for various intramolecular quenching cases. In the isotropic phase, the activation parameters for the monomer quenching are independent of the alkyl chain length. In cholesteric phases, they vary greatly with chain length. In the smectic phase of BCCN, very little quenching of P3P or P22P fluorescence occurs. In the nematic phase, the activation parameters for quenching increases with increasing chain length. From an analysis of the results, it has been concluded that the nature of the influence of mesophase order on the hydrocarbon chain bending depends in a subtle way upon the length of the chain. In isotropic solvents, bending efficiency is solvent dependent but solute chain independent.

Subramanian et al.[101] have examined the luminescence behavior of dodecylcyanobiphenyl as a probe for liquid crystals. The fluorescence intensity exhibited discontinuities at temperatures which correspond to phase transitions: crystalline ↔ smectic ↔ isotropic. The results have been interpreted in terms of monomer and excimer emissions. Earlier there were brief reports of similar applications of pyrene (excimer/monomer) emission intensity ratio in liquid crystals such as MBBA and cholestanyl-benzoate.[102–104]

7.3.3 Aspects of Organic Photoreactions

Photodimerization of Acenaphthylene[91,92]

Earlier discussions on the micellar photochemistry pointed out that the stereospecificity of photocycloaddition reactions is influenced by the mode

TABLE 7.3. Activation Parameters for Intermolecular and Intramolecular Quenching of Pyrene Fluorescence in Liquid Crystalline Solvents

Property	Phase	Intramolecular					Intermolecular	
		Pyrene—$(CH_2)_n$—Pyrene				Pyr—$(CH_2)_3$—DMA	Pyrene/pyrene	Pyrene/Cholesteric DMA
		$n = 3$	10	13	22			
E_{act} (kcal/mol)	cholesteric	10.5	5.7	11.5	16.1	9.7	9.3	9.9
	isotropic	11.4	11.7	11.8	12.9	7.8	7.0	5.3
A (10^{13} sec^{-1})	cholesteric	3.8	5.6	6.8	3.1	0.24	3.6 M^{-1}	21 M^{-1}
	isotropic	1.2	7.5	9.9	2.3	24.0	1.7 M^{-1}	0.018 M^{-1}
	nematic	5.9	7.1					
	smectic		8.1					
S (eu)	cholesteric	1	−16	3	15	3	5	5
	isotropic	4	3	4	6	−4	−2	−10
	nematic	−11	−11					
	smectic	−9	−9	−9	−9			
Reference		97,100	100	100	100	95	94	92

of solubilization and by the nature of the microenvironment around the solute. Photodimerization of acenaphthylene is one such reaction:

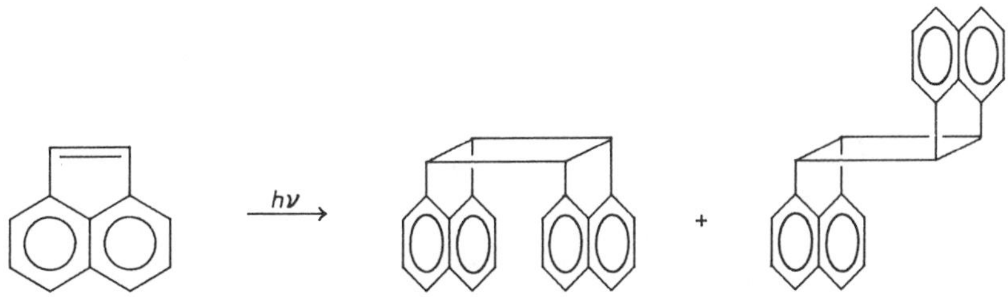

In liquid crystalline solvents, the preferred mode of solubilization (alignment) affects the solute motion, and dramatic differences in the quantum efficiency of the process are observed. Figure 7.7 illustrates the temperature dependence of quantum yields for the dimerization in neat toluene (Φ_T), n-butyl stearate (Φ_{BS}) and in cholesteric mixture CM (1 : 1 w/w mixture of cholestanyl acetate and cholestanyl nonanoate) (Φ_{CM}).

In isotropic solvents such as toluene, the quantum yield is very low and shows a small increase with temperature ($\Phi_T = 0.0105$ and 0.0127 at 10 and 60°C, respectively). The quantum yield in the liquid crystalline medium BS varies slightly as a function of temperature and remains near Φ_T. (n-Butyl stearate exhibits a smectic phase below 25°C, transforming to an isotropic phase above this temperature). In the cholesteric mixture CM at 55–80°C, one has the isotropic mesophase and Φ_{CM} remains approximately at 2.5Φ_T. However, in the cholesteric phase (10–45°C), Φ_{CM} increases dramatically to 21–25Φ_T. The quantum yields in the CM are also strongly dependent on the acenaphthylene concentration. Highest Φ_{CM} is observed in the cholesteric phase at [acenaphthalene] = 0.02 M at 35°C.

A priori, we expect acenaphthylene to be oriented similarly in the smectic, nematic, and cholesteric phases with the long axis parallel to the long axis of the solvent molecules and, hence, the photochemical consequences to be same. Presumably, the very large enhancements observed in the cholesteric phase are due to the increase in the fraction of those solute collisions which have coplanar or parallel plane orientations and not to due increases in the total number of collisions per unit time. Nematic and smectic phases appear to enhance to a much lesser extent the fraction of preferred collisions.

Photolysis of Dibenzylketones (Norrish Type I process)[98]

Photolysis of dibenzylketones has been a prototype reaction to study the dynamics of photogenerated radical pairs in a wide variety of microheterogeneous media. Photolysis leads to the generation of a singlet radical pair which can undergo either recombination or decarbonylate to generate a new radical pair. The new radical can either separate to free radicals or recombine

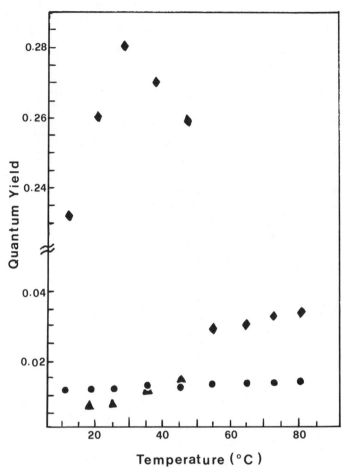

FIG. 7.7. Temperature dependence of the dimerization quantum yield of acenaphthylene (0.08 M) in toluene ($\bullet$), n-butylstearate ($\blacktriangle$) and in a 1:1 mixture of 5α-cholestan-3β-yl nonanoate and 5α-cholestan-3β-yl acetate ($\blacklozenge$). (From Nerbonne and Weiss.[91] Copyright 1979 American Chemical Society.)

to form products:

$$A—CO—B \xrightarrow{h\nu} [A—CO\cdot \ \cdot B] \xrightarrow{-CO} [A\cdot \ \cdot B] \longrightarrow (A\cdot) + (B\cdot)$$
$$A = Ph—CH_2 \qquad\qquad\qquad\qquad AB \qquad AA + BB + AB$$

The relative yields of products AA, AB, and BB depend largely on the dynamics of the radical pairs. The efficiency of recombination processes is assessed using the cage effect relation:

$$F = [AB - (AA + BB)]/[AB + (AA + BB)] \qquad (7.18)$$

TABLE 7.4. Effect of Phase Type of Liquid Crystalline Solvents on the Cage Effect Fraction F_c and Percentage Conversion in the Photolysis of 1-(4-Methylphenyl)-3-phenylpropan-2-one.[a]

Solvent[b]	Temperature (°C)/Phase[c]	F_c	% Conversion
CCl/CN (35:65)(w/w)	51(c)	0.196 ± 0.018	60
	56(c)	0.184 ± 0.008	65
	65(c)	0.143 ± 0.005	60
	70(i)	0.163 ± 0.012	35
	79(i)	0.138 ± 0.010	80
	86(i)	0.133 ± 0.007	70
BCCN	46(s)	0.088 ± 0.004	20
	61(n)	0.080 ± 0.004	25
	82(i)	0.106 ± 0.008	26
BS	9(k)	0.150 ± 0.010	22
	17(s)	0.192 ± 0.011	35
	24(s)	0.085 ± 0.010	68
	28(i)	0.021 ± 0.037	75
	31.5(i)	0.007 ± 0.008	n.d
	35(i)	-0.001 ± 0.009	57
	45(i)	0.005 ± 0.005	60

[a] Initial concentration of the ketone 1 wt%. (From Horvat et al.[98])
[b] Abbreviations: CCl = 5-cholestan-3-yl chloride; CN = 5-cholestan-3-yl nonanoate; BCCN = trans,trans-4-n-butylbicyclohexane-4-carbonitrile, and BS = n-butyl stearate.
[c] c = cholesteric; s = smectic; i = isotropic; n = nematic and k = gel/crystalline phases.

Photolysis of dibenzylketones ($Ph—CH_2—CO—CH_2—Ph$) in various liquid crystalline media (cholesteric and isotropic phases of a cholesteric mixture, smectic, and nematic phases of BCCN and BS) have shown that the cage effect fraction F depends strongly on the solvent phase in a complex manner depending both on the solvent order (phase type) and diffusion (local viscosity). Table 7.4 presents a summary of these results.

References

1. G. L. Gaines, Jr., "Insoluble Monolayers at Liquid–Gas Interfaces." Wiley (Interscience), New York, 1966.
2. G. L. Gaines, Jr., in "Surface Science and Colloids" (M. Kerker, ed.), MTP International Reviews in Science, Vol. 7, pp. 1–24. Butterworth, London, 1972.
3. H. Kuhn, D. Möbius, and H. Bücher, in "Physical Methods of Chemistry" (A. Weissberger and B. W. Rossiter, eds.), Vol. 1, Part III, pp. 577–701. Wiley (Interscience), New York, 1972.
4. I. Langmuir, J. Am. Chem. Soc. 39, 1848 (1917).
5. K. B. Blodgett, J. Am. Chem. Soc. 57, 1007 (1935).
6. K. B. Blodgett and I. Langmuir, Phys. Rev. 51, 964 (1935).

7. K. B. Blodgett, *Phys. Rev.* **55**, 391 (1939).
8. H. Kuhn and D. Möbius, *Angew. Chem. Int. Ed. Engl.* **10**, 620 (1971).
9. D. Möbius, *Ber. Bunsenges. Phys. Chem.* **82**, 848 (1978).
10. D. Möbius, *Acc. Chem. Res.* **14**, 63 (1981).
11. D. Möbius and H. Kuhn, *Isr. J. Chem.* **18**, 375 (1979).
12. H. Kuhn, *Pure Appl. Chem.* **51**, 341 (1979); **53**, 2105 (1982).
13. D. Möbius, *Mol. Cryst. Liq. Cryst.* **96**, 319 (1983).
14. D. G. Whitten, D. W. Eaker, B. F. Horsey, R. H. Schmehl, and P. R. Worsham, *Ber. Bunsenges. Phys. Chem.* **82**, 858 (1978).
15. D. G. Whitten, *Angew. Chem. Int. Ed. Engl.* **18**, 440 (1979).
16. D. G. Whitten, J. C. Russell, and R. H. Schmehl, *Tetrahedron* **38**, 2455 (1982). (a) J. Sagiv, *J. Am. Chem. Soc.* **102**, 92 (1980); *Isr. J. Chem.* **18**, 339, 346 (1979). (b) E. E. Polymeropoulos and J. Sagiv, *J. Chem. Phys.* **69**, 1836 (1978). (c) L. Netzer and J. Sagiv, *J. Am. Chem. Soc.* **105**, 674 (1983). (d) R. Elbert, T. Folder, and H. Ringsdorf, *ibid.*, **106**, 7687 (1984).
17. K.-P. Seefeld, D. Möbius, and H. Kuhn, *Helv. Chim. Acta* **60**, 2608 (1977).
18. W. D. Bellamy, G. L. Gaines, Jr., and A. G. Sweet, *J. Chem. Phys.* **39**, 2528 (1963); **41**, 2068 (1964).
19. A. G. Sweet, G. L. Gaines, Jr., and W. D. Bellamy, *J. Chem. Phys.* **40**, 2596 (1964).
20. W. Sperling and B. Ke, *Photochem. Photobiol.* **5**, 857, 865 (1966).
21. T. Trosper, R. B. Park, and K. Sauer, *Photochem. Photobiol.* **7**, 451 (1968).
22. S. M. B. de Costa, J. R. Froines, J. M. Harris, R. M. LeBlanc, B. H. Orger, and G. Porter, *Proc. R. Soc. London, Ser. A* **326**, 503 (1972).
23. H. Heithier, H.-J. Galla, and H. Möhwald, *Z. Naturforsch.*, C **33C**, 382 (1978).
24. H. Heithier and H. Möhwald, *Z. Naturforsch.*, C **38C**, 1003 (1983).
25. H. Heithier, K. Ballschmitter, and H. Möhwald, *Photochem. Photobiol.* **37**, 201 (1983).
26. J. A. Bardwell and J. R. Bolton, *Photochem. Photobiol.* **39**, 735 (1984).
27. J. A. Bardwell and J. R. Bolton, *Photochem. Photobiol.* **40**, 319 (1984).
28. T. L. Penner and D. Möbius, *J. Am. Chem. Soc.* **104**, 7407 (1982).
29. J. Aghion, S. S. Brodye, and S. S. Brody, *Biochemistry* **7**, 3120 (1969).
30. S. S. Brody, *Z. Naturforsch.*, B **26B**, 134 (1971); **28C**, 397 (1973); **30C**, 318 (1975).
31. P. Chin and S. S. Brody, *Biochemistry* **14**, 1190 (1975).
32. P. Chin and S. S. Brody, *Z. Naturforsch.*, C **31C**, 569 (1976).
33. S. M. B. de Costa and G. Porter, *Proc. R. Soc. London, Ser. A* **341**, 167 (1974).
34. A. F. Jansen, J. R. Bolton, and M. J. Stillman, *J. Am. Chem. Soc.* **101**, 6337 (1979).
35. O. Gonen, H. Levanon, and L. K. Patterson, *Isr. J. Chem.* **21**, 271 (1981).
36. K. Chandrasekaran, C. Gionnotti, K. Monserrat, J. P. Otruba, and D. G. Whitten, *J. Am. Chem. Soc.* **104**, 6200 (1982).
37. G. Sprintschnik, H. W. Sprintschnik, P. P. Kirsch, and D. G. Whitten, *J. Am. Chem. Soc.* **98**, 2337 (1976); **99**, 4947 (1977).
38. G. Sprintschnik, H. W. Sprintschnik, P. P. Kirsch, and D. G. Whitten, *J. Am. Chem. Soc.* **99**, 877 (1977).
39. S. J. Valenty and G. L. Gaines, Jr., *J. Am. Chem. Soc.* **99**, 1285 (1977).
40. A. Harriman, *J.C.S. Chem. Commun.* p. 777 (1977).
41. T. Loughran, M. D. Hatlee, L. K. Patterson, and J. J. Kozak, *J. Chem. Phys.* **72**, 5791 (1980).
42. J. Teissie, J. F. Tocanne, and A. Baudras, *Eur. J. Biochem.* **83**, 77 (1978).
43. D. G. Whitten, *J. Am. Chem. Soc.* **96**, 594 (1974).
44. F. H. Quina, D. Möbius, F. A. Carroll, F. R. Hopf, and D. G. Whitten, *Z. Phys. Chem. (Wiesbaden)* **101**, 151 (1976).
45. F. H. Quina and D. G. Whitten, *J. Am. Chem. Soc.* **97**, 1602 (1975).
46. E. E. Polymeropoulos and D. Möbius, *Ber. Bunsenges. Phys. Chem.* **83**, 1215 (1979). (a) M. Morin, R. M. Leblanc, and I. Gruda, *Can. J. Chem.* **58**, 2038 (1980). (b) D. A. Holden, H. Ringsdorf, V. Deblauwe, and G. Smets, *J. Phys. Chem.* **88**, 716 (1984).

47. P. R. Worsham, D. W. Eakers, and D. G. Whitten, *J. Am. Chem. Soc.* **97**, 277 (1975); **100**, 7091 (1978).
48. F. R. Hopf and D. G. Whitten, *J. Am. Chem. Soc.* **98**, 7422 (1976).
49. J. A. Mercer-Smith and D. G. Whitten, *J. Am. Chem. Soc.* **101**, 6620 (1979).
50. B. E. Horsey and D. G. Whitten, *J. Am. Chem. Soc.* **100**, 1293 (1978).
51. W. F. Mooney, P. E. Brown, J. C. Russell, S. M. de Costa, L. G. Pedersen, and D. G. Whitten, *J. Am. Chem. Soc.* **106**, 5659 (1984).
52. K. S. Schanze, T. F. Maddox, and D. G. Whitten, *J. Am. Chem. Soc.* **104**, 1773 (1982). (a) D. A. Holden, H. Ringsdorf, and M. Haubs, *J. Am. Chem. Soc.* **106**, 4531 (1984).
53. L. K. Patterson, J. E. McCarthy, and J. J. Kozak, *Chem. Phys. Lett.* **89**, 435 (1982).
54. T. Watanabe, T. Miyasaka, A. Fujishima, and K. Honda, *Chem. Lett.* p. 443 (1978); *J. Am. Chem. Soc.* **100**, 6657 (1978).
55. P. J. Reucroft and W. H. Simpson, *Photochem. Photobiol.* **10**, 79 (1969).
56. M. J. Villar and M. R. Wurmser, *C. R. Hebd. Seances Acad. Sci., Ser. A* **275**, 861 (1972).
57. R. C. Nelson, *J. Chem. Phys.* **27**, 864 (1957). (a) A. F. Jansen and J. R. Bolton, *J. Am. Chem. Soc.* **101**, 6342 (1979).
58. M. F. Lawrence, J. P. Dodelet, and M. Ringnet, *Photochem. Photobiol.* **34**, 393 (1981).
59. M. F. Lawrence, J. P. Dodelet, and L. H. Dao, *J. Phys. Chem.* **88**, 950 (1984).
60. R. Jones, R. H. Tredgold, and J. F. O'Mullane, *Photochem. Photobiol.* **32**, 223 (1980).
61. H. Ti Tien, "Bilayer Lipid Membranes (BLM): Theory and Practice." Dekker, New York, 1974.
62. H. Ti Tien, *in* "Photosynthesis in Relation to Model Systems" (J. Barber, ed.), p. 115. Elsevier, Amsterdam, 1979.
63. M. K. Jain, "The Bimolecular Lipid Membrane, A System." Van Nostrand-Reinhold, New York, 1972.
64. H. Ti Tien, ed., *Photochem. Photobiol.* **24**, 95 (1976).
65. D. S. Berns, *Photochem. Photobiol.* **24**, 117 (1976).
70. H. M. Ullrich and H. Kuhn, *Z. Naturforsch., B* **24B**, 1342 (1969); *Biochim. Biophys. Acta* **266**, 584 (1972).
71. H. W. Trissel and P. Läuger, *Biochim. Biophys. Acta* **282**, 40 (1972).
72. J. S. Huebner, *J. Membr. Biol.* **39**, 97 (1978).
73. F. T. Hong, *Adv. Chem. Ser.* No. 188, 211 (1980).
74. F. T. Hong, *Bioelectrochem. Bioenerg.* **5**, 425 (1978).
75. J. S. Huebner, *Photochem. Photobiol.* **15**, 1035 (1979).
76. J. S. Huebner and W. E. Varnadore, *Photochem. Photobiol.* **39**, 97 (1982).
77. J. S. Huebner, R. T. Arrieta, I. C. Arrieta, and P. M. Pachori, *Photochem. Photobiol.* **39**, 191 (1984).
78. N. Alamuti and P. Läuger, *Biochim. Biophys. Acta* **211**, 363 (1970).
79. A. Steinemann, N. Alamuti, U. Brodmann, O. Marshall, and P. Läuger, *J. Membr. Biol.* **4**, 284 (1969).
80. G. W. Pohl, *Biochim. Biophys. Acta* **288**, 248 (1972).
81. P. Fromherz and W. Arden, *J. Am. Chem. Soc.* **102**, 6211 (1980); *Ber. Bunsenges. Phys. Chem.* **82**, 871 (1978).
82. R. C. Young and S. W. Feldberg, *Biophys. J.* **27**, 237 (1979).
83. S. W. Feldberg, G. H. Armen, J. A. Bell, C. K. Chang, and C. B. Wang, *Biophys. J.* **34**, 149 (1981). (a) A. Losev and D. Mauzerall, *Photochem. Photobiol.* **38**, 355 (1983).
84. F. D. Saeva, ed., "Liquid Crystals: The Fourth State of Matter." Dekker, New York, 1983.
85. S. Chandrasekar, "Liquid Crystals." Cambridge Univ. Press., London and New York, 1977.
86. P. G. DeGennes, "The Physics of Liquid Crystals." Oxford Univ. Press, London and New York, 1974.
87. J. J. Wolken and G. H. Brown, "Liquid Crystals and Biological Systems." Academic Press, New York, 1980.

88.   G. H. Brown, *J. Chem. Educ.* **60,** 900 (1983).
89.   J. M. Nerbonne and R. G. Weiss, *J. Am. Chem. Soc.* **100,** 5952 (1978).
90.   J. M. Nerbonne and R. G. Weiss, *J. Am. Chem. Soc.* **100,** 2571 (1978).
91.   J. M. Nerbonne and R. G. Weiss, *J. Am. Chem. Soc.* **101,** 402 (1979).
92.   V. C. Anderson, B. B. Craig, and R. G. Weiss, *J. Am. Chem. Soc.* **103,** 7169 (1981).
93.   J. P. Otruba and R. G. Weiss, *Mol. Cryst. Liq. Cryst.* **80,** 165 (1982).
94.   V. C. Anderson, B. B. Craig, and R. G. Weiss, *J. Am. Chem. Soc.* **104,** 2972 (1982).
95.   V. C. Anderson, B. B. Craig, and R. G. Weiss, *J. Phys. Chem.* **86** 4642 (1982).
96.   J. P. Otruba and R. G. Weiss, *J. Org. Chem.* **48,** 3448 (1983).
97.   V. C. Anderson and R. G. Weiss, *J. Am. Chem. Soc.* **106,** 6628 (1984).
98.   D. A. Horvat, J. H. Liu, N. J. Turro, and R. G. Weiss, *J. Am. Chem. Soc.* **106,** 5291 (1984).
99.   J. M. Nerbonne and R. G. Weiss, *Isr. J. Chem.* **18,** 266 (1979).
100. V. C. Anderson, B. B. Craig, and R. G. Weiss, *Mol. Cryst. Liq. Cryst.* **97,** 351 (1983).
101. R. Subramanian, L. K. Patterson, and H. Levanon, *Chem. Phys. Lett.* **93,** 578 (1983).
102. A. K. Soutar, H. J. Pownall, A. S. Hu, and L. C. Smith, *Biochemistry* **13,** 2828 (1974).
103. Y. Tomkiewicz and A. Weinrub, *Chem. Phys. Lett.* **3,** 229 (1969).
104. T. J. Novak, R. A. Mackay, and E. J. Poziomek, *Mol. Cryst. Liq. Cryst.* **20,** 213 (1973).

Chapter 8

Photoprocesses in Polymers, Polyelectrolytes and Ion-Exchange Membranes

Earlier chapters have been concerned with various types of molecular aggregated systems in water composed of simple surfactants or lipids. In this and subsequent chapters we will consider larger macromolecular host systems formed by polymeric molecules. As before, we would like to probe their static and dynamic properties using some of the well-established photophysical and photochemical methods. First, let us consider long-chain macromolecules composed of organic polymers. Photophysical and chemical studies of polymers have a long history, starting much earlier than before the recent interest in the photophysical studies of organized assemblies, and, hence, the literature is rather extensive. Our discussions will be very selective, attempting to highlight how various photophysical and photochemical studies bring about a better understanding of dynamical properties of these systems. We will consider studies of synthetic polymers with built-in chromophores (labelled at a few or more places, selectively or randomly) and polymer solutions containing free (unbound) molecules. For the most part, we will restrict ourselves to flexible polymers in dilute solutions and exclude studies of polymer solids and thin films. For lack of space, we also exclude photochemical studies of photopolymerization, photodegration, and studies of biopolymers.

Closely related to the neutral polymers are their ionic counterparts, the polyelectrolytes and their cross-linked resins, the ion-exchange membranes. Currently there is a growing interest to examine these systems using various physical methods. In later sections of this chapter, we review photophysical

255

and photochemical studies in these systems. We start our discussions with an overview of the general features of polymers in solutions and certain points of importance in polymer photophysics in solution.

8.1   General Aspects of Polymers in Solution[1-4] and of Polymer Photophysics[5-20]

Photoprocesses in polymer media are unique in several aspects, differing significantly from those of homogeneous solutions and some of the micro-heterogeneous systems encountered earlier. The attachment of a chromophore or quencher to a polymer chain renders their mobility strongly controlled by the polymer dynamics in solution. In extent of interchain and intra-chain interactions between the labelled chromophores and/or quencher in polymer solution depend largely on the configuration of the polymer chain in solution. This, in turn, depends on several factors; polymer molecular weight (average chain length), concentration, temperature, and the nature of the solvent. Good solvents in this context are those in which the attraction between polymer segments and solvent molecules is greater than that between the polymer segments themselves. Consequently there is a swelling of the polymer in good solvent solutions. For poor solvents, the reverse situation is true, and the polymer takes a more compact, nested (shrinked) configuration. The indifferent or θ solvents are those that have no effect. The scheme below illustrates schematically possible configurations of the polymer in good, poor and indifferent solvents:

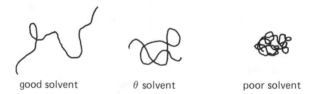

good solvent θ solvent   poor solvent

(This effect of the solvent in controlling the volume occupied by the adjacent segments of the polymer leads to the so-called excluded volume effect.)

The swelling or shrinking of the polymer chains drastically affects the average distance between the excited-state species and the quenchers/traps and, hence, the efficiencies of the photoprocesses. Polymer concentration and molecular weight affect the processes in a similar manner. Usually studies are carried out in so-called dilute solutions where the concentration of the polymer is low enough to make intermolecular forces unimportant, and polymer-coil entangling is reduced to a minimum. These effects become important at higher concentrations (as in semi-dilute solutions). In polymers with pendant chromophores and/or quencher groups, the extent of labelling and the mode of labelling (tacticity) are also important in affecting the efficiency of the photoprocesses. (Until recently, the tacticity of the polymers was not controlled in most of the studies. The effect of tacticity on polymer photophysics was not realized until it was discovered that in polystyrene the efficiency of excimer formation depended strongly on the tacticity and crystallinity of the polymer.)

Labelled polymers are broadly classified into two main types, A and B. In type A polymers, isolated chromophores are attached to a polymer chain as an end group or as minor component of a copolymer. Type B polymers are those in which the repeat units carry an absorbing (and emitting) unit. Polystyrene is a typical example of type B polymer where the pendant aromatic groups are arranged along the backbone in a random (atatic, I_a), alternating (syndiotactic, I_b) or regular (isotactic, I_c) fashion:

$$-C-C-C-C-C-C-C-C-C-C-C-C-C-C- \qquad \text{Poly(styrene)-atactic (PS-}I_a\text{)}$$

$$-C-C-C-C-C-C-C-C-C-C-C-C-C-C- \qquad \text{Poly(styrene)-syndiotactic (PS-}I_b\text{)}$$

$$-C-C-C-C-C-C-C-C-C-C-C-C-C-C- \qquad \text{Poly(styrene)-isotactic (PS-}I_c\text{)}$$

The efficiency of bimolecular excited-state processes such as excimer or exciplex formation, electron transfer, or energy transfer, measures, in essence, the probability of encounters of the excited-state chromophore and the quenchers:

$$P^* + Q \longrightarrow [PQ]^* \longrightarrow P + Q \qquad (8.1)$$

In polymers this depends on the initial distribution of the reactants and on the polymer dynamics, namely, configurations and mobility of the polymer in solution. Polymer dynamics thus amounts to a discussion of the cyclization phenomena, the process by which two ends or segments of a polymer chain (or substituents attached to these) come to close proximity. The distance of nearest approach to qualify for being at close proximity can be short (5–10 Å as in electron transfer reactions between the chain ends or segments labelled with donors and acceptors) or large (> 40 Å as in energy transfer processes by

Förster mechanisms). Similarly, the sampling time for the mobility is larger for the triplet-state reactions than for the singlet-state reactions.

For open-chain polymers, we can visualize three distinct topologies for the cyclization process:

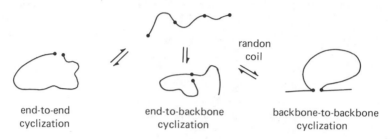

| end-to-end | end-to-backbone | backbone-to-backbone |
| cyclization | cyclization | cyclization |

For this reason, in an atactic polymer, for example, in addition to the intermolecular interactions, we need to consider each of the above intra-molecular interactions, each occurring with different efficiencies! Extensive problems encountered in the interpretation of excimer and exciplex yields and decay profiles have led to the realization that the conformational requirements play a very important role in controlling reactions in bichromophoric and polychromophoric systems.

Interpretation of various photophysical processes in terms of kinetic schemes developed for reactions in homogeneous solutions to polymers and polyelectrolytes brings in several problems. Some of these are reminiscent of those encountered earlier with organized assemblies: inhomogeneous inter-iors and exteriors (with the presence of charged interfaces in polyelectrolytes) and inhomogeneous distribution of probes and quencher molecules. Controlled synthesis of specifically labelled polymers can alleviate to some extent the latter problem. In fact, great surges in our understanding of polymer photophysics have been made possible only through elegant synthesis and study of specifically labelled polymers. Conformational/stereochemical consequences of orientation and mobility of chromophores on excited-state processes is an area where studies of polymer photophysics have provided new insights.

The dependence of the mobility of a given chromophore label on the backbone (connectivity) is a difficult problem to handle. The polymer interior approximates neither that of fluid solutions (characterized by an uniform distribution of solutes and diffusion controlled kinetics) nor that of a rigid matrix (characterized by reactants present at known/fixed distances) but an intermediate behavior. Extensive labelling with chromophores along the backbone of a polymer leads to the presence of several chromophores at close proximity. In polymers it has been found that excitation energy migrates to distances much larger than is allowed in between two isolated chromophores, via transfer from one to another to another and so on!

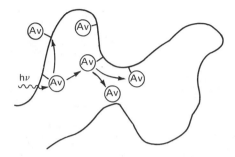

This rapid migration of the excitation energy along the backbone and across influences considerably bimolecular reactions of these excited chromophores (processes such as fluorescence quenching, fluorescence polarization, and excimer and exciplex formation). Determination of the existance of an energy migration pathway and its consequences has been a major topic in each of the polymer systems studied to date.

In solutions of polymers, the space between the polymer chains which is available for unhindered diffusion of small molecules can be quite large. In this case, the microscopic viscosities will not even approximate the high macroscopic viscosities, and processes can occur as in dilute polymer-free solutions. We can have an inverse situation where reactions are hindered or slowed. The microviscosity of polymers, thus, is an intriguing but important property to measure.

8.1.1   Singlet-State Reactions: Fluorescence Probe Analysis and Quenching

The extent of polymer coiling and solvation can be probed by examining the emission of medium-sensitive probes (attached to the polymer) in various solvents. A comparative study on the fluorescence of 5-(dimethylamino)-1-naphthalenesulfonate labelled on styrene-divinylbenzene copolymer (S-DVB) has been made in various good and poor solvents.[21] The responses of the probe to changes in the local environment parallel the gel swelling of the polymer. In poor solvents, the probe environment resembles that of S-DVB while in good solvents, the probe emission resembles the pure solvent rather than the pure polymer matrix.

Dansyl group fluorescence has been used similarly to study polymer complexes formed on mixing solutions of poly(acrylic acid) and poly(oxyethylene).[22] Dansyl group fluoresces much more strongly in organic solvents than in water. Aqueous solutions of dansyl-labelled PAA exhibit large increases in fluorescence on complexation with poly(oxyethylene), thus reflecting local exclusion of water molecules. Kinetic studies have also been made during the addition of an excess of unlabelled PAA.

Fluorescence properties of N,N-dialkylaminobenzylidene malononitrile have also been examined in the polymer matrices of alkyl methacrylates and alkyl acrylates.[23] In the presence of the polymer, the emission maximum of the probe shows blue shifts, an increase in k, and a decreasing quantum yield with chain length, results which indicate a chain-length-dependent location of the probe.

Dynamic quenching of the fluorescence of probe molecules by quenchers is a common and useful method to study the mobility and extent of penetration of molecules. Depending on the experimental conditions, we can envisage several possible quenching scenarios for the excited-state quenching processes:

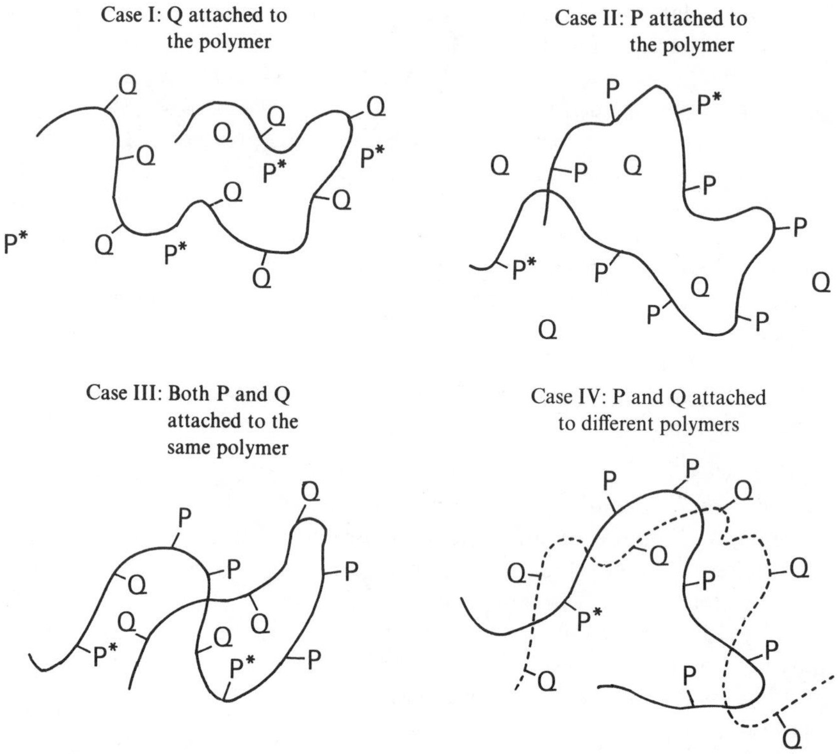

Case I: Q attached to
the polymer

Case II: P attached to
the polymer

Case III: Both P and Q
attached to the
same polymer

Case IV: P and Q attached
to different polymers

As a result of the inhomogeneous distribution of the probe P*, the quencher Q, or both, we often observe marked deviation from linearity in the Stern–Volmer plots. There have been only a few quantitative studies of these cases.

Weill and co-workers[24–26] studied the quenching of small fluorescing species in a polymer containing quenchers attached at random along the polymer backbone (case I) and in solutions containing small quencher molecules and polymers labelled with chromophores (case II, e.g., bound diphenylanthracene fluorescence quenched by N,N-dimethyltoluidine molecules). In the former case, the system may be thought of as consisting of two regions, one representing the polymer domains with a high quencher concentration and the other outside the polymer coils with no quenchers. The dependence of the fluorescence intensity on the polymer and quencher concentrations may be interpreted in terms of critical concentrations at which the polymer chains begin to be sufficiently entangled so that the system has a uniform quencher concentration. In the second case, deviation from Stern–Volmer plots are due to a difference between the solvation of fluorescing polymer by the principal solvent and its solvation by the quenching agent. The fluorescence data could be interpreted in terms of the extent to which the quencher is concentrated in the polymer domain or excluded from it.

If the polymer chain carries both the fluorescing and quenching species (case III), then the intramolecular quenching is governed by the rate of conformational transitions of the polymer, which allows the fluorescing and quenching species to diffuse toward each other. Experiments of this type were reported first by Kirsch et al.[27] The experimental results were analyzed in terms of an effective local concentration C_{eff} within the polymer coils. This is rather misleading, since an averaging of the polymer segment concentrations within some effective polymer volume of the randomly coiled polymer would lead to a decreasing values of C_{eff} with increasing length of the polymer. A treatment taking into account the connectivity of the chain must lead to C_{eff} rapidly approaching asymptotic value.

In a solution containing two similar polymers one carrying the fluorescing groups and the other the quenching species (case IV), the change in the fluorescence intensity as the concentration of the quenching polymer is changed reflects the ability of polymer coils to interpenetrate each other. Experiments have shown that interpenetration in flexible-chain molecules is negligible until a critical polymer concentration has been reached. Since no such threshold has been observed in interpolymer reactions characterized by a high activation energy, the effects observed by Kirsch et al.[27] are apparently restricted to diffusion-controlled reactions involving species with very short lifetimes such as excited-state species. From these discussions, it is obvious that in polymer solutions, the quenching rate constant k_q in general is dependent on the molecular weight and concentration of the polymer and the thermodynamic character of the solvent. Rapid progress in quantitative analysis is being made in the triplet-state quenching process, which we discuss later.

8.1.2 Singlet-State Reactions: Fluorescence Polarization

Information on the mobility (rotational diffusion) in macromolecular systems can be derived from the measurements of depolarization of fluorescence of probe molecules imbedded in them. The degree of polarization measured in the steady state is a quantity that yields a gross picture (estimate) on the rigidity (microviscosity?) experienced by the probe in the host medium. Detailed descriptions on the dynamics of the motion executed by the probe require time-resolved measurements of the fluorescence anisotropy decay. The type of motion is usually deduced by fitting various autocorrelation functions associated with different types of motion. In earlier discussions on micelles (Section 2.3) and lipid vesicles (Section 6.6), we have seen several applications of these concepts.

Fluorescence polarization in polymers such as poly(methyl methacrylate) and poly(styrene) has been determined for the probes attached to the backbone or at the end of the chain.[28-30] It was found that the rotational relaxation times for groups fixed at the interior of the chain were less than one fifth of those of same groups at the end of the polymer, reflecting the constraints to molecular motion imposed by the chain. In an examination of the polarization of fluorescence of 9,10-diphenylanthracene attached to polystyrene, Wahl concluded that the emission anisotropy could be well described by a model for polymer chain resembling a necklace of rotating ellipsoids![31]

Valeur, Monnerie, and co-workers have considered in detail various segmental motions of a polymer chain and their consequences on the observed fluorescence anisotropy.[32-36] The form of the emission anisotropy curve for the processes leading to a diffusion of orientations along the chain (three-bond motion, crankshaft, etc.) in a tetrahedral lattice has been derived to be

$$r(t) = r(0) \exp(t/\rho) \, \text{erfc} \sqrt{t/\rho} \qquad (8.2)$$

where erfc is the error function component and ρ the correlation time containing parameters descriptive of the diffusive jump frequency and conformational structure of the chain. Attempts to fit the experimental $r(t)$ with the above form were not satisfactory. When both valence angles and internal rotational angles were allowed to vary from those associated with the ideal lattice, an additional term is introduced and the expression becomes

$$r(t) = r(0) \exp(-t/\theta) \exp(t/\rho) \, \text{erfc} \sqrt{t/\rho} \qquad (8.3)$$

where θ is the relaxation time reflecting the relaxation time with respect to the ideal lattice.

Equation (8.3) was found to adequately describe emission anisotropy decay for anthracene labelled to polystyrene in the backbone (PS—Anthr—PS) and also at the end of the chain (Anthr—PS—Anthr). Wahl also found fitting his data to the above equation to be satisfactory. Analysis of ρ and θ indicate that the relaxation observed for the end-labelled polymer are much faster than that in the backbone labelled case. The mean relaxation time $\langle\sigma\rangle$ is about four times larger—results which are in good agreement with those of Soutar et al.[28–30] ($\langle\sigma\rangle$ = 9.6 nsec for PS—Anthr—PS and 2.3 nsec for Anthr—PS—Anthr). Anufrieva and Gotlieb[8] have described in detail extensive studies of polymers in solution using polarized luminescence (mostly in the russian literature).

8.1.3 Dynamics of Excimers

For a wide variety of polymers, the observation of excited dimer (excimer) luminescence along with the normal monomer fluorescence is a common feature. There have been very intense efforts (and still continue to be) directed towards an understanding of the various factors that govern the formation of excimers in polymers. Table 8.1 provides a listing of various systems (representative cases) that have been studied.[37–77]

Most often the excimers are intramolecular in origin, being present at very low concentration of the polymers. There are several factors that complicate the analysis of excimer dynamics along the conventional schemes available for intermolecular excimers formed by small (free/unbound) molecules in solution. First, the approximation of the relative motion of the two chromophores towards each other when they are bound to the polymer chain as that of unrestricted diffusional motion of free molecules in solution is rather unrealistic. Second, it is generally accepted in polymers that singlet energy

TABLE 8.1. Studies of Intramolecular Excimer Formation in Polymers

Chromophore	Polymer	Abbreviation	Reference
Phenyl	Poly(styrene)	PS	37–39
1-Naphthyl	Poly(1-vinylnaphthalene)	P1VN	40–42
2-Naphthyl	Poly(2-vinylnaphthalene)	P2VN	43–46
1-Naphthyl	Copolymer of 1VN and methyl methacrylate	P(1VN-co-MMA)	47,48
2-Naphthyl	Copolymer of 2VN and methacrylate	P(2VN-co-MA)	49
2-Naphthyl	Copolymer of 2VN and styrene	P(2VN-co-S)	50
2-Naphthyl	Naphthyl labelled poly(methylmethacrylate)	P(Naph-MMA)	51–55
1-Pyrenyl	Pyrenyl labelled poly(ethyleneoxide)	P(Pyr-EO)	56
1-Pyrenyl	Pyrenyl (end) labelled poly(styrene)	Pyr—PS—Pyr	57–60
1-Pyrenyl	Pyrenyl (end) labelled poly(dimethylsiloxane)	Pyr—PDMS—Pyr	61–62
9-Phenanthryl			63
N-Carbazolyl	Poly(N-vinylcarbazole)	PNVC	64–74
Naphthyl, pyrenyl, and anthryl	Polyesters labelled with napthyl, pyrenyl, or anthryl groups		75–77

TABLE 8.2. Studies of Intramolecular Excimer Formation Bichromophoric Model Compounds

Chromophore	Model compound	Reference
Phenyl	$Ph(CH_2)_3Ph$	78
Phenyl	$Ph—CH_2—X—CH_2—Ph, X = CH_2, O, \overset{+}{N}H_2$	79
1-Pyrenyl	$Pyr—(CH_2)_n—Pyr, n = 1-16, 19, 22$	80–84
1-Pyrenyl	$Pyr—CH_2—O—CO—(CH_2)_n—CO—O—CH_2—Pyr,$ $n = 1-4, 6, 8, 10, 14, 18, 22$	85, 86
1-Pyrenyl	$CH_3—CH(Pyr)—O—CH(Pyr)—CH_3$ (racemic, meso)	87
Naphthyl	$Ar—(CH_2)_3—Ar, Ar = 1\text{-naphthyl}, 2\text{-naphthyl}$	88
Naphthyl	$Ar—(CH_2)_3—Ar, Ar = 4(OMe)\text{-1-naph}, 4(OH)\text{-1-naph}$	89
Naphthyl	$CH_3—CH(Naph)—O—CH(Naph)—CH_3$ (racemic, meso)	90
Naphthyl	$Ar—CH_2—O—CH_2—Ar, Ar = \text{naphthyl}$	90a
Anthryl	$Ar—(CH_2)_3—Ar, Ar = 9\text{-anthryl}$	91
Anthryl	$Ar—(SiO)_n—Ar, Ar = 9\text{-anthryl}$	92
Anthryl	$Ar—CH_2—O—CH_2—Ar, Ar = 9\text{-anthryl}$	93
Phenanthryl	$Ar—(CH_2)_3—Ar, Ar = n\text{-phenanthryl}$	94, 95
N-Carbazolyl	$Ar—(CH_2)_3—Ar, Ar = N\text{-carbazolyl}$	96–98
N-Carbazolyl	$CH_3—CH(Ar)—CH_2—CH(Ar)—CH_3, Ar = N\text{-carbazolyl}$	65, 72, 73

migration among the chromophores significantly affects the extent of excimer formation. Third, only in polymers has the occurrence of more than one resolvable (structurally or kinetically) excimer emission been noted, e.g., poly-(N-vinylcarbazole) and poly(1-vinylnaphthalene).

The first problem has been attacked via synthesis and the study of various model compounds—bichromophoric and polychromophoric compounds where the chromophores are covalently linked together via methylene, ether, or amide linkages of different length. Table 8.2 presents a listing of some of these model compounds.[78-98] Efforts in this area largely helped solve some of the anamolous emission and the complex excimer kinetics observed in polymers. As mentioned earlier, assessment of energy migration channels in polymers involves a comparative analysis of various photophysical processes as they occur in a given polymer. Due to these inherent problems, steady-state analysis of excimer formation efficiencies under different conditions alone is very inadequate to provide a complete picture of the processes involved.

The kinetics of intermolecular excimers of small, unbound molecules in solution is adequately described by the following scheme:[99]

$$M \xrightarrow{hv} {}^1M^* \underset{k_{MD}}{\overset{k_{DM}[M^*]}{\rightleftharpoons}} {}^1D^*$$

with k_{IM}, k_{FM}, k_{ID}, k_{FD} leading to M, M + hv, 2M, 2M + hv′ respectively.

This conventional scheme yields the following expression for the intensity ratio of excimer to monomer emission at constant T:

$$I_D/I_M = kk_{DM}[M] \qquad (8.4)$$

where k is a composite term of rate coefficients, k_{DM} is the rate constant for the excimer formation and $[M]$, the concentration of the fluorophore. Applied to polymers, the concentration dependence can be described in terms of a function characteristic of the number of potential excimer sites in the polymer and a function which expresses the extent of energy migration that leads to the population of the excimer sites.

Steady-state fluorescence experiments have resulted in the generation of a number of expressions which describe the dependence of excimer formation on the intramolecular concentration distribution. For example, in systems containing relatively weakly interacting chromophores such as phenyl and naphthyl, adequate description of the intensity ratio I_D/I_M is given by

$$I_D/I_M = kl_a f_{aa} \qquad (8.5)$$

where f_{aa} is the fraction of aromatic pairs within the polymer (assumed to characterize the concentration of potential excimer sites) and l_a is the average sequence length of aromatic species in the polymer. The latter term, derived from fluorescence anisotropy measurements, describes the extent to which energy migration within the polymer takes place. Behavior conforming to this expression has been found in copolymers containing vinyl naphthalenes or styrenes as chromophores.

The general approach outlined above has been extended to other co-polymers in which either the excimer site population is not expected to result from nearest-neighbor interactions or the extent of energy migration is not described in terms of l_a. In cases where the excimer to monomer ratio is proportional to f_{aa}, either energy migration is sufficiently fast to ensure equal sampling of all potential excimer sites or it is not important at all in the population of the excimer sites. Quantitative analysis of the excimer formation and decay kinetics are considerably facilitated if time-resolved excitation techniques are employed. Phillips, Soutar, and co-workers have used a modified scheme that adequately explains the photophysics of excimers in polymers:[5,6]

In this scheme, M_1 and M_2 are two spectrally identical, but kinetically distinguishable monomer species, differing only in the extent to which they can participate in energy migration and thereby form excimers.

In the photophysics of excimers found in polymers, experimental analysis has been hampered by two problems. These are the assignments of the distinct excimer bands, the unsatisfactory fit of the monomer, excimer emission growth, and decay curves to the conventional Birks scheme. Recall that in the conventional scheme (with reversible formation of excimers), the monomer decay is described by a sum of two exponentials and the excimers as the difference of the same two exponentials. For a wide variety of polymers it was found that the double exponential description was far from satisfactory. It is commonly observed in these systems that the monomer fluorescence decay can be numerically resolved by a sum of *three* exponentials and the excimer emission requiring fit with functions consisting of a linear combination of more than two exponentials.

One can envisage several possible situations giving rise to such results. The invalidity of the dissociative excimer–monomer concept has led to strategies which try to correlate the number of exponentials observed with the number of interacting states in the coupled kinetic schemes. The modified kinetic scheme involving two distinct monomer states is one such case. This model is one simplification of a general kinetic scheme for a system involving three distinct excited-state species capable of interverting, fluorescing, or undergoing radiationless transitions during their excited-state lifetimes. *A priori*, it is not possible to indicate the molecular nature of these kinetically distinguishable species. Two possibilities can be envisaged: (1) two kinetically different monomer species M_1^* and M_2^* and one excimer (scheme, p. 265) or one monomer and two kinetically distinguishable excimers D_1^* and D_2^*. The first possibility was suggested by Morawetz who emphasized the importance of distinguishing different types of rotational conformers for the excited monomer state. Conformers from which a single hindered rotation can lead to excimer formation were distinguished from those that must undergo a more complex rearrangement before the excimer alignment can be achieved. Similar ideas have been advanced by DeSchryver. The second alternative has been shown to operate in the excimer kinetics of poly(N-vinylcarbazole) (PNVC).

The evolution of the assignments for excimer states in PNVC illustrates most of the problems often encountered such that it is worthwhile to review its case.[64–74] Poly(N-vinylcarbazole) in solution exhibits a broad structureless excimer band (in addition to the normal monomer fluorescence) which can be assigned to two spectrally distinct species. The excimer decay kinetics is complex and the relative yields of the two excimer bands appear to vary in a complex manner with the nature of the sample preparation and temperature.

Ng and Guillett examined the decay curves of PNVC in benzene and found that they can be fitted to a linear combination of three exponentials.[65] The lifetimes of the two excimers were estimated to be approximately 2.9 and 12.8 nsec, respectively, at room temperature.

Earlier, in 1976 Itaya et al.[71] in their studies of PNVC excimers made some very important observations concerning the possible origin of the two excimer bands. It was found that the fluorescence of PNVC depended on the mode of polymerization: cationic polymerization favored the normal excimer emission (emission maximum at 410 nm) whereas radically polymerized samples exhibit high-energy excimer fluorescence (emission maximum at 380 nm). The emission lifetimes at 77 K were 28 and 4 nsec for the cation-derived polymer and radical-derived polymer, respectively. It was also proposed that the differences observed are due to the predominant syndiotactic configuration in radical PNVC whereas cationic polymers are believed to contain up to 50% isotactic sequences. From molecular models, it can be deduced that the 1,3-sandwich configuration responsible for the long-wavelength emission can easily be formed at isotactic sequences. Weakly overlapping structures can be formed without difficulty in syndiotactic regions, and this possibly can give rise to the high-energy emission.

Confirmation on the assignments to the two excimer bands has come with the synthesis of appropriate diastereoisomeric model compounds, namely, racemic and meso forms of 2,4-di(N-carbazolyl)pentane.[65,72,73] (Interestingly, 1,3-di(N-carbazolyl)propane is not a good model. It exhibits a broad excimer band red shifted to the normal excimer bands of PNVC, possibly due to the absence of steric constraints found in the polymer.)[96−98] Both the meso and racemic forms exhibit typical carbazole monomer emission together with the excimer emission characteristic for each isomer: meso isomer at 420 nm (isotactic diad) and racemic isomer at 370 nm (syndiotactic diad) (Fig. 8.1). It was also shown that the two emissions arise independently from the respective diads and rule out the possibility of the normal excimer and the high-energy excimer in equilibrium via a direct pathway, as proposed by Ng and Guillett.

The behavior observed with poly(vinylnaphthalenes) appears to be similar to that found in PNVC.[39,46] With these polymers also any fit of the excimer kinetics as a difference of two exponentials is far from satisfactory. In studies of substituted dinaphthylpropanes, Itagaki et al. found two structureless emissions derived from two different types of excimers.[39] It was proposed that a sandwich-type structure gives the normal excimers (emission maximum at 410 nm) while partially overlapping structures cause the high-energy emission at 370 nm. In the studies of bis(α-naphthylmethyl) ethers, Todesco et al.[90a] found that intramolecular excimer formation does not follow any simple kinetic schemes. The existence of two different excimers (endo and exo) was

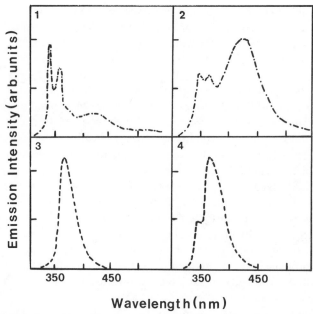

FIG. 8.1. Emission spectra of meso and racemic isomers of 2,4-di(N-carbazolyl)pentane, a model compound for poly(N-vinylcarbazole). (1) Meso isomer in isooctane at 197 K; (2) meso isomer in isooctane at 293 K; (3) racemic isomer in isopentane at 133 K; and (4) racemic isomer at 293 K. (From DeSchryver et al.[72] Copyright 1982 American Chemical Society.)

deduced (Fig. 8.2). Time-dependent fluorescence spectra and excimer decay curves for diastereoisomeric model compounds meso and racemic 1,1'-di(naphthyl)diethyl ether have shown the existence of two isomers.

The role of relative configurations in influencing excimer formation has been nicely demonstrated with a series of bis(9-phenanthryl)propanes.[94,95] Zachariasse et al. studied the formation of intramolecular excimers in a series of 1-(m-phenanthryl)-3-(n-phenanthryl)propanes, where m and n are 1, 2, 3, and 9 (abbreviated as mnP3P).[94] The monomer fluorescence of 29P3P and 39P3P was not quenched while a broad excimer emission band was observed for symmetrically substituted diphenanthrylpropanes such as 11P3P, 22P3P, 33P3P, and 99P3P. Examination of the possible geometrical structures with different overlap between phenanthryl groups can be made using Büchi–Dreiding stereomodels, and they are schematically displayed in Fig. 8.3. In the case of 29P3P and 39P3P, two kinds of structures are suggested with different partial overlap while the sandwich structure is possible for the case where $m = n$. By combining this geometric information with the excimer emission results, it can be concluded that only the sandwich structures are suitable for singlet excimers in phenanthrene systems.

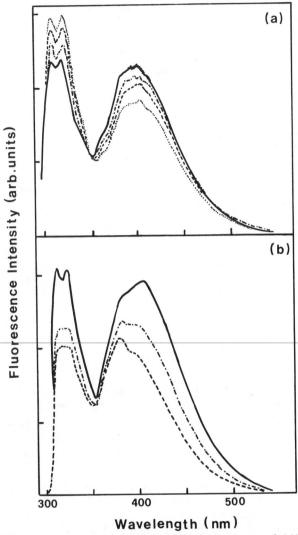

FIG. 8.2. Emission spectra of (a) racemic and (b) meso isomers of 1,1'-di(1-naphthyl) diethylether, a model compound for poly(1-vinylnaphthalene). (a) Racemic isomer in isooctane at 190, 201, 222, and 239 K, respectively. (b) Meso isomer in isooctane: (– – –), 196 K; (–·–·–·), 212 K and (——), 232 K. (From DeSchryver et al.[90] Copyright 1983 American Chemical Society.)

End-to-End Cyclization Dynamics

Excimer formation finds interesting applications in the study of cyclization dynamics in polymers. Intramolecular cyclization reactions have been classified as diffusion controlled or conformationally controlled, depending on

11P3P   22P3P   33P3P   99P3P   29P3P   39P3P

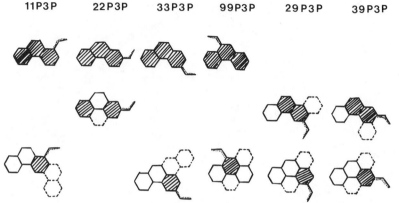

FIG. 8.3. Possible geometrical structures of some diphenanthrylpropanes. Shaded parts represent the overlap of two phenanthryl groups. (From Tamai *et al.*[95] Copyright 1983 American Chemical Society.)

whether the corresponding bimolecular reaction between the unbound reactants occurs at a diffusion-controlled rate or much slower ($k_{model} \ll k_{diff} \Rightarrow$ reaction conformationally controlled).

For the latter case, the equilibrium constant for the intramolecular encounter pair formation k_{cv} is given by k_1/k_{-1} and is proportional to the cyclization probability. In the former case, the observed rate will itself be the forward rate k_1 and, hence, the efficiency of cyclization. Excimer formation and triplet–triplet annihilation reactions usually occur at diffusion-controlled rates and hence are suitable for probing cyclization dynamics.

Cuniberti and Perico[56] first explored the utility of pyrene excimers for the study of end-to-end cyclization dynamics by examining poly(ethylene oxide) end-labelled with pyrene. Winnik and co-workers have largely extended these investigations using end-labelled polymers such as pyrene—polystyrene— pyrene (P—PS—P), Pyrene—poly(dimethylsiloxane)—pyrene (P— PDMS—P), and pyrene—poly(ethylene oxide)—pyrene(P—PEO—P).[57–62] (Fig. 8.4) The scheme below outlines the reactions of interest:

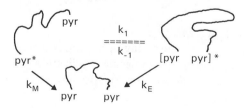

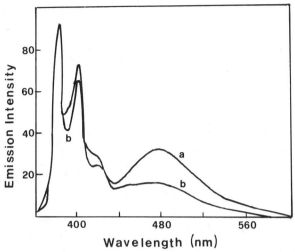

FIG. 8.4. Fluorescence spectra of pyrene end-labelled polymers in solution. (a) Pyrene—poly(dimethylsiloxane)—pyrene in toluene (2×10^{-6} M) at 22°C. (b) Pyrene—poly(styrene)—pyrene in cyclohexane at 34.5°C. (Both chains have their pyrene chromophores separated by $\bar{N} = 175$ bonds). (From Svirskaya et al.[62])

The forward reaction (occurring at a diffusion-controlled rate (k_1) describes the mean rate of cyclization k_{cy}. Excimer dissociation to locally excited pyrene is described by k_{-1} and its decay (radiative and nonradiative) is described by k_E. The corresponding monomer decay k_M is determined by reference to a model compound containing only one pyrene (e.g, methyl ester of pyrene butyric acid for P—PDMS—P). Its exponential decay is set as $k_{model} = k_M$. Excimer dissociation becomes more important for short chains than for long chains since a large fraction of very short chains cyclize during the excited-state lifetime of pyrene ^{1}Pyr*. According to the scheme, the monomer fluorescence intensity decay $I_M(t)$ should be described by a sum of two exponentials for short chains and by a single exponential for long chains if cyclization is by a single relaxation of the polymer chains. The experimental data are in accord with these predictions.

For polymers with mean segment length $N < 300$, k_1 values were determined from the exponential decay times using the relation

$$\tau - \tau_{model} = k_1 \qquad (8.6)$$

In all cases, the cyclization rate k_1 was typically in the range of 10^4 sec^{-1}. The dynamic rate constant for cyclization k_1 decreases with increasing mean chain length as one would expect, and the dependence of k_1 on the mean chain length N is adequately described by the relation:

$$k_1 = k_{cy}^{intra} \propto N^{-1.5} \qquad (8.7)$$

(The exponent values for the polymers were typically in the range of $(1.4-1.6)$ a relation predicted theoretically by Wilemski and Fixman.[194]) The polymer PDMS cyclized somewhat faster (more than two times faster) than PS chains of the same length in solvents of similar viscosity and solvating power. The result emphasizes the conclusion that k_1 does indeed measures the dynamic flexibility of polymer chains.

Zachariasse et al. have also made some early studies of end-to-end cyclization dynamics using dipyrenylalkanes, $Pyr—(CH_2)_n—Pyr$, $n = 2-16$ and 22, as model compounds.[80-82] The excimer to monomer intensity ratio I_E/I_M is a sensitive function of the number of methylene groups (chain length) separating the terminal chromophores. For $n = 2, 4, 5, 6, 8$, and 9, not only is the excimer formation inhibited but the excimer emission is blue shifted with respect to the model compound Me—Pyr. Only for the case of $n = 3$ is the excimer fluorescence normal and intense—a result which confirms early observations on diphenylalkanes. Kanaya et al.[86] in their studies of bis(1-pyrenylmethoxy)carbonylalkanes ($n = 1-22$) have found that the rate constant for excimer formation (end-to-end cyclization) was approximately proportional to -1.5 times the power of the chain length above $-60°C$ $[\log(I_E/I_M) \propto N^{-1.5}]$.

Backbone Cyclization dynamics

There have also been a few studies of excimers polymers with pendant pyrenyl groups in the backbone of the polymer (e.g., polystyrene containing evenly spaced pyrenyl groups (approximately every 25–30 monomer units) mixed with various concentrations of the unlabelled polymer. In a θ solvent such as cyclopentane, hydrodynamic screening causes a decrease in the rate of excimer formation at elevated polymer concentrations. In toluene (a good solvent), hydrodynamic screening effects are offset by the excluded volume screening so that intramolecular excimer formation is much less affected than in cyclopentane. An important observation has been made that the changes in the cyclization rates induced by increased polymer concentration are an order of magnitude smaller than those corresponding increases due to shear viscosity of the solution. Pyrene emission intensity ratios I_E/I_M in backbone-labelled polymers have also been used to obtain Hildebrand's solubility parameter δ_H of soluble polymers at infinite solution.

8.1.4   Aspects of Exciplexes

Extensive studies of the intra- and intermolecular interactions leading to excimers and exciplexes in polymers have been made by Tazuke et al. using fluorescence as an analytical tool. A wide variety of polymers containing pendant carbazolyl, 1-pyrenyl, and 9-anthryl groups on poly(esters), poly(amides) (composed of terephathalate, malonyl or urethane units), poly(methylmethacrylate), or poly(vinylcarbazole) have been synthesized:

$$[-O-(CH_2)_n-OCO-(CH_2)_n-CO-]_m \qquad [-NH-(CH_2)_n-NHCO-(CH_2)_n-]_m$$
$$\text{polyester} \qquad\qquad\qquad\qquad \text{polyamide}$$

Exciplex formation have been observed during the quenching of polymer-pendant chromophore by small molecules such as dicyanobenzene, dimethyl-terephthalate, and dimethylaniline[100-112] (Table 8.3). Structural effects, solvent effects, molecular weight dependence, and thermodynamic parameters with respect to polymer effects have all been elucidated. As outlined earlier, examination of the model compounds and systems has been extremely useful in unraveling the origin of various emission bands.[113-121]

An interesting observation in these studies has been the detection of triple exciplexes, e.g., terephthalate quenching of poly(N-vinylcarbazole) emission [carbazole—carbazole—terephthalate]*. Recall that PNVC exhibits dual excimer emission in addition to the normal monomer fluorescence. These excimers are sufficiently long lived to undergo excited-state quenching in the presence of electron donors and form triple exciplexes in addition to the normal (two-component) exciplex. Only using diastereoisomeric model compounds such as 2,4-di(N-carbazolyl)pentane (DNCP) has it been possible to unambiguously assign the various emission bands[121] (Table 8.4 contains typical results).

TABLE 8.3. Studies of Intramolecular Exciplexes in Polymers and in Model Compounds

Chromophore (acceptor)	Quencher (donor)	Compound	Reference
		Polymers	
N-Carbazolyl	Terephthalate TPT	Poly(N-vinylcarbazole)	101–104
N-Carbazolyl	Dimethylaniline DMA	Poly(N-vinylcarbazole)	105
N-Carbazolyl	Dicyanobenzene DCNB	Poly(N-vinylcarbazole)	106
9-Anthryl	Dimethylaniline DMA	Polyesters	107
1-Pyrenyl	Dimethylaniline DMA	Polyesters	108–110a
1-Pyrenyl	Dimethylaniline DMA	Poly(methyl methacrylate)	111,112
		Model compounds	
Phenyl	Dimethylamine	$Ph-X-N(CH_3)_2$, $X=C_8H_6$ (bicyclooctyl)	113
Phenyl	Dimethylamine	$Ph-(CH_2)_n-N(CH_3)_2$	114,115
1-Naphthyl	Dimethylamine	$Naph-(CH_2)_n-N(CH_3)_2$	116,117
2-Naphthyl	Dimethylamine	$Naph-(CH_2)_n-N(CH_3)_2$	118
9-Anthryl	Dimethylaniline DMA	$Anthr-(CH_2)_n-DMA$	119,120
N-Carbazolyl	Dicyanobenzene DCNB	2,4-di(N-carbazolyl)pentane + DCNB (intermolecular)	121

TABLE 8.4. Emission Maxima of Various Excimer and Exciplex Species during the Quenching of Mono and Bis-carbazolyl Model Compounds by m-Dicyanobenzene (DCNB)[a]

Compound[b]	Without DCNB	With DCNB
N-Isopropylcarbazole	347 nm (monomer)	410 nm exciplex (monomer + DCNB)
Racemic 2,4-DNCP	347 nm + 370 nm (monomer) (II excimer)	460 nm exciplex (II excimer + DCNB)
meso-2,4-DCNP	347 nm + 420 nm (monomer) (normal excimer)	510 nm exciplex (normal excimer + DCNB)

[a] From Masuhara et al.[121]
[b] DNCP = Di(N-carbazolyl)pentane.

Polymer chains in addition to producing local concentration effects also affect the solvation, mobility, and association of bound chromophores. In polymers with potential exciplex-forming groups (donor–acceptor pairs such as carbazolyl–terephthalate, aryl–dimethylanilino groups) there is a marked tendency to form ground-state association pairs. This is in distinct contrast to excimer-forming polymers such as those with pendant aryl groups alone. It has been proposed that polymer chains assist in the zipping effects which amplify very weak donor–acceptor ground-state interactions, and the process requires a regular arrangment of exciplex-forming pairs. Random copolymers of p-cyanostyrene and p-dimethylaminostyrene showed no ground-state association and only an intrapolymer exciplex.[110a]

Masuhara and Mataga[12] have studied the primary photoprocesses in polymers with pendant aromatic groups and free electron acceptors by time-resolved photolysis techniques. Interesting dynamic behaviors characteristic of polymers have been observed; some of these are:

(1) intense laser excitation produces several fluorescent chromophores in one polymer chain and their interactions lead to efficient intrapolymer singlet–singlet annihiliation;

(2) addition of electron acceptors converts more than one fluorescent chromophore into an ion radical, resulting in the formation of a transient polyelectrolyte;

(3) the yields of ionic photodissociation of polymers by quenching with electron donors or acceptors decrease as mean degree of polymerization increases;

(4) the absorption spectra of a polymer cation and their decay dynamics are different from those of its monomeric model compounds.

Detailed analyses have shown that the observed effects have nothing to do with excimer states in the polymer but are consistent with the general effects due to chromophore aggregation. The above results have been obtained with the following polymer systems: poly(N-vinylcarbazole), polyesters containing 1-pyrenyl groups and poly(urethane), poly(acrylate), and poly(methacrylate) carrying 1,3-dicarbazolylcyclobutane.

8.1.5 Energy Transfer and Migration Phenomena[122-132]

An intriguing process in the functioning of the green plant photosynthetic machinery is the efficient migration of the excitation energy among the antenna chlorophyll molecules of the chloroplast. Electronic excitation of any one of the 200–300 chlorophyll molecules present leads to a series of energy transfers until the excitation energy is trapped by the reaction centers. It is believed that the efficiency of this process is as high as 75–94%. It has been a great challenge to chemists to duplicate this energy migration process *in vitro*. There are growing evidences that suggest the migration of excitation energy among the chromophores attached to a polymer chain to be efficient, Guillett *et al.* in particular have made some elegant studies in this area, examining singlet energy transfer in homopolymers, copolymers, and terpolymers and pendant aryl groups such as naphthyl, 9-anthryl, or 9-phenanthryl [e.g., P(naph—MMA—co-anthr—MMA), P(9-phen—MMA), P(9-phen—MMA—co-9-anthr—MMA), P(9-phen—MMA—co-9-anthr—MMA—co-MMA). Intrachain singlet energy transfer in 2.06% anthr—MMA polymer content polymer P(naph—MMA—co-anthr—MMA) in THF at 25° is as high as 84%. Studies of the energy transfer efficiencies in different solvents and solvent mixtures indicate that the energy transfer process is very sensitive to coil dimensions in much the same way as the extent of the excimer formation and fluorescence quenching. On going from good solvents such as THF or dioxane to poor solvents such as methanol, a pronounced increase in the efficiency is observed due to a decrease in the average donor–acceptor distance produced by the contraction of the individual polymer coils in poor solvents.

Time-resolved fluorescence studies of the acceptor (trap) fluorescence in these systems have provided information on the mechanism and time scales of singlet energy transfer. It has been proposed that energy migration involves two processes, as illustrated schematically in Fig. 8.5. The first is a one-step dipole–dipole transfer from the donors to the acceptors. This Förster-type transfer occurs when the excited-state donor lies within a sphere of radius R_0 characteristic of dipole–dipole interaction between two chromophores. (This spherical picture is, in fact, a simplification in that there is an angular dependence of rate of transfer in Förster mechanisms). The second process involves singlet energy migration between donor chromophores followed by Förster transfer to an acceptor, once the excitation resides on a chromophore at a distance of R_0 from the trap. It has been proposed that this energy

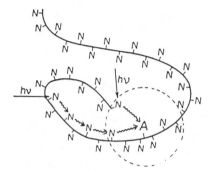

FIG. 8.5. Conceptual illustration of excitation energy transfer process in antenna polymers. (From Guillett.[14])

migration can be described by a random walk of discrete hops of the excitation. The chromophore separation over which D–D transfer occurs is consistent with those predicted by Förster mechanism.

Kinetic treatments of energy transfer quenching processes can be considered under two limiting cases:

(1) the donor–acceptor distance remains fixed during the quenching process as in rigid media; here the long range quenching via dipole–dipole mechanisms of Förster or the short-range quenching via Perrin's model are applicable;

(2) the probe and/or the quencher diffuses freely in which case diffusional models and Stern–Volmer kinetics apply. The above cases can be defined in terms of molecular energy transfer distance r,

$$r = [2(D + \Lambda)\tau_0]^{1/2} \qquad (8.8)$$

where D is the relative molecular diffusion coefficients of donors and acceptors, Λ the donor excitation energy migration coefficient, and τ_0 the donor luminescence lifetime. For intermediate cases, two theories due to Voltz et al.[133] and Yokoto and Tanimoto[134] have been developed. The former is derived from the diffusion theory while the latter is developed from Förster theory. Stern–Volmer kinetics are realizable when complete statistical mixing of the excited donor D* and the acceptor A occurs owing to material diffusion and/or excitation migration. This results in an enhanced rate of exponential decay of the donor fluorescence. Förster kinetics describe systems in which both the donor D* and A remain effectively stationery during transfer, and as a result, the donor decay is nonexponential.

8.1.6 Triplet-State Characteristics and Quenching Studies

Properties of the lowest triplet excited state of poly(2-vinylnaphthalene) (T–T absorption spectra, extinction coefficients, lifetimes, and yields) have been determined in benzene for different samples with varying degree of polymerization.[135] As compared to the model compound, τ_T decreases slightly but triplet yields are significantly low in polymers [$\tau_T = 10$–14 μsec versus 60 μsec and Φ_T (0.6–2.5) $\times$ 10^{-2} versus 0.8]. Intracoil singlet–singlet and singlet–triplet annihilation are believed to be responsible for the low yields observed.

Studies of Cyclization Dynamics Using Triplets

Mita, Schnabel, and co-workers have used the triplet state of anthracene in anthracene end-labelled polymers such as anthr—polystyrene—anthr (A—PS—A) to study the end-to-end cyclization process.[136–138a] Anthracene triplets were produced by irradiation with 25-nsec laser pulses of 347 nm light. At the end of the pulse, a great portion of the polymer carried two triplets of anthracene. The triplet decay via triplet–triplet annihilation was followed by either triplet–triplet absorption or delayed fluorescence:

$$\text{An—PS—An} \xrightarrow{2h\nu} \underset{(T)}{^3\text{An}^*}\text{—PS—}\underset{(T)}{^3\text{An}^*} \xrightarrow[\text{intra}]{k_{TT}} \underset{(S_1)}{^1\text{An}^*}\text{—PS—}\underset{(S_0)}{\text{An}} \qquad (8.9)$$

The triplet decay in benzene solution was exponential for the polymer labelled on one end only (A—PS). The polymer labelled on both ends (A—PS—A) showed an initial fast component, attributable to intramolecular triplet–triplet annihilation.

Analysis of the triplet decay as a function of solvent and mean segment length N showed that the rate constants for cyclization k_{cy}, as with pyrene excimer case mentioned earlier, were in the range of 10^3–2×10^5 sec^{-1}, decreasing with increasing chain length ($k_{TT} \propto N^{-n}$). The exponent value n is 1.0 ± 0.06 in benzene for $N = 110$–3000. In θ solvents such as cyclohexane $n = 1.5$ for $N < 300$ but $n = 1.0$ for $N > 300$ as illustrated in Fig. 8.6. The delayed fluorescence decay exhibited three distinct slopes, attributable, in order, to the inter- and intramolecular T–T annihilation and the first-order decay of triplets. The exponent n in the relation $k_{cy} \propto N^{-n}$ was $\simeq 0.6$ (butanone) and $\simeq 0.53$ (benzene), being determined at much higher polymer concentrations than those used for T–T absorption studies (10^{-5}–10^{-4} M versus 10^{-6}–10^{-5} M).

In a related study, Mita *et al.* measured rates of intramolecular photo-dimerization rates in A—PS—A spectrophotometrically and obtained similar dependence of k_{cy} on the chain length. The exponent value was even lower, about 0.3. Photodimerization reactions of anthracenes, however, go through singlet excimers. This means that the reaction occurs only for those polymers in which the end-to-end distance is such that the two ends can encounter each during the short singlet lifetime.

The intramolecular phosphorescence quenching reaction of alkenyl esters of benzophenone-4-carboxylic acids studied by Mar *et al.*[139] illustrates

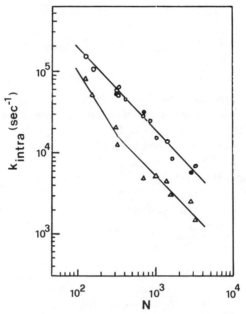

FIG. 8.6. Dependence of intramolecular rate constant k_{intra} for end-to-end cyclization on the degree of polymerization N of anthr—poly(styrene)—anthr in benzene ($\bigcirc$) and in cyclohexane ($\triangle$) determined using triplet absorption decay. Included are similar data in benzene measured using delayed fluorescence ($\ominus$). (From Horie *et al.*[138] Copyright 1981 American Chemical Society.)

end-to-end cyclization occurring under conformational control:

$$C_6H_5-CO^*-C_6H_4-CO-O-(CH_2)_n-CH\!=\!CH_2 \underset{k_{-1}}{\overset{k_1}{\rightleftarrows}} \left[\begin{array}{c} CH_2\!=\!CH-\!\!\smile \\ (CH_2)_n \\ \overset{O^*}{\underset{\|}{}} \\ Ph-C-C_6H_4-COO-\!\!\smile \end{array} \right]$$

$$\xrightarrow{k_2} C_6H_5-CO-C_6H_4-CO-O-(CH_2)_n-CH\!=\!CH_2$$

In these molecules, the end-group monosubstituted alkene acts as the quencher of the ketone triplet, a reaction that occurs 100–1000 times slower than the diffusion-controlled rate. The k_q values derived showed a sharp increase with increasing chain length, reaching a maximum at $n = 12$, indicating stereochemical demands on the reaction.

Triplet Quenching

Dynamic aspects of polymers can be probed by triplet quenching studies in various forms. Triplet–triplet annihilation considered above is one particular case. Let us consider the case of small-molecule excited-state donors being quenched by polymer-bound acceptors. Typical examples are quenching of the triplet of unbound (free) aromatic molecules, such as biphenyl, benzophenone, and biacetyl, by 1-naphthyl or 2-naphthyl groups pendant on poly(1-vinyl or 2-vinylnaphthalenes) (Table 8.5).[140-147]

For small-molecule, polymer-pendant side group reactions, it has been shown that the bimolecular quenching rate constant k_q depends on the intrinsic viscosity η of the polymer and on the polymer molecular weight $\bar{M}$ (or equivalently on the polymer mean segment length N):

$$k_q \propto \eta[\bar{M}] \tag{8.10}$$

In the descriptions of polymer dynamics, the interrelationship between these polymer properties is given by the Mark–Howink relation:

$$\eta = KM^\alpha \tag{8.11}$$

where K and α are constants that depend on the polymer, solvent, and

TABLE 8.5. Studies of Triplet Excited-State Quenching Reactions in Polymer Solutions [a]

Triplet chromophore	Quencher	Polymer	Reference
Biphenyl	Naphthyl-	P2VN	135
Benzil-	Anthryl-	Mixture of anthryl-PS and benzil-PS	136,138
Anthryl-	Anthryl-	Anthryl-PS-anthryl	138a–c
Phenylketone-	Phenylketone-	P(VPK)	140a
Benzophenone	Naphthyl-	P2VN	140,141
Benzophenone-	Naphthalene, benzophenone-, dimethylhexadiene	Mixture of PMMA-PVN and PMMA-PVBP	140b
Benzophenone-	Naphthyl-	Copolymer of PVBP-PMMA-P2VN	140c
Phenylketone-	Napthyl-	Copolymer of PVPK-P2VN	140d
Benzophenone	Polystyrene?	PS	140e
Naphthyl-	Piperylene	P2VN	141
Arylketone-	Arylketone-	Copolymer of PTVK-PMVK	142
Phenylketone-	Naphthyl-	Copolymer of PVPK-PMMA-P2VN	143
Arylketone-	Arylketone-	PMAP	144
Biphenyl-	Naphthyl-	Mixture of P2VN and PVBPh	145
Benzophenone, benzil, and biacetyl	Polystyrene?	PS and its substituted styrene analogs	146
Benzophenone, benzil, and biacetyl	Naphthyl-	Copolymer of PMMA-P2VN	147

[a] Polymer pendant chromophores or quenchers are indicated by a hyphen. Abbreviations used for polymers: P2VN is poly-2-vinylnaphthalene; PS, polystyrene; PVPK, polyvinylphenylketone; PMMA, polymethylmethacrylate; PVBP, polyvinylbenzophenone; PTVK-PMVK, poly(o-tolylvinylketone-methyl vinylketone), and PVBPh, polyvinylbiphenyl.

temperature. Hence, the above dependence of rate constant k_q can also be recast as:

$$k_q \propto N^{(\alpha - 2)/3} \tag{8.12}$$

where N is the mean polymer segment length. It has been found that the α values derived from triplet quenching experiments are in good agreement with those derived independently through viscometric measurements on the polymer. For example, in the case of quenching of various aromatic triplets by pendant biphenyl on poly(4-vinylbiphenyl), α determined by triplet quenching is 0.45, which compares well with 0.46 determined viscometrically.

Encinas *et al.* have shown that the average chain length dependence of the rate constant can also be derived by applying the Smoulouchowski equation to a volume of an equivalent quenching sphere, $k_q = 4\pi NrD$, where r is the radius of the equivalent quenching sphere. This inverse exponential dependence of k_q with polymer segment length has also been observed in the intracoil triplet quenching reactions as well, e.g., polymer-pendant triplet benzil quenched by polymer-pendant, end-labelled anthracene (PS—B—$*$PS by A—PS).

Webber *et al.* have investigated triplet sensitization of polymer-bound chromophores by various small-molecule donors present in the solution:

$$\text{\textasciitilde}\text{-A} \quad + \quad D^* \quad \text{------} \rightarrow \quad \text{-A}^* \quad + \quad D \tag{8.13}$$

For acceptor-labelled polymers poly(1-vinylnaphthalene) or poly(2-vinylnaphthalene), the dependence of the sensitizer triplet quenching rate k_q was examined for various sensitizers with different triplet energies. The results indicate two types of acceptor sites on the polymer: a low-energy one at energies comparable to that of unbound model compound and a high-energy one tentatively assigned to end-labelled groups!

With acceptors such as biphenyl, the triplet-state sensitization of polymer-bound species is also accompanied by the delayed fluorescence of biphenyl. From numerical solutions of a kinetic model and general physical considerations, it has been argued that intracoil triplet–triplet annihilation yields the delayed fluorescence. The sensitization rate k_q for triplet of poly(2-vinylnaphthalene) in benzene has been found to vary with the degree of polymerization according to

$$k_q \propto N^{-0.41} \qquad \text{for exothermic sensitizers } (\Delta E_T > 3 \text{ kcal/mole}) \tag{8.14}$$

$$k_q \propto N^{-1} \qquad \text{for others } (\Delta E_T < 3 \text{ kcal/mole}) \tag{8.15}$$

where ΔE_T is the energy difference between the triplets of the sensitizer and naphthalene.

8.2 Photoprocesses in Polyelectrolytes

8.2.1 General Features of Polyelectrolytes and Gegenion Binding[148-152]

There are many macromolecules that exist in aqueous solution (or other polar solvents) in the form of macroions. Included in this group are the synthetic polyelectrolytes such as polystyrenesulfonate (PSS) and poly-ethylenesulfonate (PES) and biopolymers such as proteins and nucleic acids. Thus, these molecules carry several ionic groups all along the polymer backbone:

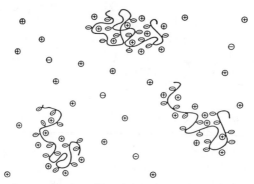

To preserve electrical neutrality, polyelectrolyte solutions always contain enough small ions to neutralize the charge on the macroion. Thus, structurally polyelectrolytes bear some resemblance to ionic micelles formed by surfactants. As in micelles, a large fraction of the total counterion concentration is tightly bound within the polymer domain, moving along with the macroion as a single kinetic entity in diffusion and transference measurements (Fig. 8.7).

The electrostatic forces due to the high local charges have important effects on the configuration of the macroions and on their chemical interactions with other ions. As in all electrolyte solutions, the counterion binding and exchange processes are influenced considerably by the ionic strength of the medium. It is

FIG. 8.7. Schematic representation of ion distribution in dilute solutions of polyelectrolytes.

believed that polyelectrolytes would, in their uncharged form and at very high ionic strength, have the random coil configuration typical of ordinary uncharged polymers such as polystyrene. The polyion–gegenion binding interactions are not always electrostatic in origin, especially for large hydrophobic counterions such as quarternary ammonium salts. It has been proposed that the interactions of quartenary ammonium cations with carboxylic polyelectrolytes are predominantly of the hydrophobic type. Here we can visualize the formation of thermodynamically stable, but structurally dynamic microheterogeneous systems composed of a hydrophobic interior core covered by a hydrophilic layer. When electrostatic forces dominate, the association of macroion–gegenion can lead to either a structurally loose (atmospheric binding) or a more stoichiometric and structurally specific binding (site binding). Thus different structures are expected depending on whether hydrophobic or electrostatic forces dominate in the binding mechanism.

8.2.2   Luminescence Probe Analysis

The luminescence of five fluorescence probes (pyrene, pyrene-3-carbox-aldehyde, dinapthylpropane, a surfactant derivative of indole 6-In-11$^+$, and a dicarbazolylpropane) have been used to investigate the behavior of poly(ethylene oxide)–poly(propylene oxide) water-soluble copolymers in aqueous solution.[153-155] The various fluorescence probe parameters measured at fixed probe concentrations (for pyrene: fluorescence lifetime and vibronic band intensity ratio (III/I); for pyrenecarboxaldehyde: emission maxima; for 6-In-11$^+$: fluorescence polarization; and for dinaphthylpropane and dicarbazolylpropanes: monomer-to-excimer intensity ratios) were found to depend on the concentration of the copolymer, in three distinct concentration domains. At low concentrations (region 1), probes such as Py—CHO and 6-In-11$^+$ are not solubilized, and they stay in water. At middle concentrations (region 2), the former probe shows increasing blue shifts and the latter increasing microviscosities of the probe environment, suggesting solubilization in a hydrophobic environment. In high concentrations (region 3), both probes show similar but very pronounced changes. The results have been interpreted as due to the formation of monomolecular (intra) polymer micelles in region 2 and further aggregation of these to form polymolecular (inter) polymer micelles.

The presence of polyelectrolytes such as poly(sodium acrylate) PAA, or poly(sodium ethanesulfonate) PES, enhance considerably the fluorescence intensity of Tb^{3+} ions in aqueous solution, due to the binding of Tb^{3+} ions to the asymmetrically arranged polyanions.[156] With poly(sodium methacrylate) PMA, and poly(styrenesulfonate) PSS however, the fluorescence intensity of Tb^{3+} did not increase appreciably. When methyl or phenyl groups are attached to the polymer chains, the polymer becomes less flexible and multidentate complex formation becomes more difficult. When Tb^{3+} solutions were treated with Li–, Na–, K–, and Rb–PAA salts, and Tb^{3+} fluorescence increased sharply over a small range of PAA and reached a constant value at higher concentrations. Binding constants calculated for Tb^{3+}–PAA complexes in the presence of Li^+, Na^+, K^+, and Rb^+ ions, are 300, 390, 500, and 600 M^{-1} respectively. This order reflects the competition between Tb^{3+} and alkali metal ions for binding to the polyelectrolyte. Addition of large amounts of NaCl increased the Tb^{3+} fluorescence intensity by reducing the surface charge and promoting the binding of fluorescent ions. The addition of EDTA to the PAA–Tb^{3+} complex causes the fluorescence intensity to decrease drastically, due to the stronger complexation of Tb^{3+} with EDTA than with PAA.

Various features of pyrene monomer fluorescence (lifetime, (III/I) ratio, intensity, etc.) can be used to follow the incorporation of the probe in poly(methacrylate) as a function of pH.[157,158] Pyrene is sprangly soluble in water but may be solubilized readily at low pH in the presence of PMA. Sharp changes in pyrene fluorescence, however, occur at pH around 5 as shown in Fig. 8.8. The decrease in fluorescence intensity, (III/I) ratio, and excimer fluorescence from aggregated pyrene molecules all indicate an opening of the polymer coils at this pH and ejection of pyrene into the aqueous phase. Similar abrupt changes in the polymer configuration have also been observed during pH changes of a solution of copolymer of methacrylic acid and 1-pyrenyl acrylic acid.[158] Solubilization in the low pH region provides sufficient protection to the probes that with molecules such as 3-bromopyrene we can readily observe room-temperature phosphorescence (as in micellar solutions). Fluorescence changes of pyrene are so pronounced during the phase changes that they can be used in a stopped-flow experiment to study the kinetics of polymer unfolding. Solubilization of pyrene in the polymer coils reduces the access to it by other molecules such as I^-, Tl^+, and CH_3NO_2. This is indicated by a decrease in the quenching rate constants in the presence of the polymer. Studies in two similar polymers poly(acryclic acid) PAA and poly(methacrylic acid) PMA show the effects to be much smaller in PAA. A comparative study of pyrene photophysics has been made when the probe is attached to PMA and PA in three different ways: (1) as a guest molecule in simple solubilization, (2) covalently attached to the polymer ends, and (3) randomly bound to the polymer backbone.[158]

Addition of poly(vinylsulfate) PVS to aqueous solutions (pH 2.0) of UO_2^{2+} leads to a reduction of uranyl ion fluorescence intensity. The high local concentration in the vicinity of the polyelectrolyte causes self-quenching.[159] At pH 3.0, uranyl ion binding produced even greater reduction of the fluorescence intensity and a change in the fluorescence spectrum similar to that observed when the acidity of the medium is increased. Addition of Fe^{2+}

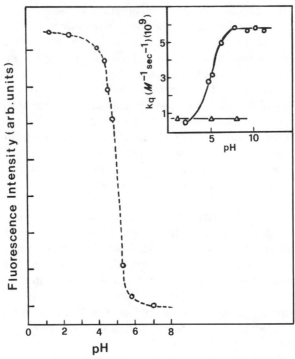

FIG. 8.8. Dependence of fluorescence intensity of pyrene in poly(methacrylate) (0.1 M PMA) solutions as a function of pH. (From Chen and Thomas.[157]) Insert: pH dependence of quenching rate constant k_q of monomeric Ru(bpy)$_3^{2+}$ ($\triangle$) and poly(acrylic acid)-pendant Ru(bpy)$_3^{2+}$ by methyl viologen ($\bigcirc$). (From Kurimura et al.[188] Copyright 1982 American Chemical Society.)

ions to UO$_2^{2+}$ solutions leads first to a rapid quenching due to the presence of both the ions in the polymer. This is followed by a pronounced increase in fluorescence intensity as Fe^{2+} ions replace uranyl ions in the polyelectrolyte binding sites. Eventually, the fluorescence decreases again due to quenching of free UO$_2^{2+}$ ions in solution. The experiments also indicate that the exchange of bound and free counterions is negligible during the lifetime of the UO$_2^{2+*}$ excited state.

A fluorescence probe study has been made on poly(styrene sulfonate) PSS using a cationic indole fluorescence probe 6-In-11$^+$:

$$CH_3-(CH_2)_5$$

$$(CH_2)_{11}-\overset{+}{N}(CH_3)_3Br^-$$

6—In—11$^+$

Measurements of fluorescence maxima and lifetimes and a comparison to various ionic micellar solutions indicate that the probe is strongly associated with PSS in a region of high hydrophobicity. The microviscosity estimates ($\simeq 15$ cP) are similar to that observed in ionic micellar assemblies.

The emission spectrum of $Ru(bpy)_3^{2+}$ undergoes little change in the presence of polyelectrolytes such as poly(vinyl sulfate) PVS.[160] The emission decay (at 600 nm) and the bleaching recovery of ground-state absorption (at 450 nm), however, indicate multi(three)-component decay.[161] A fast process ($t \leqslant 10$ nsec) accounted for the decay of a significant fraction of the excited $Ru(bpy)_3^{2+}$ species. This is followed by mixed first- and second-order decays, which remained after the end of the fast process. The results have been interpreted in terms of triplet–triplet annihilation reactions caused by the local concentration effects in the vicinity of the polyelectrolyte. When less than four $Ru(bpy)_3^{2+}$ ions were present on an average in each polyelectrolyte molecule, then the only reaction observed was a natural decay of $Ru(bpy)_3^{2+}$.

The interactions of $Ru(bpy)_3^{2+}$ with the polyelectrolyte PSS can also be studied by examining the elution behavior of $Ru(bpy)_3^{2+}$–PSS mixtures on passing through a cation exchange resin Dowex 50W-X-8.[162] In the absence of PSS, cations such as $Ru(bpy)_3^{2+}$, $Co(phen)_3^{2+}$, and Cu^{2+} are strongly adsorbed on the Dowex. With increasing concentrations of PSS, quantitative elution of $Ru(bpy)_3^{2+}$ can be achieved. The elution, however, is suppressed on introduction of increasing concentrations of NaCl in the $Ru(bpy)_3^{2+}$–PSS mixtures. The luminescence intensity of $Ru(bpy)_3^{2+}$ was found to increase, with a small red shift (~ 4 nm) in the presence of PSS. Since the addition of urea nullified the intensity enhancements, the observed effects have been interpreted as due to hydrophobic interactions of the dye with PSS.

The perfluorosulfonate constituent of Nafion membranes can be extracted out under reflux in 50:50 $EtOH$–H_2O as the solvent and the alcohol subsequently removed by heating. The luminescence of probes such as $Ru(bpy)_3^{2+}$, Auromine O cation (AO^+), and a surfactant indole, 6-In-11$^+$, have been examined in Nafion polyelectrolyte solutions.[163] As with PSS, the indole probe shows a large (18-nm) blue shift. The emission intensity decreases on initial addition of PES, followed by an increase with further addition. The luminescence intensity of $Ru(bpy)_3^{2+}$ and AO^+ show a gradual increase with PES addition, levelling off at stoichiometric equivalent point. Since the divalent cation Mg^{2+} does not effectively displace $Ru(bpy)_3^{2+}$, it has been argued that the observed effects are due to combined electrostatic and hydrophobic interactions.

8.2.3 Dynamics of Excimers

Formation of intra- and intermolecular excimers by bichromophoric molecules such as bis(1-napthylmethyl)ammonium chloride DNMA and monochromophore 1-naphthylmethylammonium chloride NMA:

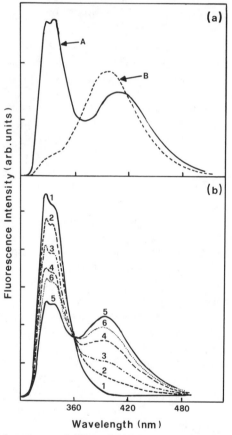

are strikingly enhanced in the presence of hydrophobic polyelectrolyte PSS as shown in Fig. 8.9. For DNMA, studies of excimer lifetimes indicate formation of both inter- and intramolecular excimers, with lifetimes of 45 ± 5 nsec and

FIG. 8.9. (a) Emission spectra of di(1-naphthyl)methylammonium chloride in the absence (A) and in the presence (B) of 10^{-4} M of poly(2-acrylamido-2-methyl-1-propanesulfonate). (b) Emission spectra of 1-naphthylmethylammonium chloride (NMAC) in the presence of poly(styrene sulfonate)(PSS). PSS = (1) 0 M, (2) 1.5 × 10^{-5} M, (3) 3.0 × 10^{-5} M, (4) 4.0 × 10^{-5} M, (5) 8 × 10^{-5} M, and (6) 1.6 × 10^{-4} M. (From Turro and Pierola.[165,166] Copyright 1983 American Chemical Society.)

23 ± 3 nsec, respectively.[164-168] The enhanced excimer formation for DNMA and NMA in the presence of PSS has been suggested to result from cooperative interactions (hydrophobic and hydrophilic) between the probes and the macroion. Addition of NaCl and $Co(NH_3)_6Cl_3$ retards excimer formation of DNMA presumably via an electrostatic shielding effect. Application of high pressures (1–2610 bars) increased both the emission intensity and lifetime of the excimers of DNMA bound to PSS. The cationic monochromophoric compound NMA exhibits a strikingly enhanced intermolecular excimer formation relative to that in water. Kinetic analysis indicates that intermolecular excimer rate is enhanced by approximately three orders of magnitude when NMA is bound to PSS.

The emission maxima of these probes are also influenced by complexation with PSS. Interestingly, we observe similar changes on change of pH of the probe solutions in the absence of the polyelectrolyte. A transition of the probe structure from a tetrahedral one (stable at low pH) to a pyramidal one characteristic of polyelectrolyte association (unstable in basic solutions) has been proposed:

tetrahedral   R = 1-naphthyl   pyrimidal

The fluorescence decay in all cases consists of two first-order components, whose relative contributions to the overall intensity depended on pH or polyelectrolyte concentration. A study of excimer formation in different sulfonic polyelectrolytes (PSS, PES, and PAMPS) has revealed that the probe conformation in the ground, and excited states depends on the type of interaction (hydrophobic or electrostatic) that dominates the binding strength.

8.2.4 Singlet Energy Transfer

Nonradiative singlet energy transfer has been studied in aqueous solutions containing mixtures of naphthyl- and anthryl-labelled neutral or charged polymers (copolymers and terpolymers derived from dimethylacrylamide, vinylpyrrolidone and acrylic acid).[169] An enhanced efficiency of energy transfer is observed when the donor and acceptor chromophores are attached to polymers carrying opposite charges. This effect decreases with increasing ionic strength of the medium. In mixtures of anionic copolymer of dimethylacrylamide and a cationic copolymer of vinylpyrrolidone, however, both hydrogen bonding and hydrophobic interactions contribute to the stability of the polymer complex. Studies of energy transfer in donor- and acceptor-labelled polyacrylic acid at different pH values indicated that the probability of encounter between polyanion-bound donors and acceptors is not affected to any demonstrable degree by their mutual repulsion.

Efficiency of singlet energy transfer (^{1}naphthalene* to anthracene) in a copolymer of acrylic acid polymer terminated with one anthracene per chain was found to increase drastically (12–70%) on going from dioxane solvent to 0.02 M NaOH.[170] Substantial amounts of naphthalene excimer emission were also observed in aqueous medium although it was almost absent in dioxane. Apparently, in aqueous medium, naphthalene—MMA units aggregate. The process could involve micellization of aggregated NMMA units within individual coils or multimerization, viz., NMMA units from several chains aggregating with hydrophobic groups functioning as transient ties between the macromolecules.

8.2.5 Phosphorescence Quenching and Analysis of Gegenion Binding

Halonaphthalene derivatives such as n-(4-bromonaphthoyl)alkyltrimethylammonium chlorides ($n = 1$, 5, and 10) phosphoresce readily in degassed aqueous solutions with a maximum around 570 nm:

$$Br{-}\underset{}{\bigcirc\!\!\bigcirc}{-}CO(CH_2)_n{-}\overset{+}{N}(CH_3)_3Br^- \qquad BNK{-}n^+$$

The phosphorescence can be readily quenched by ionic quenchers such as nitrite. The quenching process can be adapted to determine the rate constants for the association/dissociation of the probe molecules to the host aggregate/assembly.[171,172] The scheme below outlines the reactions of interest:

$$
\begin{array}{ccccc}
BN^* & + & P & \underset{k_-}{\overset{k_+[P]}{\rightleftharpoons}} & [BN^*...P] \\
k_M \Big\downarrow \;\; \Big\downarrow k_q[Q] & & & & k_P \Big\downarrow \;\; \Big\downarrow k'_q[Q] \\
BN & & & & [BN...P]
\end{array}
$$

It is based on the competitive quenching of a polyelectrolyte-bound probe by a quencher that has the same sign of charge as the polyelectrolyte. A cationic probe associated with a polyanion will be protected from quenching by anionic species by electrostatic repulsion effects of the polyion, as long as it is associated with the polyelectrolyte during its excited-state lifetime. Thus, the extent of scavenging by solution-confined quenchers will depend on the relative rate constants for probe association (k_+) dissociation (k_-) and quenching (k_q). Note that $k'_q \ll k_q$. Further discussions on the method can be found under ionic micelles (cf. Section 3.3).

For the quenching of BNK-1$^+$ by nitrite in the presence of PSS, the probe association and dissociation constants with the polyelectrolyte have been evaluated to be $6.8 \times 10^9 \ M^{-1} \ sec^{-1}$ and $3.7 \times 10^4 \ sec^{-1}$, respectively. The magnitude of k_+ is close to diffusion controlled, suggesting that the probe association to the polyelectrolyte is one of atmospheric binding and not of site binding type. The magnitude of k_+ and the equilibrium constant K decreases with NaCl addition or with increasing polymer concentration while k_- decreases. NaCl is known to shield the electrostatic attractions between the polyanion and the probe cation and, hence, decreases the association rate. The dissociation process, on the other hand, depends on the extent of attraction and is enhanced by the shielding effects.

A comparative study has also been made of the phosphorescence quenching of these probes by $Co(NH_3)_6^{3+}$ when the two reactants are in aqueous solution containing either simple electrolytes (e.g., NaCl, BaCl$_2$, or LaCl$_3$) or anionic polyelectrolytes such as PSS, PES, or PA. In aqueous solution, the quenching is very inefficient ($k_q = 2.6 \times 10^7 \ M^{-1} \ sec^{-1}$). Simple electrolytes increase rates slightly and the rate enhancements are explained quantitatively with Brønsted–Bjerrum–Debye–Hückel theory. Polyelectrolytes however show pronounced rate enhancements especially at low concentrations due to the large electrostatic potential fields of polyions. The k_q values show a maximum close to diffusion-controlled rate, decreasing at high polyelectrolyte concentrations. The $k_{q,0}$ values for BNK-5$^+$ are $5.3 \times 10^7 \ M^{-1} \ sec^{-1}$ (water), $1.5 \times 10^8 \ M^{-1} \ sec^{-1}$ (NaCl), $1.53 \times 10^{10} \ M^{-1}$ (PES), and $2.53 \times 10^9 \ M^{-1} \ sec^{-1}$ (PSS), respectively.

8.2.6 Luminescence Quenching and Photoredox Processes

A major application that uses the presence of charged interfaces in ionic micellar systems has been in the optimization of charge separation yields in photoredox processes (Section 3.5). The microheterogeneous character and the presence of strong electrostatic potential fields in polyelectrolytes should also facilitate the charge separation phenomena. It has been known for quite some time that the presence of potential fields in polyelectrolytes affects the rates of thermal reactions between ionic species. Such effects have also been observed in the quenching of excited states of polyion-bound proble molecules.

The luminescence quenching of Ru(bpy)$_3^{2+}$* by metal ions[173–175] such as Cu^{2+} or Fe^{3+} is enhanced by several orders of magnitude in the presence of poly(vinyl sulfate) (PVS). Both the metal complex and quencher ions are attracted to the vicinity of the polyelectrolyte field. In principle, the quenching process can be static occurring among the pre-existing ion pairs in the polymer domain or dynamic occurring via mutual diffusion of the two reactants toward one another in the polymer field. Time-resolved studies have established that for the Cu^{2+} quenching case, the process is indeed dynamic.

Interestingly, at high concentration of the quencher, the enhanced quenching efficiency vanishes. The quenching reversal process involves displacement of $Ru(bpy)_3^{2+}$ by inert cations (or by quenchers themselves). Earlier we mentioned a similar occurrence during the Fe^{2+} quenching of UO_2^{2+} emission. Such displacement studies also allow estimation of the number of binding sites and the binding constants for the probe and quenchers. The Fe^{3+} quenching exhibits a complex behavior and it has been proposed that the iron crosslinks the PVS network.

The presence of PVS affects significantly the rate of quenching of $Ru(bpy)_3^{2+}*$ by neutral species (e.g., ferric nitriloacetate (FeNTA), Co (acac)$_3$, Ru(acac)$_3$, nitrobenzene, O_2).[175] Only with FeNTA and Co(acac)$_3$ have photoproducts been observed, and the redox quantum yields increase substantially on addition of PVS. These two acceptors on photoreduction undergo chemical reactions (the former protonation and the latter ligand substitution) and that the polyelectrolyte effect on back reactions is minimal. Assistance of polyelectrolytes in the separation of photoredox products has been examined during $Fe(CN)_6^{3-}$ quenching of the neutral complex $Ru(bpy)_2(CN)_2$ excited state in cationic polybrene solutions:[176]

$$\{[-(CH_2)_6-N(CH_3)_2-(CH_3)_3-N(CH_3)_2-]^{2+}(Br^-)_2\}_n \qquad \text{polybrene (cationic)}$$

Low cage escape yield ($\leqslant 0.1$) in aqueous solutions increases markedly in the presence of PVS (0.93). The rate of disappearance of the redox products exhibits a composite decay (mixed first and second order) in the pH range 2–12.2.

Photosensitizers such as $Ru(bpy)_3^{2+}$ have also been covalently attached to various polyelectrolytes and the efficiency of charge separation in photoredox processes assessed.[177–182] Kurimura et al.[177] have described the synthesis and luminescence of a copolymer-bound $Ru(bpy)_3^{2+}$ derivative [Ru(bpy)$_2$(bpy— COOH) labelled onto p(aminostyrene-N-vinylpyrrolidone)]. The effect of Cu^{2+} quenching in this polyion-bound system, however, is similar to the free/unbound species mentioned earlier. Morishima et al., however, found interesting effects in amphiphilic phenanthrene-labelled polyanions (carrying SO_3^- groups) and polycations (carrying quarternary ammonium groups NR_3^+):

$$-(CH_2-CH)_{0.09}-(CH_2-CH)_{0.91}-$$
Phenan   CO
                    |
                    NH
                    |
          CH(CH$_3$)$_2$—CH$_2$SO$_3^-$
(APh)$^-$

$$-(CH_2-CH)_{0.14}-(CH_2-C(CH_3))_{0.86}-$$
Phenan   CO
                    |
                    NH
                    |
          (CH$_2$)$_3$Ň(CH$_3$)$_3$
(APh)$^+$

Ru complexes have been synthesized using the pendant phenanthryl ligands and effect of the polyelectrolyte on the photoredox quenching with two viologen acceptors [methyl viologen, MV^{2+}, and propylviologensulfonate (PVS)] studied.[178] In the case of the anionic polymer–MV^{2+} system, strong coulombic interactions led to extensive ground-state interactions and low yields of MV^+;

$$Ru(phen)_3^{2+} + MV^{2+} \underset{}{\overset{hv}{\rightleftharpoons}} Ru(phen)_3^{3+} + MV^+ \qquad (8.16)$$

The highest yields of MV^+ have been obtained for the polycation–MV^{2+} system, which also showed a slowing down of the reverse electron transfer ($k_b = 3.2 \times 10^8\ M^{-1}\ sec^{-1}$). Most effective for the retardation of the reverse reaction was the polyanion–PVS pair ($k_b = 8.7 \times 10^7\ M^{-1}\ sec^{-1}$) though the cage escape yield in this system is low.

The photoredox quenching of a polymer-pendant $Ru(bpy)_3^{2+}$ (pendant on a copolymer of 4-methyl-4'-vinyl-2, 2'-bipyridine and acrylic acid) and MV^{2+} is subject to significant pH effects. As shown in the insert of Fig. 8.8, the quenching rate constant k_q shows a remarkable pH dependence and is an order of magnitude larger than that of monomeric $Ru(bpy)_3^{2+}$:[182]

$$-(CH_2-CH)_{0.024}-(CH_2-CH)_{0.037}-(CH_2-CH)_{0.939}$$
$$\quad\quad Ru(bpy)_2(bpy)\quad\quad\quad (viologen)\quad\quad\quad COOH$$

The results can be understood in terms of electrostatic effects of polyanion produced by the dissociation of carboxylic groups of the acrylic acid moiety.

8.3   Photoprocesses in Ion-Exchange Membranes

Analogous to the simple polyelectrolytes are their cross-linked analogs, the ion exchange membranes. These also adsorb fairly large quantities of ions from solution and provide a microheterogeneous environment. Studies in ionic micelles have amply demonstrated how one can manipulate hydrophobic and hydrophilic domains to achieve better charge separation in photoredox processes. The goals of studies in ion exchange membranes are on similar lines. Thorton and Laurence[183] were the first to explore the influence of adsorption of $Ru(bpy)_3^{2+}$ on the cationic exchange membrane Sephadex SP on the quenching of its MLCT excited state by metal ions such as Cu^{2+} or Fe^{3+}. Cation exchange resins such as Dowex or Sephadex adsorb divalent cations such as $Ru(bpy)_3^{2+}$ readily from aqueous solution. At concentrations well below their exchange capacity, the adsorbed cations are not easily eluted out. Co-adsorption of quencher ions such as Cu^{2+} leads to an apparent increase in the quenching efficiency. Though the adsorbed ions are in close proximity, it has been shown that the quenching process is a dynamic one, involving mutual diffusion of the two species towards one another.

Lee and Meisel[184] have made a more detailed study of such processes in perfluorosulfonate membranes (known commercially as Nafion). Adsorption of $Ru(bpy)_3^{2+}$ onto Nafion membranes produces a small blue shift (~ 3 nm) in the ground-state absorption spectrum, but more pronounced effects are observed in the emission spectrum — 12-nm blue shifts and ~ 10-nm narrowing of the bandwidth as compared to that in aqueous solution. $Ru(bpy)_3^{2+}$ does not show any such shifts either in anionic sodium lauryl sulfate (SDS) micelles or in the presence of polyelectrolytes such as poly(vinyl sulfate). We can observe similar shifts, however, on solubilization in perfluorooctanoate micelles. Hence, it is likely that in Nafion membranes, $Ru(bpy)_3^{2+}$ interacts with the fluorocarbon chains rather than with the sulfonate head groups.

Excited-state quenching reactions with metal ions such as Ni^{2+}, Cu^{2+}, Cr^{3+} or Eu^{3+} — all inefficient in aqueous solutions — are not influenced significantly by the adsorption of the dye to the membrane. For relatively fast reactions such as quenching by methyl viologen ($k_q = 4.5 \times 10^8 \, M^{-1} \, sec^{-1}$), the observed quenching behavior is different. The Stern–Volmer plots are nonlinear and the decay of $Ru(bpy)_3^{2+*}$ exhibits a biphasic character — a fast component whose contribution and rate increases with increasing quencher concentration and a slow component which is independent of the quencher and rate the same as that of unquenched decay. Recall that this behavior is reminescent of case I of the quenching scenarios discussed under ionic micelles (Section 2.5) and, hence, is treated in an analogous manner. Slama-Schwok et al.[185] have also studied the electron transfer quenching of Sephadex-bound $Ru(bpy)_3^{2+}$ by various species [Cu^{2+}, Fe^{3+}, FeNTA, O_2, benzophenone, $Cr(acac)_3$]. In all cases, the quenching process has been confirmed to be dynamic. Except for the case of O_2 and benzophenone, electron transfer products have been observed with quantum yields ranging from 0.1 for FeNTA and $Cr(acac)_3$ to 1.0 for Fe^{3+}.

The fluorescence intensity of Eu^{3+} ions was found to be enhanced on their binding to Nafion membranes[186] and this effect is observable at Eu^{3+} concentrations as low as $10^{-5} \, M$. Presumably, the asymmetric binding of Eu^{3+} ions in the potential field of polyelectrolytes affects the photophysical properties. In Nafion membranes with co-adsorbed Eu^{3+} and UO_2^{2+} ions, the fluorescence intensity of Eu^{3+} (at 616 nm) is greatly enhanced when UO_2^{2+} bands are irradiated. The results indicate that ion clusters exist in Nafion membranes where the Eu^{3+} and UO_2^{2+} ions are at close proximity to facilitate energy transfer:

$$UO_2^{2+*} + Eu^{3+} \longrightarrow UO_2^{2+} + Eu^{3+*} \tag{8.17}$$

When Eu^{3+} and Co^{2+} are present together in the Nafion, very efficient quenching of Eu^{3+} emission is observed. When Nafion membranes containing Eu^{3+} and UO_2^{2+} are heated, the fluorescence intensity of Eu^{3+} decreases sharply around 130°C, indicating thermal deformation of the clusters in the membrane. At high loading levels, the fluorescence of Eu^{3+} exhibits concentration (self) quenching, but thermal treatment of Nafion membranes leads to a redistribution of the europium ions as evidenced by the enhancement of emission intensities.

Nafion membranes incorporated with semiconductor crystallites of CdS and one or more redox catalysts have been examined as potential "integrated chemical systems" for the photocatalytic production of H_2/water cleavage.[186a-c] On Nafion/CdS/Pt films, the efficiency of photoproduction of H_2 in sulfide containing aqueous solutions (S^{2-}/H_2O) is greater than those commonly observed with unsupported colloidal or powdered semiconductors under similar conditions.

The efficiency of the photoredox reaction between $Ru(bpy)_3^{2+}$ and MV^{2+} when these reactants are co-adsorbed on cation exchange membranes Dowex 50-W-X8 or on an iminodiacetic acid type chelate resin Diaion CR 150 has been investigated.[187,188] Irradiation of the latter membranes in deaerated aqueous solutions led to a rapid development of blue color (characteristic of MV^{+}) on the resin beads. Apparently, the resin constitutent iminodiacetic acid itself acts as an electron donor (sacrificial) to reduce one of the photoproducts $Ru(bpy)_3^{3+}$:

$$Ru(bpy)_3^{2+} + MV^{2+} \overset{h\nu}{\rightleftharpoons} Ru(bpy)_3^{3+} + MV^{+} \tag{8.18}$$

$$Ru(bpy)_3^{3+} + IDA \longrightarrow Ru(bpy)_3^{2+} + IDA^{+} \tag{8.19}$$

On sulfonato membranes such as Dowex, we do not observe such buildup of MV^{+}. The photoredox reactions are reversible, but we do observe net photoreduction of MV^{2+} in the presence of donors such as EDTA or triethanolamine. Molecular oxygen (O_2) also competes effectively with MV^{+} in the reverse reaction of (8.18) to give superoxide radicals O_2^{-}:

$$MV^{+} + O_2 \longrightarrow MV^{2+} + O_2^{-} \tag{8.20}$$

$$O_2^{-} + H^{+} \longrightarrow HO_2 \cdot \longrightarrow H_2O_2 \tag{8.21}$$

In aqueous solutions, superoxide radicals however are readily protonated to yield H_2O_2 as a product. The utility of Dowex 50W-D1 adsorbed $Ru(bpy)_3^{2+}$ as a photosensitizer for the generation of singlet oxygen has been explored.[189] The limiting quantum yield for the production of singlet oxygen ($^1O_2$) in methanol is indeed quite high (0.9), comparable to a homogeneous system using a dye rose bengal but more efficient than other known heterogeneous photosensitizers.

References

1. H. Morawetz, "Macromolecules in Solution," 2nd Ed. Wiley, New York, 1975.
2. C. Tanford, "Physical Chemistry of Macromolecules." Wiley, New York, 1975.
3. F. W. Billmeyer, "Textbook of Polymer Science," 3rd Ed. Wiley, New York, 1984.

4.   P. C. Hiemenz, "Polymer Chemistry: Basic Concepts." Dekker, New York, 1984.
5.   S. W. Beaven, J. S. Hargreaves, and D. Phillips, *Adv. Photochem.* **11**, 207 (1979).
6.   K. P. Ghiggino, A. J. Roberts, and D. Phillips, *Adv. Polym. Sci.* **40**, 69 (1981).
7.   H. Morawetz and I. Z. Steinberg, eds., *Ann. N.Y. Acad. Sci.* **366** (1981).
8.   E. V. Anufrieva and Y. Y. Gotlieb, *Adv. Polym. Sci.* **40**, 1 (1981).
9.   Y. Nishijima, *Prog. Polym. Sci.* **6**, 199 (1973).
10. M. Kaneko and A. Yamada, *Adv. Polym. Sci.* **55**, 1 (1984).
11. D. Phillips and A. J. Roberts, eds., "Photophysics of Synthetic Polymers." Sci. Rev., Northwood, England, 1982.
12. H. Masuhara and N. Mataga, *J. Lumin.* **24/25**, 511 (1981).
13. M. A. Winnik, *Chem. Rev.* **81**, 491 (1981).
14. J. E. Guillett, *Pure Appl. Chem.* **52**, 285 (1980).
15. H. Morawetz, *Pure Appl. Chem.* **52**, 277 (1980).
16. S. Farid, F. A. Martic, R. C. Daly, D. R. Thompson, D. P. Specht, S. E. Hartman, and J. L. R. Williams, *Pure Appl. Chem.* **51**, 241 (1979).
17. W. Schnabel, *Pure Appl. Chem.* **51**, 2373 (1979).
18. M. A. Winnik, *Acc. Chem. Res.* **10**, 173 (1977).
19. H. Morawetz, *Science* **203**, 405 (1979). (a) G. Wilemski and M. Fixman, *J. Chem. Phys.* **58**, 4009 (1973); M. Fixman, *ibid.* **69**, 1527, 1528 (1978).
20. F. C. DeSchryver, N. Boens, and J. Put, *Adv. Photochem.* **10**, 359 (1977).
21. K. J. Shea, Y. Okahata, and T. K. Dougherty, *Macromolecules* **17**, 296 (1984).
22. H. L. Chen and H. Morawetz, *Macromolecules* **15**, 1445 (1982).
23. K. Y. Law and R. O. Loutfy, *Polymer* **24**, 439 (1983).
24. G. Duportail, D. Froelich, and G. Weill, *Eur. Polym. J.* **7**, 711 (1971).
25. L. Moldvan and G. Weill, *Eur. Polym. J.* **7**, 1023 (1971).
26. J. P. Benty, J. B. Beryl, G. Beinert, and G. Weill, *Eur. Polym. J.* **11**, 495 (1975).
27. Y. E. Kirsch, N. R. Pavlova, and V. A. Kabanov, *Eur. Polym. J.* **11**, 495 (1975).
28. A. M. North and I. Soutar, *J.C.S. Faraday I* **68**, 1101 (1972).
29. G. J. Kettle and I. Soutar, *Eur. Polym. J.* **14**, 895 (1978).
30. K. Brown and I. Soutar, *Eur. Polym. J.* **10**, 433 (1974).
31. P. Wahl, G. Meyer, and J. Parrod, *Eur. Polym. J.* **6**, 585 (1970).
32. B. Valeur and L. Monnerie, *C. R. Hebd. Seances Acad. Sci., Ser. C* **280**, 57 (1975).
33. B. Valeur and L. Monnerie, *Eur. Polym. J.* **14**, 11, 29 (1976).
34. B. Valeur, L. Monnerie, and J. P. Jarry, *Eur. Polym. J.* **13**, 675 (1975).
35. J. L. Viovy, L. Monnerie, and J. C. Brochon, *Macromolecules* **16**, 1845 (1983).
36. N. Kasparyan-Tardiveau, B. Valeur, L. Monnerie, and I. Mita, *Polymer* **24**, 205 (1983).
37. K. P. Ghiggino, R. P. Wright, and D. Phillips, *J. Polym. Sci., Polym. Phys. Ed.* **16**, 1499 (1978).
38. D. Phillips, A. J. Roberts, G. Rumbles, and I. Soutar, *Macromolecules* **16**, 1198 (1983).
39. K. Itagaki, K. Horie, I. Mita, M. Washio, S. Tagawa, and Y. Tabata, *J. Chem. Phys.* **79**, 3996 (1983).
40. K. P. Ghiggino, R. P. Wright, and D. Phillips, *Chem. Phys. Lett.* **53**, 552 (1978).
41. D. Phillips, A. J. Roberts, and I. Soutar, *Polymer* **22**, 427 (1981).
42. A. Gupta, R. Liang, J. Mocanin, D. Kliger, R. Goldenbeck, J. Horwitz, and V. M. Miskowski, *Eur. Polym. J.* **17**, 485 (1981).
43. K. Demeyer, M. Van der Auweraer, L. Aerts, and F. C. DeSchryver, *J. Chim. Phys.* **77**, 493 (1980).
44. F. C. DeSchryver, K. Demeyer, M. Van der Auweraer, and E. Quanten, *Ann. N.Y. Acad. Sci.* **366**, 93 (1981).
45. M. Irie, T. Kanijo, J. Aikawa, T. Takemura, H. Hayashi, and H. Baba, *J. Phys. Chem.* **81**, 1571 (1977).

46. A. J. Roberts, D. V. O'Connor and D. Phillips, *Ann. N.Y. Acad. Sci.* **366**, 109 (1981).
47. D. Phillips, A. J. Roberts, and I. Soutar, *J. Polym. Sci., Polym. Phys. Ed.* **18**, 2401 (1980).
48. D. Phillips, A. J. Roberts, and I. Soutar, *J. Polym. Sci., Polym. Phys. Ed.* **20**, 411 (1982).
49. D. Phillips, A. J. Roberts, and I. Soutar, *Polymer* **22**, 293 (1981).
50. S. Ito, M. Yamamoto, and Y. Nishijima, *Polym. J.* **13**, 791 (1981).
51. D. A. Holden and J. E. Guillett, *Macromolecules* **13**, 289 (1980).
52. D. A. Holden, P. Y. K. Wang, and J. E. Guillett, *Macromolecules* **13**, 295 (1980).
53. L. Merle-Aubry, D. A. Holden, Y. Merle, and J. E. Guillett, *Macromolecules* **13**, 1138 (1980).
54. D. A. Holden and J. E. Guillett, *Macromolecules* **15**, 1475 (1982).
55. H. Itagaki, A. Okamoto, K. Horie, and I. Mita, *Eur. Polym. J.* **18**, 885 (1982).
56. C. Cuniberti and A. Perico, *Eur. Polym. J.* **16**, 887 (1980).
57. A. E. C. Redpath and M. A. Winnik, *Polymer* **24**, 1286 (1983).
58. M. A. Winnik, A. E. C. Redpath, K. Paton, and J. Danhelka, *Polymer* **25**, 91 (1984).
59. M. A. Winnik, X. B. Li, and J. E. Guillett, *Macromolecules* **17**, 699 (1984).
60. X. B. Li, M. A. Winnik, and J. E. Guillett, *Macromolecules* **17**, 992 (1984).
61. M. A. Winnik, A. E. C. Redpath, and P. Svirkaya, *Polymer* **24**, 473 (1983).
62. P. Svirskaya, J. Danhelka, A. E. C. Redpath, and M. A. Winnik, *Polymer* **24**, 319 (1983).
63. A. Itaya, K. Okamomoto, and S. Kusabayashi, *Bull. Chem. Soc. Jpn.* **49**, 2086 (1976); **50**, 52 (1977).
64. K. P. Ghiggino, D. A. Archibald and P. J. Thistlewhite, *J. Polym. Sci., Polym. Lett.* **18**, 673 (1980).
65. D. Ng and J. E. Guillett, *Macromolecules* **14**, 405 (1981).
66. C. E. Hoyle, T. L. Nemzek, A. Mar, and J. E. Guillett, *Macromolecules* **11**, 429 (1978).
67. A. Ikaya, K. Okamoto, and S. Kusabayashi, *Bull. Chem. Soc. Jpn.* **49**, 2082 (1976).
68. A. J. Roberts and D. Phillips, *Macromolecules* **15**, 678 (1982).
69. A. J. Roberts, C. G. Cureton, and D. Phillips, *Chem. Phys. Lett.* **72**, 554 (1980).
70. A. J. Roberts, D. Phillips, F. A. Abdul-Rasoul, and A. J. Ledwith, *J. C. S. Faraday I* **77**, 2725 (1981).
71. A. Itaya, K. Okamoto, H. Masuhara, N. Ikeda, N. Mataga, and S. Kusabayashi, *Macromolecules* **15**, 1213 (1982).
72. F. C. DeSchryver, J. Vandendriessche, S. Toppet, K. Demeyer, and N. Boens, *Macromolecules* **15**, 406 (1982).
73. F. Evers, K. Kobs, R. Memming, and D. R. Terrell, *J. Am. Chem. Soc.* **105**, 5988 (1983).
74. H. F. Kaufman, W.-D. Weixelbaumer, J. Bauerbaumer, A.-M. Schmoltner, and O. F. Olaj, *Macromolecules* **18**, 104 (1985).
75. S. Tazuke and F. Banba, *Macromolecules* **9**, 451 (1976).
76. S. Tazuke and N. Hayashi, *Polym. J.* **10**, 443 (1978).
77. S. Tazuke, H. Ooki, and K. Sato, *Macromolecules* **15**, 400 (1980).
78. F. Hirayama, *J. Chem. Phys.* **42**, 3163 (1965).
79. M. Goldenberg, J. Emert, and H. Morawetz, *J. Am. Chem. Soc.* **100**, 7171 (1978).
80. K. A. Zachariasse and W. Kühnle, *Z. Phys. Chem. (Wiesbaden)* **101**, 267 (1976).
81. K. Zachariasse and W. Kühnle, *Ber. Bunsenges. Phys. Chem.* **78**, 1254 (1974).
82. K. A. Zachariasse, W. Kühnle, and A. Weller, *Chem. Phys. Lett.* **59**, 375 (1978).
83. M. J. Snare, P. J. Thistlewhite, and K. P. Ghiggino, *J. Am. Chem. Soc.* **105**, 3328 (1983).
84. K. A. Zachariasse, G. Duveneck, and R. Busse, *J. Am. Chem. Soc.* **106**, 1045 (1984).
85. M. Yamamoto, K. Goshiki, T. Kanaya, and Y. Nishijima, *Chem. Phys. Lett.* **56**, 333 (1978).
86. T. Kanaya, K. Goshiki, M. Yamamoto, and Y. Nishijima, *J. Am. Chem. Soc.* **104**, 3580 (1982).
87. A. Collart, K. Demeyer, S. Toppet, and F. C. DeSchryver, *Macromolecules* **16**, 1390 (1983).
88. E. A. Chandross and C. J. Dempster, *J. Am. Chem. Soc.* **92**, 3586 (1970).

89. H. Itagaki, N. Obukata, A. Okamoto, K. Horie, and I. Mita, *J. Am. Chem. Soc.* **104**, 4469 (1982).
90. F. C. DeSchryver, K. Demeyer, and S. Toppet, *Macromolecules* **16**, 89 (1983). (a) R. Todesco, J. Gelan, H. Martens, J. Put, and F. C. DeSchryver, *J. Am. Chem. Soc.* **103**, 7304 (1981).
91. M. Itoh, K. Fuke, and S. Kusabayashi, *J. Chem. Phys.* **72**, 1417 (1980).
92. J. Ferguson, A. Castellan, P. Desvargne, and H. Bonas-Laurent, *Chem. Phys. Lett.* **78**, 446 (1981).
93. F. C. DeSchryver, K. Demeyer, J. Huybrechts, H. Bonas-Laurent, and A. Castellan, *J. Photochem.* **20**, 341 (1982).
94. K. A. Zachariasse, R. Busse, U. Schrader, and W. Kühnle, *Chem. Phys. Lett.* **89**, 303 (1982).
95. N. Tamai, H. Masuhara, and N. Mataga, *J. Phys. Chem.* **87**, 4461 (1983).
96. G. E. Johnson, *J. Chem. Phys.* **61**, 3002 (1974); **63**, 404 (1975).
97. W. Klöpfer, *Ber. Bunsenges. Phys. Chem.* **74**, 6931 (1970).
98. H. Masuhara, N. Tamai, N. Mataga, F. C. DeSchryver, J. Vandendriessche, and N. Boens, *Chem. Phys. Lett.* **95**, 471 (1983).
99. W. Klopfer, *in* "Organic Molecular Photophysics" (J. B. Birks, ed.), Vol. 1, p. 357. Wiley, New York, 1973.
100. S. Tazuke and Y. Matsuyama, *Macromolecules* **8**, 280 (1975).
101. Y. Matsuyama and S. Tazuke, *Polym. J.* **8**, 481 (1976).
102. S. Tazuke and Y. Matsuyama, *Macromolecules* **10**, 215 (1977).
103. U. Lachish, R. W. Anderson, and D. J. Williams, *Macromolecules* **13**, 1143 (1982).
104. U. Lachish and D. J. Williams, *Macromolecules* **13**, 1322 (1982).
105. H. Masuhara, H. Shioyama, N. Mataga, T. Inoue, N. Kitamura, T. Tanake, and S. Tazuke, *Macromolecules* **14**, 1738 (1981).
106. H. Masuhara, S. Ohwada, N. Mataga, A. Itaya, K. Okamoto, and S. Kusabayashi, *Chem. Phys. Lett.* **59**, 188 (1978); **70**, 276 (1980).
107. S. Tazuke, K. Sato, and F. Banba, *Chem. Lett.* p. 1321 (1975); *Macromolecules* **10**, 1224 (1977).
108. S. Tazuke, H. L. Yuan, Y. Iwaya, and K. Sato, *Macromolecules* **14**, 267 (1981).
109. S. Tazuke and H. L. Yuan, *Polym. J.* **14**, 45, 695 (1982).
110. S. Tazuke and H. L. Yuan, *Polym. J.* **15**, 125 (1983). (a) S. Tazuke and H. L. Yuan, *J. Phys. Chem.* **86**, 1250 (1982).
111. S. Tazuke, Y. Iwaya, and R. Hayashi, *Photochem. Photobiol.* **35**, 621 (1981).
112. Y. Iwaya and S. Tazuke, *Macromolecules* **15**, 396 (1982).
113. R. S. Davidson, R. Bonneau, J. Joussot-Dubien, and K. J. Toyne, *Chem. Phys. Lett.* **63**, 269 (1979).
114. M. Vander Auweraer, A. Gilbert, and F. C. DeSchryver, *Nouv. J. Chim.* **4**, 153 (1980).
115. M. Vander Auweraer, A. Gilbert, and F. C. DeSchryver, *J. Am. Chem. Soc.* **102**, 4007 (1980).
116. E. A. Chandross and H. T. Thomas, *Chem. Phys. Lett.* **9**, 393 (1971).
117. G. S. Beddard and R. S. Davidson, *J. Photochem.* **1**, 491 (1972–1973).
118. F. Meens, M. Vander Auweraer, and F. C. DeSchryver, *Chem. Phys. Lett.* **74**, 218 (1980).
119. K. Gnadig and K. B. Eisenthal, *Chem. Phys. Lett.* **46**, 339 (1977).
120. N. Migita, T. Okada, N. Mataga, Y. Sakata, S. Misumi, N. Nakashima, and K. Yoshihara, *Bull. Chem. Soc. Jpn.* **54**, 3304 (1981).
121. H. Masuhara, J. Vandendriessche, K. Demeyer, N. Boens, and F. C. DeSchryver, *Macromolecules* **15**, 1471 (1982).
122. D. A. Holden and J. E. Guillett, *Macromolecules* **13**, 289 (1980).
123. C. E. Hoyle and J. E. Guillett, *J. Polym. Sci., Polym. Lett.* **16**, 185 (1978).
124. J. S. Aspler, C. E. Hoyle, and J. E. Guillett, *Macromolecules* **11**, 925 (1978).
125. T. Nakahira, C. Minami, S. Iwabushi, and K. Koyima, *Makromol. Chem.* **180**, 2245 (1979).
126. D. Ng and J. E. Guillett, *Macromolecules* **15**, 724, 728 (1982).

127. D. A. Holden and J. E. Guillett, *Macromolecules* **15**, 1475 (1982).
128. D. A. Holden, S. E. Shepherd, and J. E. Guillett, *Macromolecules* **15**, 1481 (1982).
129. D. Ng, K. Yoshiki, and J. E. Guillett, *Macromolecules* **16**, 568 (1983).
130. D. A. Holden, X. X. Ren, and J. E. Guillett, *Macromolecules* **17**, 1500 (1984).
131. J. S. Hargreaves and S. E. Webber, *Macromolecules* **15**, 424 (1982).
132. N. Kim and S. E. Webber, *Macromolecules* **15**, 430 (1982).
133. R. Voltz, G. Laustrait, and A. Cocher, *J. Chim. Phys.* **63**, 1255 (1973).
134. M. Yokoto and O. Tanimoto, *J. Phys. Soc. Jpn.* **22**, 779 (1967).
135. R. V. Bensasson, J. C. Ronfard-Haret, E. L. Land, and S. E. Webber, *Chem. Phys. Lett.* **68**, 438 (1979). (a) S. Tagawa, N. Nakashima, and K. Yoshihara, *Macromolecules* **17**, 1167 (1984).
136. K. Horie, K. Tomamune, and I. Mita, *Polym. J.* **11**, 539 (1979).
137. I. Mita, K. Horie, and M. Masuda, *Polym. Bull.* **4**, 369 (1981).
138. (a) K. Horie, W. Schnabel, I. Mita, and H. Ushiki, *Macromolecules* **14**, 1422 (1981). I. Mita, K. Horie, and M. Takeda, *Macromolecules* **14**, 1428 (1981). (b) H. Ushiki, K. Horie, A. Okamoto, and I. Mita, *Polymer J.* **13**, 191 (1981). (c) H. Ushiki, K. Horie, A. Okamoto, and I. Mita, *Polym. Photochem.* **1**, 303 (1981).
139. A. Mar, S. Fraser, and M. A. Winnik, *J. Am. Chem. Soc.* **103**, 4941 (1981).
140. J. F. Pratte, W. A. Noyes, and S. E. Webber, *Polym. Photochem.* **1**, 3 (1981). (a) J. Kiwi and W. Schnabel, *Macromolecules* **8**, 430 (1975). (b) J. Kiwi and W. Schnabel, *Macromolecules* **9**, 468 (1976). (c) W. Schnabel, *Macromol. Chem.* **180**, 1487 (1979). (d) K. Hayashi, M. Irie, J. Kiwi, and W. Schnabel, *Polym. J.* **9**, 41 (1977). (e) I. Mita, T. Takagi, K. Horie, and Y. Shindo, *Macromolecules* **17**, 2256 (1984).
141. J. F. Pratte and S. E. Webber, *Macromolecules* **15**, 417 (1982).
142. J. P. Bays, M. V. Encinas, and J. C. Scaiano, *Polymer* **21**, 283 (1980).
143. P. K. Das and J. C. Scaiano, *Macromolecules* **14**, 693 (1981).
144. J. C. Selwyn and J. C. Scaiano, *Polymer* **21**, 1365 (1980).
145. J. F. Pratte and S. E. Webber, *Macromolecules* **16**, 1188, 1193 (1983).
146. A. F. Oleas, M. V. Encinas, and E. A. Lissi, *Macromolecules* **15**, 111 (1982).
147. M. V. Encinas, E. A. Lissi, L. Gargello, D. Radic, and A. F. Olea, *Macromolecules* **17**, 2261 (1984).
148. S. A. Rice and M. Nagasawa, "Polyelectrolyte Solutions." Academic Press, New York, 1961.
149. E. Selegny and A. Rembaum, eds., "Polyelectrolytes and Their Applications." Reidel, Dordrecht, Netherlands, 1975.
150. F. Oosawa, "Polyelectrolytes." Dekker, New York, 1971.
151. N. Ise, T. Okubo, and S. Kanugi, *Acc. Chem. Res.* **15**, 171 (1982).
152. G. S. Manning, *Acc. Chem. Res.* **12**, 443 (1979).
153. N. J. Turro, B. H. Baretz, and P.-L. Kuo, *Macromolecules* **17**, 1321 (1984).
154. N. J. Turro and T. Okubo, *J. Am. Chem. Soc.* **104**, 2985 (1982).
155. N. J. Turro and C.-J. Chung, *Macromolecules* **17**, 2123 (1984).
156. I. Nagata and Y. Okamoto, *Macromolecules* **16**, 749 (1983).
157. T. Chen and J. K. Thomas, *J. Polym. Sci., Polym. Chem.* **17**, 1103 (1979).
158. D.-Y. Chen and J. K. Thomas, *Macromolecules* **17**, 2142 (1983).
159. I. A. Taha and H. Morawetz, *J. Am. Chem. Soc.* **93**, 829 (1971).
160. D. Meisel and M. S. Matheson, *J. Am. Chem. Soc.* **99**, 6577 (1977).
161. S. Kelder and J. Rabani, *J. Phys. Chem.* **85**, 1637 (1981).
162. Y. Kurimura, H. Yokata, K. Shigehara, and E. Tsuchida, *Bull. Chem. Soc. Jpn.* **55**, 55 (1982).
163. N. E. Prieto and C. R. Martin, *J. Electrochem. Soc.* **131**, 751 (1984).
164. N. J. Turro, T. Okubo, C.-J. Chung, J. Emert, and R. Catena, *J. Am. Chem. Soc.* **104**, 4799 (1982).

165. N. J. Turro and I. F. Pierola, *Macromolecules* **16**, 906 (1983).
166. N. J. Turro and I. F. Pierola, *J. Phys. Chem.* **87**, 2420 (1983).
167. N. J. Turro and T. Okubo, *J. Phys. Chem.* **86**, 1485 (1982).
168. P. Ander and M. K. Mahmoudhagh, *Macromolecules* **14**, 87 (1981).
169. I. Nagata and H. Morawetz, *Macromolecules* **14**, 87 (1981).
170. D. A. Holden, W. A. Rendall, and J. E. Guillett, *Ann. N.Y. Acad. Sci.* **366**, 11 (1981).
171. T. Okubo and N. J. Turro, *J. Phys. Chem.* **85**, 4034 (1981).
172. N. J. Turro and T. Okubo, *J. Phys. Chem.* **86**, 1535 (1982).
173. D. Meisel, J. Rabani, D. Meyerstein, and M. S. Matheson, *J. Phys. Chem.* **82**, 985 (1978).
174. C. D. Jonah, M. S. Matheson, and D. Meisel, *J. Phys. Chem.* **83**, 257 (1979).
175. D. Meyerstein, J. Rabani, M. S. Matheson, and D. Meisel, *J. Phys. Chem.* **82**, 1879 (1978).
176. R. E. Sassoon and J. Rabani, *J. Phys. Chem.* **84**, 1319 (1980).
177. Y. Kurimura, N. Shinozaki, F. Ito, Y. Uratani, K. Shigehara, E. Tschuida, M. Kaneko, and A. Yamada, *Bull. Chem. Soc. Jpn.* **55**, 380 (1982).
178. Y. Morishima, Y. Itoh, S. Nozakura, T. Ohno, and S. Kato, *Macromolecules* **17**, 2264 (1984).
179. M. Furue, K. Sumi, and S. Nozakura, *Chem. Lett.* p. 1349 (1981).
180. M. Kaneko, S. Nemato, A. Yamada, and Y. Kurimura, *Inorg. Chim. Acta* **44**, L289 (1980).
181. T. Matsuo, T. Sakamoto, K. Takuma, K. Sakura, and T. Ohsako, *J. Phys. Chem.* **85**, 1277 (1981).
182. M. Kaneko, A. Yamada, E. Tsuchida, and Y. Kurimura, *J. Phys. Chem.* **88**, 1061 (1984).
183. A. T. Thornton and G. S. Laurence, *J.C.S. Chem. Commun.* p. 408 (1978).
184. P. C. Lee and D. Meisel, *J. Am. Chem. Soc.* **102**, 5477 (1980).
185. A. Slama-Schwok, Y. Feitelson, and J. Rabani, *J. Phys. Chem.* **85**, 2222 (1981).
186. I. Nagata, R. Li, E. Banks, and Y. Okamoto, *Macromolecules* **16**, 903 (1983). (a) M. Krishnan, J. R. White, M. A. Fox, and A. J. Bard, *J. Am. Chem. Soc.* **105**, 7002 (1983). (b) A. W. H. Mau, C. B. Huang, N. Kakuta, A. J. Bard, A. Campion, M. A. Fox, J. M. White, and S. E. Webber, *J. Am. Chem. Soc.* **106**, 6537 (1984). (c) D. Meissner, R. Memming, and B. Kastening, *Chem. Phys. Lett.* **96**, 34 (1983).
187. Y. Kurimura and K. Katsumata, *Bull. Chem. Soc. Jpn.* **55**, 2560 (1982).
188. Y. Kurimura, M. Nagashima, K. Takato, E. Tsuchida, M. Kaneko, and A. Yamada, *J. Phys. Chem.* **86**, 2432 (1982).
189. S. L. Buell and J. N. Demas, *J. Phys. Chem.* **87**, 4675 (1983).

Chapter 9

Photoprocesses in Molecular Inclusion Complexes

9.1 Introduction

An interesting development in the coordination chemistry of inorganic systems in recent years has been the synthesis of novel multidentate, macrocyclic compounds such as crown ethers, cyclodextrins, and zeolites. These large, permanent molecular systems are capable of trapping (or complexing) small molecules in their cavities. This inclusional process is analogous to the dynamic solubilization of molecules in organized molecular aggregated systems such as micelles discussed earlier. A distinguishing feature of these host systems, however, is the existence of finite size cavities or pores that lead to shape selectivity on the molecules they include. Consequently, there is intense interest in the characterization and exploitation of the host–guest complex chemistry. In this chapter we focus our attention on the photophysical and photochemical studies of these molecular inclusion complexes. It should be pointed out that, unlike the classical clathrate compounds that exist only in the solid state, these complexes are present in homogeneous solutions at room temperature. In some aspects, these host systems can also be considered as inorganic polymeric systems with a certain repeat unit (cyclodextrins as cyclic polysugars; crown compounds as cyclic polyethers, and zeolites as framework structure aluminosilicates). The origins of interest in these systems are numerous. Foremost of them is in catalysis and in the design of biomimetic systems (enzyme catalysis, ionophore mediate membrane transport, etc.).

9.2   Photoprocesses in Cyclodextrins

9.2.1   General Features of Cyclodextrins and Their Inclusion Complexes[1-5]

Cyclodextrins are cyclic oligosaccharide molecules consisting of six, seven, or eight glucose units. Isolated as degradation products of starch as early as 1891 by Villiers,[6] they were characterized as cyclic polysugars by Schardinger in 1904.[7] (Hence, the attribute to these compounds as "Schardinger dextrins".) The common hexamers, heptamers, and octamers are identified as α-, β-, and γ-cyclodextrins. The glucose monomers are arranged in a torus, and the coupling of glucose moieties gives cyclodextrins a rigid conical structure with a hollow interior of a specific volume. X-ray analysis of crystalline cyclodextrins (abbreviated hereafter as CDs) indicates that the glucose rings are in chair conformation and NMR and ORD studies have shown that the glucose residues retain their chair conformation in solution as well. Due to the apparent lack of free rotation about the glucosidic bond that links the glucose

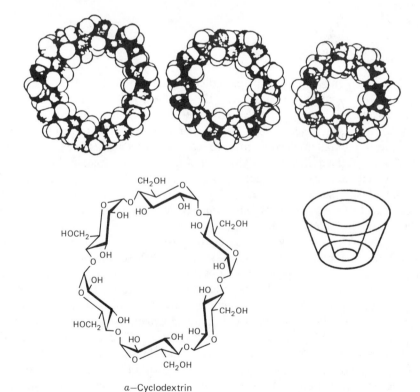

α—Cyclodextrin

Fig. 9.1. Schematic representation of structural features of cyclodextrins (including CPK models of α-, β-, and γ-cyclodextrins).

TABLE 9.1. Solubility and Structural Parameters of Various
Cyclodextrins in Aqueous Solutions [a]

Cyclodextrin	Number of glucose units	Solubility in water (g/100 ml)	Diameter (Å)	
			Cavity	External
α	6	14.5	4.7–5.2	14.6 ± 0.4
β	7	1.85	6.0–6.4	15.4 ± 0.4
γ	8	23.2	7.5–8.3	17.5 ± 0.4

[a] From Saenger.[5]

units, the cyclodextrins are not perfectly cylindrical but are somewhat cone shaped. The 6-hydroxyl face is the narrow side while the 2,3-hydroxyl face is somewhat wider. Figure 9.1 presents certain structural features of cyclodextrins. Table 9.1 presents data on the solubility of CDs in water and estimates on the cavity and external diameters based on Corey–Pauling–Koltun (CPK) models.

The most remarkable property of CDs is their ability to form inclusion complexes with a variety of molecules which apparently must satisfy one condition, namely, they must fit entirely, or at least partially, into the CD cavity. If the substrate is too small, it will pass in and out of the pore with little apparent binding. Larger molecules simply cannot fit into the cavity. The host and guest associate with each other apparently without any specific interactions. A great majority of molecules have been shown to form 1:1 complexes with CDs. However there are increasing evidences that suggest inclusion of more than one guest molecule in large-cavity CDs such as γ-CD at the same time, thus forming 1:1:1 or 2:1 complexes. The interior of toruses of CDs is relatively nonpolar and, therefore, is capable of including aliphatic and aromatic compounds, gases, etc. The association constants K for most molecules are in the range of 10^3 M/L, characteristic of weak intermolecular interactions. Other than the size and shape criteria, there is no obvious correlation between the chemical properties of guests and the K values. Kinetic studies with a series of substituted azo dyes (all having similar K_D values $\simeq 10^3$ M/L) have shown that the rate constants for the inclusion and dissociation processes vary by a factor of 10^5 as the volume of the substituent increases.[8] The hydrophobic and steric control of the inclusion process drastically influences the mode and reactivity of guests in the cavity in a manner much resembling the way enzymes bind substrates and carry out stereospecific enzymatic catalytic processes.

9.2.2 Fluorescence Probe Analysis of Cyclodextrins

A simple approach to probe the inner cavity and the inclusion process is to use medium-sensitive fluorescence probes as guest molecules. The very early application in 1967 was by Cramer *et al.*[8] who used 1,8-anilinonaphthalene-sulfonate (ANS) as a probe. ANS shows a strong fluorescence in organic solvents such as ethanol ($\Phi = 0.6$) but in water it exhibits only a very weak fluorescence. Thus, on its transfer from water to the apolar/semipolar interiors of CD, we would expect an increase in the fluorescence intensities. Indeed we observe such enhanced intensities, twofold in the smallest α-CD but as much as a factor of 10 with larger β- and γ-CDs. Concurrent with the intensity enhancement are the blue shifts in the emission maxima of ANS along the series α, β, and γ. In addition to the demonstration of the inclusion process *per se*, fluorescence techniques can be used to determine the association constants, the extent of accessibility of the included probe to other molecules, and the stereochemical mode of incorporation of the probe itself.

The driving forces for the inclusion process are mainly hydrophobic, and, hence, the host–guest binding could be enhanced by a decrease in the hydrophobic surface exposed in water. In capped cyclodextrins this is done by capping one end of the pore via covalent linkages. Tabushi *et al.* have demonstrated such enhanced binding of ANS via fluorescence measurements in β-CD capped with terephalate or diphenylmethane sulfonate units ($K_D = 6.4 \times 10^2\ M^{-1}$ and $1.3 \times 10^3\ M^{-1}$, respectively):[9]

$$H_2C \quad \text{\lbrack benzene\rbrack} - SO_2 - O \quad \text{\lbrack benzene\rbrack} - SO_2 - O \quad [\beta\text{---}CD] \qquad \text{\lbrack benzene\rbrack} \begin{matrix} CO - O \\ CO - O \end{matrix} [\beta\text{---}CD]$$

The binding is 11, 24 times stronger than in the parent β-CD.

The fluorescence intensity enhancement along with the spectral blue shifts on inclusion in the CD cavity appear to be a general feature with a wide variety of hydrophobic probes: benzene derivatives,[10] cationic surfactant indole,[11] Coumarin 540A,[12,13] and others. Quantitative data on the fluorescence enhancements for several benzene derivatives have been determined by Hoshino *et al.*[10] and some of their data are presented in Table 9.2. The fluorescence enhancements have been ascribed to an increase in the radiative rates, a decrease in the probe rotational freedom, and the elimination of water molecules surrounding the probe molecule in aqueous solution. Absorption and emission data as a function of CD concentration have been analysed according to Benesi and Hildebrand and association constants derived. Fluorescence studies indicate that dyes such as rhodamine B which aggregate readily in aqueous solution are deaggregated in cyclodextrins prior to inclusion. (The spectra are sharper with enhanced fluorescence intensities).[14] Figure 9.2 illustrates the effect of various cyclodextrins on the fluorescence of coumarin 540A and pyrene in aqueous solutions.

TABLE 9.2. Association Constants and Fluorescence Quantum Yields for Benzene Derivatives in Aqueous β-Cyclodextrin Solutions [a]

Guest	$K(M^{-1})$		Φ_{fl}		
	Absorption	Fluorescence	H_2O	EtOH	β-CD
Benzene		196 ± 10	0.0058	0.003	0.0071
Phenol		40 ± 2	0.14	0.20	0.24
Ethoxybenzene		286 ± 15	0.18	0.23	0.22
Aniline		50 ± 3	0.025	0.085	0.10
N-Methylaniline	48 ± 3	53 ± 3	0.029	0.096	0.17
N,N-Dimethylaniline	230 ± 10	217 ± 10	0.048	0.067	0.13
Diethylaniline	960 ± 40	862 ± 40	0.036	0.056	0.076

[a] From Hoshino et al.[10]

The changes in the vibronic band intensities of pyrene monomer fluorescence and in lifetimes can be used to monitor the inclusion process of the probe in CDs.[15] The peak ratio (III/I) and lifetimes increase on transfer of pyrene from water to the cavities of β-CD. Two processes have been identified, an initial fast inclusion of the arene followed by a slower aggregation of the CD–arene complexes. Addition of surfactants at concentrations below the critical micelle concentration leads to the formation of 1:1:1 [arene:CD:surfactant] complexes and an increase in the hydrophobicity of the pyrene environment. An increase of surfactant concentration well above the CMC leads to the formation of micelles and a transfer of arenes from the CD cavity to micelles! Probe accessibility studies via fluorescence quenching with I⁻ and nitromethane indicate that arenes are protected considerably by the inclusional process. It has been found that the pyrene inclusion and the subsequent aggregation of [arene:CD] complexes are accelerated by the addition of ethanol.[16] Fluorescence quenching studies on methyl-2-aminobenzoate,[17] naphthalene,[18] and methoxynaphthalene[19] by iodate (IO_3^-) and I⁻ in the presence of β-CD indicate an efficient protection against the former quencher but much less efficiency against the smaller I⁻ ions. A comparison of the data on quenching rate constants for various probe/quencher systems is presented in Table 9.3.

The dual fluorescence of dimethylaminobenzonitrile (DMABN) and diethylaminobenzonitrile (DEABN) has been used to probe the mode of incorporation of these dyes in the cavities of cyclodextrins.[19] Enhanced emission of the twisted internal charge transfer state was observed in CDs, with the maximum effect with the DMABN–β-CD system. Polarity experienced by the probes, however, did not correlate with the size of the equilibrium constants determined.

The fluorescence of methyl salicylate[20] in aqueous solutions differs dramatically from that in the presence of α- and γ-CDs but is similar in the

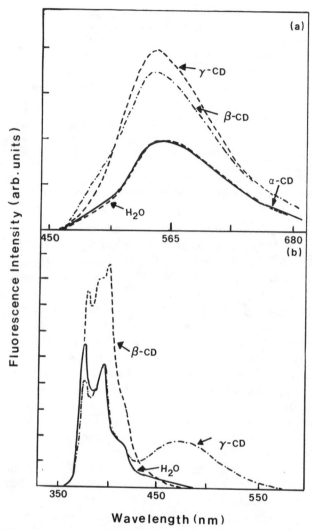

FIG. 9.2. Fluorescence spectra of (a) coumarin 540A and (b) pyrene in aqueous and three types of cyclodextrin solutions. (From Scypinski and Drake[13] and Yorozv et al.[25] Copyright 1982 and 1985 American Chemical Society.)

presence of β-CDs as illustrated in Fig. 9.3. The fluorescence intensity of β-naphthol[21] and α-naphthylacetic acid[22] in aqueous solution is slightly enhanced on addition of γ-cyclodextrin. Marked enhancement, however, is observed in the presence of both γ-CD and cyclohexanol. The role of cyclohexanol presumably is as a space regulator in narrowing the cavity of γ-CD and allows formation of stable 1:1:1 complexes. The similarity in the fluorescence lifetime of chiral-(1-naphthyl)-ethylamine in water (5.28 nsec),

TABLE 9.3. Rate Constants for the Quenching of Fluorescence of Probe Molecules in Homogeneous and Cyclodextrin Media [a]

Probe	Quencher[b]	H_2O	α-CD	β-CD	γ-CD	Reference
Pyrene	DMA	0.015	0.014	0.73	0.50	15,31,32
Pyrene	DEA	0.03	0.014	0.37	0.19	15,31,32
Pyrene	TMA	0.044	0.029	14.0	0.13	15,31,32
Naphthalene	DEA			2.1		15,31,32
Naphthalene	DEA	(17)	(25)	(102)	(36)	15,31,32
Naphthalene	TMA	(5.6)	(6.1)	(195)	(30)	15,31,32
Pyrene-1-SO_3^-	DEA	(23)	(19)	(33)	(29)	15,31,32
1-Naph-SO_3^-	DEA	(27)	(45)	(32)		15,31,32
2-Naphthol	I^-	4.4		1.6		15,31,32
Naphthalene	IO_3^-	2.4		0.45		
Naphthalene	I^-	6.0		3.9		15,31,32
2-Methoxynaphthalene	IO_3^-	2.8		0.6		15,31,32
2-Methoxynaphthalene	I^-	6.0		2.6		15,31,32

[a] Units are given $\times 10^9 \ M^{-1} \ sec^{-1}$.

[b] Abbreviations used: DMA = N,N-dimethylaniline; DEA = N,N-diethylaniline; TMA = trimethylamine. Values in parentheses are Stern–Volmer constants K_{SV} in M^{-1}.

α-CD, and γ-CD suggest lack of incorporation of the probe in these cyclodextrin molecules.[22a] In the large cavity γ-CD, the fit of the probe along its long axis is too loose to yield a stable complex, even in the presence of a cyclic alcohol spacer such as cyclohexanol. Inclusion in β-CD however occurs and the process is accompanied by an increase in τ_{fl} to about 16.6 nsec. Table 9.4 presents a comparison of the fluorescence properties of various dye molecules in homogeneous and cyclodextrin solutions.

9.2.3   Fluorescence Depolarization

Measurements of depolarization of fluorescence of probes included in the cavity provide an elegant means of evaluating the inner microviscosity η and the freedom of motion experienced by the probe. Studies using surfactant derivatives of indole, 6-In-11$^+$, have yielded η values of 80, 150, and 100 cP for the α-, β-, and γ-CD complexes, respectively.[12] In order to ascertain whether the polarization observed is due to the probe motion in the cavity or due to the CD–probe complex as a whole, the macroscopic viscosity of the medium was increased by a factor of approximately 10 by addition of methyl α-glucoside to the aqueous CD solutions. Unchanged polarization loss, coupled with the unchanged probe emission maxima and lifetimes confirm that the observed polarization is entirely due to the probe motion in the CD cavity.

Kitamura et al.[23] have examined fluorescence depolarization of two fluorescein derivatives conjugated to linear polysaccharides, amylose and dextran, one randomly labelled throughout the chain and other locally on a

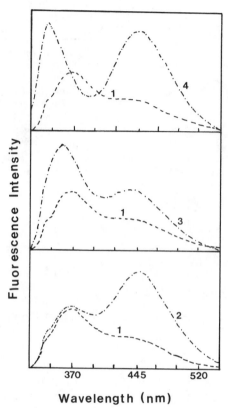

FIG. 9.3. Fluorescence spectra of methyl-1-salicylate in aqueous (curve 1) and cyclodextrin solutions (γ-CD, curve 2; β-CD, curve 3, and α-CD, curve 4). (From Cox and Turro.[20] Copyright 1984, Pergamon Press.)

terminal segment. Rotational relaxation time measurement indicate, as we expect, the terminal segments to undergo more rapid microbrownian motion than the interiors.

9.2.4 Excimer and Exciplex Dynamics

Intermolecular Excimers

When there is a good match between the cavity size and the probe size, we observe a simple 1:1 (probe:CD) complexation between the probe and the CD. However, with small molecules and large-cavity CDs, there is a potential possibility of two probe molecules to be included in the same cavity at the same time. For condensed aromatic hydrocarbons such as pyrene, this can result in high yields of excimers. Apparently such double occupancy of probes in the CD cavity appears to be a common occurrence. Ueno *et al.*[24] demonstrated such feasibility first in 1980 in their fluorescence study of α-naphthyl-

TABLE 9.4. Fluorescence Properties of Various Dye Molecules in Homogeneous and Cyclodextrin Media.[11,19–21]

Probe/property	Homogeneous solvents		α-CD	β-CD	γ-CD
	H_2O	EtOH			
Methyl salicylate[20]					
Absorption maximum (nm)	303	306	307	306	303
Fluorescence maximum (nm)	362	349	339	355	364
$[\Phi(LWE)/\Phi(SWE)]^a$	0.52	1.12	0.93	0.59	1.86
Binding constant $K(M^{-1})$	—	—	20	156	47
Dimethylaminobenzonitrile[19]					
Fluorescence maximum (nm)	553	488	453	520	
Fluorescence lifetime (nsec)	1.8	2.46	3.26	2.43	
Fluorescence quantum yield (relative)	1.0		23.2	3.5	
Binding constant $K(M^{-1})$	—	—	160	440	
Diethylaminobenzonitrile[19]					
Fluorescence maximum (nm)	552	480	513	517	
Fluorescence lifetime (nsec)	0.56	2.88	1.91	1.68	
Fluorescence quantum yield (relative)	1.0		4.9	4.2	
Binding constant $K(M^{-1})$	—	—	575	2500	
2-Naphthol[21]					
$\Phi[\text{Naph}]/\Phi[\text{Naph}^-]^b$	0.57		0.075	0.14	0.53
$\Phi[\text{Naph-CD}]/\Phi[\text{Naph}]$			1.5	1.4	1.0
Binding constant $K(M^{-1})$			79	16	190
6-Indole-11^{+11}					
Fluorescence maximum (nm)	374	353	366	360	352
Fluorescence lifetime (nsec)	15		6–9	6–9	6–9

[a] Ratio of quantum yields for the long- and short-wavelength emission bands.
[b] Ratio of fluorescence quantum yields of neutral and ionized (pH = 12) forms of 2-naphthol.

acetate included in α-, β-, and γ-CDs. The effect of increasing the CD concentration on the fluorescence of naphthalene derivative was negligible for α-CD (no insertion!) and large in the normal monomer fluorescence in the β-CD ([1:1] complex). With the large γ-CD, in addition to the normal intensity enhancements in monomer fluorescence, an additional excimer-like emission around 410 nm (broad and structureless) was observed. Presumably in γ-CDs, 1:2 [CD:probe] complexes are formed. Insertion of pyrene occurs essentially in a similar manner:[25,26] α-CD shows no effects, β-CD shows enhancements in the monomer fluorescence peak ratio (III/I), intensity, and lifetimes. With γ-CDs we observe enhanced monomer fluorescence accompanied by an excimer fluorescence (cf. Fig. 9.2). Interestingly, addition of n-butanol caused an increase in the monomer fluorescence at the expense of excimer fluorescence.[26] Apparently, on n-butanol addition, unstable 1:2 [2 pyrene:1 CD] complexes are replaced by 1:1:1 complexes of [pyrene:CD:BuOH]. Both the peak ratio (III/I) and the lifetime increase on addition of n-butanol in accordance with the above interpretations.

An accurate dynamical description of the relative motion of pyrene molecules that results in excimer formation require time-resolved studies on the growth of the excimer emission and on the changes in the ground-state absorption spectra. It has been found that the initial growth of excimer fluorescence in γ-CD is very rapid (complete within 0.8 nsec laser pulse).[25] Hence, it has been proposed that excimer-like fluorescence originates from excited-state dimers which have already been formed in the ground state. Since the ground state of pyrene excimers is dissociative, it is most likely that the pyrene molecules are located in very close proximity to allow very rapid formation of excimers with minimum relative motion. In a related study, Hamai has examined excimer fluorescence of naphthalene in the presence of β-CD.[18] Based on an analysis of the concentration and temperature dependence of absorption and emission intensities, it has been concluded that species responsible for excimer fluorescence are 2:2 [naphthalene:CD] complexes formed by the association of two 1:1 [naphthalene:CD] complexes.

Intramolecular Excimers

Complementary to the intermolecular case are those involving intra-molecular excimers formed by molecules containing two chromophoric groups covalently linked to each other (as in 1,3-diarylpropane and analogs). There have been few studies with bichromophoric systems such as bis(4-biphenylmethyl)ammonium chloride,[27] 1,3-dinaphthylpropanes,[28,29] and 2,2-bis(1-naphthylmethyl)1,3-dithiane.[30] As shown in Fig. 9.4, α- and β-CDs significantly inhibit intramolecular excimer formation of biphenylmethyl-ammonium chloride while γ-CD enhances it considerably, indicating its ability to induce a conformational change in the probe molecule during the formation of the inclusion complex.[27]

Fluorescence of two 1,3-bis(2-naphthyl)propane, 1-(1-naphthyl)propane and 3-(2-naphthyl)propane (2,2- and 1,2-DNP) in CD solutions has been probed by Itoh and Fujiwara.[29] Intramolecular excimer and ground-state complex formation were found to take place both in β- and γ-CD solutions of 1,2- and 2,2-DNP. The excimer formation is more significant in 1,2- than in 2,2-DNP. The excimer-like emission (maximum at 410 nm) showed a complex decay behavior, approximated to triple exponentials—two short lifetimes of 1–2 nsec and 8–25 nsec and a long one of 165 nsec.

Turro et al.[28] have examined the excimer formation in several 1,3-bichromophoric systems of the type R_1—CH_2—X—CH_2—R_1 [R_1 = 1-naphthyl or phenyl and X = —CH_2, —CO or —CHOH] in cyclodextrin solutions. In all cases where molecular models reveal a good correspondence between the size and shape of the eclipsed conformation of the bichromo-phoric system and the CD, a stable [CD:probe] complex is formed. In these

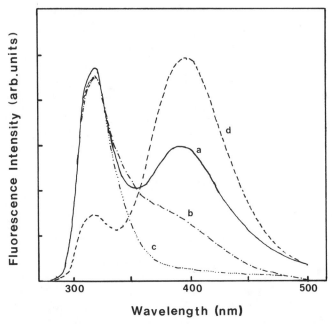

FIG. 9.4. Fluorescence spectra of bis(4-phenylmethyl)ammonium chloride (10^{-4} M) in (a) ethanol and in water containing 5×10^{-3} M of (b) α-CD, (c) β-CD, and (d) γ-CD. (From Emert et al.[27])

cases, the ratio of excimer to monomer emission intensity I_{ex}/I_{mon} is much larger than found in homogeneous solutions. Table 9.5 presents some of their data of relative intensity ratios and excimer and monomer fluorescence lifetimes.

Intermolecular Exciplexes

Quenching of aromatic hydrocarbon singlet excited states by electron donor molecules such as amines in nonpolar solvents often results in the formation of luminescent excited state complexes:

$$^{1}Ar^* + RNH_2 \rightleftharpoons [Ar^- \cdots RNH_2^+] \longrightarrow [Ar \cdots RNH_2]^*$$

$$\downarrow \qquad\qquad \downarrow \text{polar} \qquad\qquad \downarrow \text{nonpolar}$$
$$\qquad\qquad\qquad \text{solvent} \qquad\qquad\quad \text{solvent}$$

$$Ar + h\nu \qquad Ar_s^- + (RNH_2^+)_s \qquad Ar + RNH_2 + h\nu'$$

In polar solvents, the intermediate radical ion pairs dissociate to form free solvated radical ions. Given the nonpolar nature of the cyclodextrin cavity where an excited arene and an amine can find themselves, it is of interest to examine the quenching of arene fluorescence by amines and look for the exciplex formation.

TABLE 9.5. Fluorescence Properties of Excimer-Forming Probes in Cyclodextrin Media [25-30]

Probe[a]	Property	Homogeneous solvent		α-CD[b]	β-CD[b]	γ-CD[b]
		H$_2$O	Hexane			
Intermolecular excimers[25,26]						
Pyrene	τ(mon) (nsec)	200		200	370	250
Pyrene	τ(ex) (nsec)	56		66	—	77
Pyrene	(III/I) ratio	0.65		0.65	1.15	0.90
Pyrene	complex type	—	—	—	1:1	1:2
Naphthalene	τ(mon) (nsec)	40			48	
Naphthalene	τ(ex) (nsec)				68	
1-Methylnaphthalene	τ(mon) (nsec)		77			
2-NMDT	τ(mon) (nsec)		5.51	20		
Intramolecular excimers[27-30]						
2,2-Bis(NMDT)	τ(mon) (nsec)		3.75		4.4,53	
2,2-Bis(DMDT)	τ(ex) (nsec)		69.8			50–69
2,2-DCAN	τ(ex) (nsec)					14,51
1,2-DCAN	τ(ex) (nsec)					10,42
2,2-DNP	τ(ex) (nsec)				1.9,8.7 48.2	1.8,7.8 56.0
1,2-DNP	τ(ex) (nsec)				2.0,15.7 53.7	1.6,25.8 93.8
DNK	I(ex)/I(mon)	0.1	7	0.2	0.7	4
DNL	I(ex)/I(mon)	0.2	1	0.2	0.4	0.5
PNL	I(ex)/I(mon)	0.1	2	0.1	4	4
PNA	I(ex)/I(mon)	0.04	0.08	0.04	0.07	0.5

[a] PNA = 1-phenyl-2-naphthylpropan-2-ol,   PNK = 1-phenyl-2-naphthyl-propan-2-one DNL = 1,3-di(1-naphthyl)propan-2-ol,   DNK = di(1-naphthyl)propan-2-one,   NMDT = naphthylmethyldithiane, and DNP = dinaphthylpropanes.

[b] Data with more than one τ correspond to cases where the decay is multiexponential.

Hamai[18] has examined the fluorescence and its quenching in 2-methoxy-naphthalene by *o*-dicyanobenzene in aqueous β-CD solutions. Insertion of the probe is accompanied by enhanced fluorescence intensities of the probe and quenching in the presence of the donor. The probe fluorescence quenching is accompanied by a broad, structureless exciplex-like emission. Analysis of the absorption and emission intensities has suggested the formation of 1:1:2 [probe:quencher:CD] complexes.

The quenching of fluorescence of pyrene and naphthalene by various aliphatic and aromatic amines in the presence of α-, β-, and γ-CDs has been investigated by Kano *et al.*[31,32] and Kobayashi *et al.*[33] For small aliphatic amines, β- and γ-CDs accelerated the quenching process. The Stern–Volmer plots for the fluorescence lifetimes and intensities indicated that static quenching occurred in these systems, presumably due to the formation of 1:1:1 [probe:quencher:CD] complexes. On the other hand, CDs, especially the β-CD, inhibited the fluorescence quenching of pyrene by N,N'-dimethylaniline suggesting the formation of a tight 1:1 complex [CD:quencher] (or [CD:probe] ?). Surprisingly, no enhanced exciplex emission has been reported for any of the amine quenching cases examined.

Intramolecular Exciplexes

The probe 1-(1-naphthyl)-3-(dimethylamino)propane exhibits only naphthalene monomer fluorescence (emission maxima at 324, 334, and 338 nm) in aqueous solutions. In the presence of β-CD, a new broad emission (maximum at 550 nm) attributable to an intramolecular exciplex has been observed.[34] No such exciplex emission, however, is observable with α- or γ-CD solutions. The former possesses too small a cavity for association with the probe while the latter possesses too large a cavity for the probe to pass through freely without forming a stable complex. Comparison of the exciplex emission maxima and yield with those observed in organic solvents indicates that the probe environment in the cavity is similar to that of ethanol.

9.2.5 Room-Temperature Phosphorescence

One of the favorable outcomes of the organization of reactants at a microscopic level and the protective environment experienced by the probe molecules in aqueous micellar solutions is the ready observation of room-temperature phosphorescence (cf. Chapter 3). The inclusion of guest molecules in the hydrophobic cavity of the cyclodextrins is expected to provide similar effects as has been observed experimentally by Turro, Cline Love, and co-workers. In order to provide the ideal rate and yield parameters for the triplet state, studies of room-temperature phosphorescence (RTP) in aqueous CD media often employ internal and external heavy atom effects.

Intense room-temperature phosphorescence has been reported in aqueous β- and γ-CD solutions from 1-bromonaphthalene,[35] 1-chloronaphthalene,[35] and several 1-bromonaphthoyl and 4-bromonaphthoyl surfactant probe molecules.[36,37] As in the fluorescence, the size and shape selectivity criteria control whether a given guest molecule is included in the CD cavity and exhibits RTP. For naphthalene derivatives, the cavity of α-CD is too small to accommodate them and hence no RPT observed. Halonaphthalenes readily phosphoresce in N_2-purged aqueous β-CD solutions. Addition of acetonitrile was found to increase both the phosphorescence intensity and lifetime. The effect of acetonitrile presumably is similar to that of alcohols on the fluorescence of included arenes mentioned earlier, namely, to provide a more stable 1:1:1 [probe:CD:CH$_3$CN] complex. For 2-bromonaphthalene, however, no acetonitrile effect was observed. Previous work has shown that large differences (due to steric effects) exist in the ability of 1- and 2-substituted naphthalenes.

The phosphorescence of the 4-bromonaphthoyl group is readily quenched by O_2 in homogeneous solvents. However, when this probe is incorporated in the γ-CD cavity, RTP was observable even in aerated solutions.[37] Phosphorescence lifetime measurements indicate two types of [probe:CD] complexes with lifetimes of 0.6 and 3.5 msec, respectively. Oxygen completely quenched the fast decay but only partially quenched the slow decay. With β-CDs, 1:1 complexation does occur with concomitant RTP, but O_2 quenches the phosphorescence.

Nonhalogenated polynuclear aromatic hydrocarbons generally show only enhanced fluorescence on inclusion in CDs. With these, Cline Love et al. have successively applied external heavy-atom effects—addition of a third non-ionic haloalkane 1,2-dibromoethane (DBE), to the CD solutions.[38-40] Due to their hydrophobic nature, DBE readily incorporates into the cavity along with the arene to form a 1:1:1 complex and provides the necessary external heavy-atom effect. Concentration dependence studies with DBE do show that phosphorescence intensities increase and level off at 1:1 [CD:DBE] concentrations. Ionic salts such as NaBr, KBr, KCl, AgNO$_3$, or TlNO$_3$ were found ineffective in inducing RTP. The list of compounds that exhibit RTP in this manner is quite long: polynuclear aromatic compounds such as naphthalene, pyrene, phenanthrene, and chrysene; nitrogen heterocyclics such as quinoline, isoquinoline, phenanthridine, acridine, and phenazine (Fig. 9.5); and bridge compounds such as fluorene, carbazole, dibenzofuran, and dibenzothiaphene. As with bromonaphthalenes, acenaphthene included in β-CDs along with DBE phosphoresces at room temperature, even in aerated solutions—about 31% of CD–RTP remains even after aeration!

Shape selective inclusion of molecules, as monitored by RTP can have practical applications in analytical chemistry. Unsubstituted molecules and those containing bulky substituent groups can be selectively discriminated based on their emission behavior in CD–DBE solutions. An example is provided by β-CD–DBE solutions containing both naphthalene and 1-phenylnaphthalene. The former exhibits only RTP while the latter only molecular fluorescence and no RTP. (The latter is too bulky to be included into the CD cavity.) Comparison of the emission in different CDs, for the same reasoning, is equally effective. For example phenanthrene exhibits only fluorescence in α-CD but phosphoresces intensely in β-CD.

The inclusion process in the CD cavity is a dynamical process, similar to the solubilization process in micellar solutions. Thermodynamic equilibrium as represented by the binding constant gives the static picture of the inclusion process. We need to know the rate constants for the guest entry and exit in and out of the cavity for a kinetic description. Earlier discussions on micellar

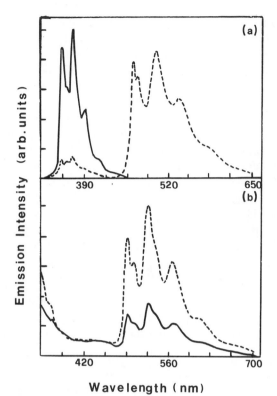

FIG. 9.5. (a) Room-Temperature phosphorescence spectra of phenanthrene in deaerated β-CD (10^{-2} M) solutions in the presence (– – –) and in the absence (——) of 0.58 M 1,2-dibromoethane: $\lambda_{\text{excit}} = 300$ nm. (b) Room-temperature phosphorescence spectra of acenaphthene in β-CD (10^{-2} M) solutions with 0.58 M 1,2-dibromoethane in the presence (——) and in the absence (– – –) of dissolved air: $\lambda_{\text{excit}} = 300$ nm. (From Scypinski and Cline Love.[12])

solubilization (Chapter 3) showed how we can utilize the long-lived phosphorescence of guest molecules to derive such kinetic data. Turro *et al.* have made detailed kinetic studies and obtained rate constants data for the inclusional process of guest molecules in cyclodextrins:[36]

$$P + CD \underset{k_-}{\overset{k_+}{\rightleftharpoons}} [P \cdot CD], \qquad K = k_+/k_- \qquad (9.1)$$

The scheme below outlines the principles of these experiments. It is based on the quenching of excited probe molecules when they are in the aqueous phase in competition with their re-entry into the CD cavity:

$$P^* + CD \underset{k_-}{\overset{k_+}{\rightleftharpoons}} [P^* \cdot CD]$$

$$\tau_{P^*} \Big\downarrow\Big\downarrow k_q[Q] \qquad \tau_{P'^*} \Big\downarrow\Big\downarrow k_q'[Q]$$

$$P \qquad\qquad [P \cdot CD]$$

When the probe resides in the cavity, it is protected against the quenchers that are restricted to the aqueous phase, as exemplified by the very long phosphorescence lifetimes (~ 3 msec). Note that the guest exclusion (or exit) is a unimolecular process (units: $\sec^{-1}$) while the inclusion (or entry) step is bimolecular, depending on the concentration of the CD molecules (units: $M^{-1} \sec^{-1}$). Available rate data for various probe molecules are collected in Table 9.6. For a series of surfactant probes, n-(4-bromonaphthoyl)alkyltrimethylammonium bromides, BNK-n^+ ($n = 1$, 5, and 10), depending on the temperature, probe, and type of CD employed, the binding constant values vary between 440 and 30000 M^{-1}. For γ-CDs, the strength of complexation increased in the order BNK-1^+ < BNK-5^+ < BNK-10^+. Importantly, all k_+ values obtained were of the order of $10^7 \ M^{-1} \ \sec^{-1}$, a value below the diffusion-controlled limit and rather insensitive to the system involved. The k_- values, on the other hand, were sensitive to the surfactant probe and the CD structure and were found to vary between 270 and 50,000 $\sec^{-1}$.

9.2.6 Energy Transfer Processes

Inclusion of more than one guest molecule in the cyclodextrin cavity at close proximity also raises significantly the probability of observing efficient transfer of excitation energy from one excited guest molecule to another:

$$[P_1^* \cdot P_2 \cdot CD] \longrightarrow P_1 \cdot P_2^* \cdot CD \tag{9.2}$$

Tabushi *et al.* have demonstrated that selective inclusion of acceptor molecules into β-CDs capped with benzophenone moieties (I) produced specific and efficient energy transfer (triplet–triplet) between the derivatives

TABLE 9.6. Rate Constants for Entry and Exit of Guest Molecules Into Cyclodextrin Cavities at 25°C.[a]

Probe[b]	Host medium	k_+ ($M^{-1} \sec^{-1}$)	k_- ($\sec^{-1}$)	K_D (M^{-1})
1-Bromonaphthalene	β-CD, 10% CH$_3$CN	8×10^7	1×10^4	1000
1-Bromonaphthalene	β-CD, 2% CH$_3$CN	2×10^7	2×10^4	1000
1-Chloronaphthalene	β-CD, 2% CH$_3$CN	1×10^7	1×10^4	1000
BNK-1^+	β-CD	3.5×10^7	5×10^4	700
BNK-5^+	β-CD	2.2×10^7	4.2×10^4	520
BNK-10^+	β-CD	2.4×10^7	4.0×10^4	590
BNK-1^+	γ-CD	1.5×10^7	4.2×10^4	360
BNK-5^+	γ-CD	1.9×10^7	4.2×10^4	440
BNK-10^+	γ-CD	1.5×10^7	1.2×10^4	1300

[a] From Turro *et al.*[36]
[b] BNK-n^+ = n-(4-bromonaphthoyl)alkyltrimethylammonium bromide

of host and the included species (1-naphthylbromide or 1-naphthylethyl-bromide):[41]

$$^3PhCOPh^* + Naph \cdot Br \quad \longleftarrow \quad ^3Naph \cdot Br^* + PhCOPh \tag{9.3}$$

No energy transfer was observable in homogeneous solvents with the open chain analog $CH_3 \cdot COO \cdot C_6H_4 \cdot CO \cdot COO \cdot CH_3$ and the acceptor molecules:

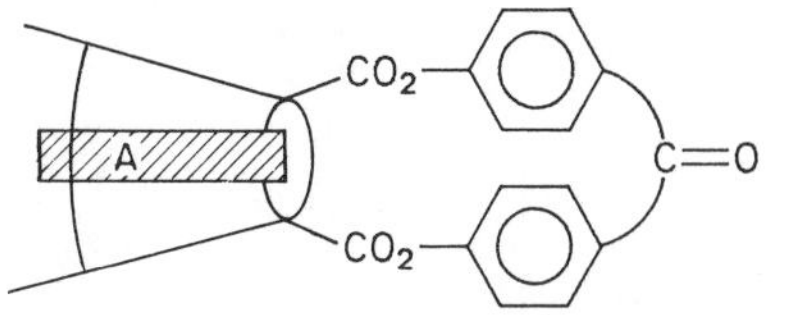

D = benzophenone

A = 1-naphthylbromide

Sensitized room-temperature phosphorescence of biacetyl upon co-inclusion with polynuclear aromatic hydrocarbons (phenanthrene, biphenyl, fluorene, etc.) in the cavity of β-CD has also been observed.[42] Though the efficiency of such energy transfer was better than that observed in homogeneous solvents, utilization of aqueous micelles to achieve molecular organization at microscopic level for these processes was much superior.

9.2.7 Organic Photochemical Reactions

Stereoselective inclusion of organic molecules in the CD cavity often leads to dramatic differences in the photochemical reactivity and selectivity in the product distribution. Organic photochemical reactions in cyclodextrin media outlined below are reminiscent of similar behavior in micellar supercages.

Photochemical Fries Rearrangement

Restricted mobility of photogenerated radical intermediates in photochemical Fries rearrangements of anilides and aryl esters [reactions (9.4) and (9.5)] in the CD cavity leads to three principal effects:[43,44] (a) increased percentage conversions; (b) migration to the para position is preferred over ortho migration, and (c) decreased yields of elimination products such as phenol or aniline. Table 9.7 contains data that illustrates some of these effects:

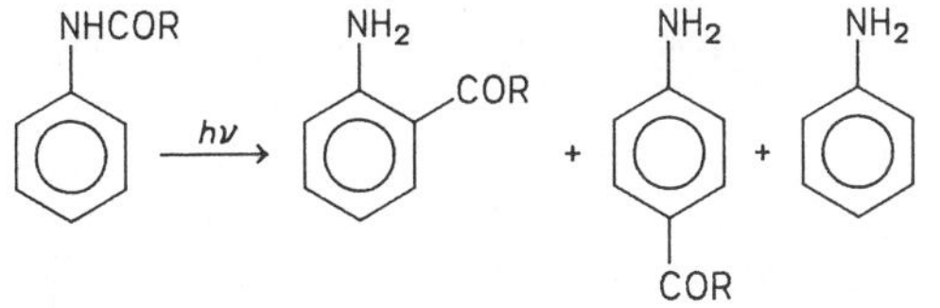

$$\tag{9.4}$$

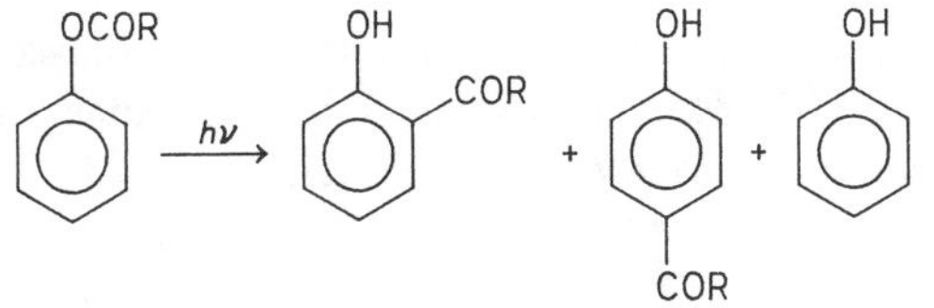

$$\tag{9.5}$$

TABLE 9.7. Influence of Cyclodextrins on the Relative Yields of Products in Photochemical
Fries Rearrangements [a]

		Relative yields (%)			(p/o)
Substrate/medium[b]	% Conversion	Ortho	Para	Others[c]	ratio
1, Homogeneous solvents	12	25.7	25.7	48.6	1
1, Methyl-α-glucopyranoside	25.2	32.4	32.4	35.4	1
1, β-CD	41	11.2	69.0	19.8	6.2
2, 2-Propanol	10	35.5	31.6	32.9	1.1
2, Water	18	74.3	19.4	6.3	3.8
2, β-CD	43	64.8	34.7	0.5	1.9
3, Ethanol	3.7	52.6	47.1	0.3	1.1
3, Water	1.3	79.8	20.2	—	4.0
3, β-CD	2.0	58.1	41.8		1.4
4, Ethanol	9.8	21.4	45.0	32.7	2.2
4, Water	15	15.0	71.9	13.1	4.8
4, β-CD	22	6.8	90.0	2.3	13.1

[a] From Ohara and Watanabe[43] and Chenevert and Plante.[44]

[b] 1 = phenylacetate, 2 = acetanilide, 3 = benzanilide and 4 = ethylphenylcarbonate.

[c] Other products for (1) phenol, for (2) aniline, for (3) benzoic acid, and for (4) phenol.

Photodimerization of Water-Soluble Anthracenes

Anthracene-1- or anthracene-2- sulfonic acid or carboxylic acids were found to form 1:2 [host:guest] complexes on inclusion in γ-CDs.[45] The complex formation greatly enhanced the photodimerization of these anthracene derivatives. The dimerization quantum yield is approximately 0.05 in water (a value comparable to that for anthracene in benzene) but in the presence of γ-CD, the quantum yield increases to 0.47 ± 0.05. The quantum yield for the reverse reaction, however, was unaffected by the presence of cyclodextrins. The α-CD showed no intermolecular interactions with the two anthracene derivatives; and β-CD showed a behavior different from that of γ-CD, namely, it formed a 1:1 complex (?) and led to enhanced photo-dimerization (not quantitative) competing with photooxidation reactions.

Cis–Trans Photoisomerization of Surfactant Stilbenes

Trans-stilbene derivative, S_6, forms 1:1 complex with carboxycyclo-amylose:[46]

The inclusion is accompanied by enhancements in fluorescence yields and a retardation in the photoisomerization. (Quantum yields are about 0.22 or about 44% of that observed in homogeneous $DMSO–H_2O$ mixtures.) Ueno *et al.*[47] have described an interesting application of the relatively stronger binding of a trans azo dye to β-CD (compared to the cis dye) to photoregulate the hydrolysis of *p*-nitrophenylacetate (NPA) catalysed by β-CD. Photoisomerization of *trans-p*-phenylazobenzoate (PAB) inserted into β-CD leads to conversion of the azo dye to the cis form and subsequent release from the CD and an inhibition of the catalytic activity:

NPA   PAB

Type II Photoelimination Reactions of Ketones

Inclusion of surfactant ketones such as $Ph—CO—(CH_2)_n—COOH$ ($n = 3$ and 10) in carboxymethylamylose affects the photoreactivity of these ketones.[48] The binding constants increase with an increase with chain length. Type II photoelimination processes which occur with unit quantum efficiency in water are retarded significantly in CD media ($\Phi_{II} = 0.75$ and 0.36 for surfactant ketones with $n = 3$ and 10 at [amylose] $= 1.81 \times 10^{-4}$ M).

9.3 Photoprocesses in Aluminosilicates—Zeolites

9.3.1 General Features of Zeolites and Their Cavities[49–52]

Silicates provide an interesting array of structural forms due to their heterocatenation property—a process where the silicon atoms share none, one, or more of the four coordinated oxygen atoms of the SiO_4 tetrahedron with adjacent silicon atoms. Based on the extent of sharing, silicates are classified broadly into four structural types: discrete anions (Si sharing none or at most one oxygen as in ortho, pryo, and meta silicates), infinite chains (Si sharing two oxygen atoms with adjacent Si as in pyroxenes and amphiboles), layers or sheets (Si sharing three of four oxygens as in clays such as kaolin and talc), and finally framework structures (total sharing of all of the four oxygen atoms with adjacent Si, as in feldspar, zeolites, and ultramarines). Aluminosilicates are compounds derived from silicates by isomorphous substitution of Si^{4+} by Al^{3+}. Zeolites are a special class of crystalline aluminosilicates with a well-defined framework structure within which there are cavities (pores or channels) of varying dimensions. Electroneutrality of structures is achieved by the presence of additional cations outside the framework structure.

Figure 9.6 illustrates schematically the structural framework of various forms of zeolites. In zeolites, the fundamental building blocks are SiO_4 and AlO_4 tetrahedra. These tetrahedra are arranged in such a way that each of the four oxygen atoms is shared with another silica or alumina tetrahedron to

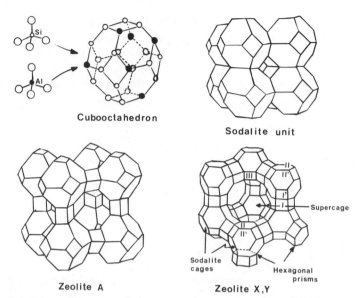

FIG. 9.6. Schematic representation of the structural features of zeolites A, X, Y and their constituent units (cubo-octahedron and sodalite unit). (From J. H. C. Van Hooff.[51a])

form a three dimensional crystal lattice. In Y- and X-type zeolites, 24 SiO_4 and/or AlO_4 tetrahedra join together to form a secondary building unit in the form of a cubo-octahedron, the so-called sodalite unit. In the sodalite unit, we can distinguish six four-membered rings (the faces of a cube) and eight six-membered rings (the faces of an octahedron).

Couping of the cubo-octahedra to form crystalline structures can occur in different ways: (1) when the cubo-octahedra are stacked so that each four membered ring is shared by two units, sodalite is formed; (2) when the cubo-octahedra are connected by bridge oxygen atoms between the four-membered rings zeolite A is formed; and (3) the cubo-octahedra are connected by bridge oxygen atoms between the six-membered rings. Based on the different ratios of (Si/Al), synthetic zeolites are classified as A, X, Y, etc. As can be seen from the figure, zeolites are a class of open structures with relatively large openings of different dimensions. Table 9.8 presents structural data on some of the commonly used zeolites.

The existence of finite size pores (or channels) confers in zeolites an ability to selectively adsorb molecules of a certain shape and size. The cavities can be occupied by large ions and water molecules, both of which have considerable freedom of movement. Thus, the distinguishing properties of zeolites are their ion-exchange capacity and reversible hydration/dehydration. In the absence of adsorbed water molecules in the cavity, zeolites behave as molecular

TABLE 9.8. Characteristic Parameters of Zeolites and Their Cavities

Zeolite	Channel/cavity dimensions	Nature of the channel/cavity	Si/Al ratio	Kinetic parameter[b]
Na^+-A	Main cavity = 11 Å channel opening = 4 Å	3-Dimensional channels with cavities	0.7–1.2	4
Na^+-X	Main cavity = 13 Å channel opening = 8 Å	3-Dimensional channels with cavities	1.0–1.5	8
Na^+-Y	Main cavity = 13 Å channel opening = 8 Å	3-Dimensional channels with cavities	1.5–3.0	8
Na^+-Mordenite	Main channels = 7 Å side channels = 3 × 6 Å	2-Dimensional channels with pertinent 1-D cavities	4.5–5.0	7
Silicalite	Main channels = 5.2 × 5.7 Å side channels 5.4 Å	3-Dimensional channels with no cavities	99% Si	6

[a] From Turro and Wan.[74]

[b] Kinetic parameter represents approximately the largest sized molecule (in angstroms) that can be accommodated in the voids of the zeolite.

sieves. Various uncharged molecules such as CO_2, NH_3, and organic compounds can be selectively adsorbed in the cavities. The efficiency of processes is controlled by the pore size and by the type of pores, that is, some pores are three dimensional. Others are more restrictive and are of lower dimensionality.

9.3.2 Photochemistry of Inorganic Ions Exchanged into Zeolites

Early photochemical studies on zeolites have been concerned with the luminescence of simple inorganic ions such as Fe^{3+}, Mn^{2+}, Eu^{3+}, Cu^{2+}, and Ag^+ ions on host lattices.[53–56] Rare earth ion Eu^{3+}-exchanged zeolites in particular have been studied by several groups. Arakawa et al. have reported on the luminescence properties in Y-zeolites measured under a variety of conditions. Thermal treatment of Eu^{3+}-zeolites leads to a reduction ($Eu^{3+} \rightarrow Eu^{2+}$) as indicated by the presence of Eu^{2+} emission bands. The order of susceptibility to reduction under evacuation at 600°C was Eu-M > Eu-Y > Eu-13X $\simeq$ Eu-A. Adsorption of water by an activated Eu-Y or Eu-M zeolite leads to evolution of H_2 gas:

$$Eu^{3+}\text{-Y} \xrightarrow[-H_2O]{\Delta} Eu^{2+}\text{-Y} \xrightarrow[+H_2O]{\Delta} Eu^{3+}\text{-Y} + H_2 + OH^- \qquad (9.6)$$

The order of the quenching process has been found to be $H_2O > D_2O > EtOH$. This reversible thermal reduction/oxidation of Eu-exchanged zeolites has led to some interest in their potential use in thermal water-splitting cycles.

Ag^+- and Ti^{3+}-exchanged zeolites have been found to be sensitive to visible light and have been adapted for photochemical and thermal water-splitting cycles.[57-59] Ag^+-zeolites release molecular oxygen (O_2) from adsorbed water on visible light photolysis along with the formation of molecularly dispersed Ag^0:

$$Ag^+\text{-}Y \xrightarrow[+H_2O]{h\nu} Ag^0\text{-}Y + H^+ + O_2 \tag{9.7}$$

The reduced zeolites are capable of thermal reduction of water to H_2 and oxidation of Ag^0 at temperatures $\geq 600°C$:

$$Ag^0\text{-}Y \xrightarrow[+H_2O]{\Delta} Ag^+\text{-}Y + H_2 + OH^- \tag{9.8}$$

The zeolite system looses the reversibility after several cycles, presumably due to the sintering of Ag or due to are loss of OH groups during the thermal treatment. Ti^{3+}-exchanged zeolites evolve H_2 upon photolysis:

$$Ti^{3+}\text{-}Y \xrightarrow[+H_2O]{h\nu} Ti^{4+}\text{-}Y + H_2 + OH^- \tag{9.9}$$

However, unlike the Ag-exchanged zeolites, thermolysis of oxygenated zeolite does not evolve O_2 in any significant amounts.

Stucky et al. have studied the phosphorescence lifetimes of several Eu-exchanged zeolites and have shown that the Eu^{3+} decay constant is linearly related to the water molecules surrounding the Eu ions in the supercages of zeolites A, X, Y, and ZSM-5.[60,61] Strome, Klier, and co-workers have utilized the luminescence in Cu-exchanged zeolites to monitor the location of cation sites in zeolites and their migration during the adsorption of gases such as CO.[62-65] Thermal activation leads to green luminescence ($\lambda_{max} = 540$ nm) attributable to monovalent copper:

$$Cu^{2+}\text{-}Y \xrightarrow[H_2O]{\Delta} Cu^+\text{-}Y \tag{9.10}$$

In the presence of CO at a pressure of 740 torr, the original emission at 540 nm disappears and one at 505 nm appears. After the CO has been pumped off, new emission maxima at 470 and 560 nm are observed. It has been proposed that different maxima correspond to the location of Cu in different sites in the zeolite cavities.

Uranyl ion (UO_2^{2+})-exchanged zeolites have also received some scrutiny for potential applications in solar energy conversion.[66,67] The luminescence of UO_2^{2+} in amorphous zeolite A and untreated zeolite ZSM-5 is very broad as in aqueous solutions, suggesting the lack of incorporation of the ions in zeolite. In thermally treated ZSM-5, Y-, X-zeolite, and Mordenite, fine-structure emission of UO_2^{2+} (as in the solid state) are observed. Lifetime studies indicate two different lifetimes (one in the range of 20–50-μsec and the other 80–300-μsec range). The lifetimes vary from zeolite to zeolite and also after dehydration. Efficient energy transfer between UO_2^{2+} and Eu^{3+} has also been observed (Fig. 9.7). The efficiency depended on (1) the nature of pore–channel system in the zeolite (A > Y > ZSM-5 > Mordenite) and (2) the mode of incorporation of the two ions (simultaneous > sequential).[67a]

In related studies isopropanol has been found to quench the luminescence of incorporated UO_2^{2+} ions. Stern–Volmer analysis indicated a dual quenching pattern corresponding to quenching at different sites in the zeolite[67] (external and internal regions). Bulk photolysis in isopropanol lead to the formation of molecular H_2 and acetone after an induction period of approximately 10 min. The photoassisted catalytic oxidation of isopropanol to acetone and H_2 is sustained for periods of over 300 hours!

$$(CH_3)_2CHOH \xrightarrow[UO_2^+\text{-zeolite}]{hv} (CH_3)_2CO + H_2 \tag{9.11}$$

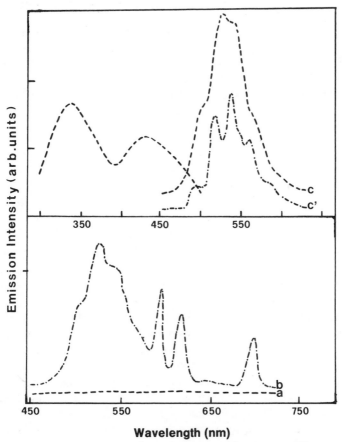

FIG. 9.7.   Emission Spectra of (a) Eu^{3+}, (b) Eu^{3+} and UO_{2+}^2, and (c) UO_2^{2+} ions exchanged into Y-zeolites. (From Suib et al.[66,67] Copyright 1984 American Chemical Society.)

9.3.3   Photochemistry of Inorganic Complexes

There have been a few studies on the photochemistry and photoreactivity of transition metal complexes in a bound state on zeolites. $Ru(bpy)_3^{2+}$ complex has been prepared in the large cavities of Y-zeolites by allowing 2,2′-bipyridine to react within a $Ru(NH_3)_6$-Y zeolite.[68] The complex is characterized by absorption and emission bands similar to that observed in homogeneous solutions. With increasing concentrations of the Ru complex, self-quenching of the luminescence is observed, due to the high local concentrations of the complex within the zeolite cavities. Dexter-type resonance energy transfer has been suggested as a possibility. Molecular oxygen is an efficient quencher of the excited state of $Ru(bpy)_3^{2+}$ in Y-zeolite. The effect of the adsorbed water is a complex one, in that the luminescence is both promoted and quenched, depending on the concentrations of the $Ru(bpy)_3^{2+}$ and the amount of water present in the lattice.

The luminescence of $Ru(bpy)_3^{2+}$ and its redox quenching by electron donors such as TMPD ($N,N,N′,N′$-tetramethyl-p-phenylenediamine) and 10-phenylphenothiazine in X-zeolites has also been examined.[69] As with the Eu-exchanged zeolites, at least two different modes of quenching were deduced from lifetime measurements. The formation of electron transfer prodducts has been identified demonstrating that zeolites are capable of separating the redox products of light-induced electron transfer reactions:

$$Ru(bpy)_3^{2+} + D \xrightleftharpoons{h\nu} Ru(bpy)_3^{+} + D^{+} \qquad (9.12)$$

The photoaquation of the rhodium(III) complex, $Rh(NH_3)_5I^{2+}$, in fully and partially hydrated Y-zeolites has been found to be inhibited.[70] The aquation quantum yield drops from 1.06 in water to 0.18 and 0.13 in hydrated and partially hydrated zeolites, respectively.

9.3.4   Photochemistry of Organic Molecules

Pyrene-3-carboxaldehyde is a fluorescence probe molecule whose emission properties are highly sensitive to the environmental polarity and, hence, have been used to obtain estimates of polarity at the micelle–water interfaces (cf. Chapter 2). On adsorption in Y-zeolites, with increasing surface coverage the probe shows a growth in the excimer emission (maximum at 560 nm).[71] As with other probe molecules, inclusion in zeolites leads to locally high concentrations as has been observed in γ-CD cavities (multiple occupancies). The fluorescence maxima for both monomer and excimer indicated exposure of the probe to polar environments. On protiated zeolites (H^+-Y), the absorption spectrum shows a large bathochromic shift ($\lambda_{max} \simeq 420$ nm) indicating oxidation of the pyrene chromophore. The emission spectrum also shows different maxima arising from the pyrene radical cations.

The dynamics of intermolecular excimer formation with pyrene in the supercages of zeolites has been the subject of a recent investigation.[72] Evidence has been presented for the ground-state association of pyrene, deposition of microcrystallites on external zeolite surfaces, and formation of monomers and excimers in the large supercages of large-pore-sized zeolites. The preparation of materials controls what types of zeolites are dehydrated first. Presence of organic solvents of differing polarity also affects the monomer fluorescence. Halonaphthalenes and Cs^+ ions can be used to dissociate the ground-state associated species. Luminescence decays for monomers and excimers were found to exhibit double exponential behavior in most of these systems. Table 9.9 presents some of the data on the pyrene monomer vibronic intensity ratio (III/I), monomer and excimer emission intensity ratios I_{ex}/I_{mon}, and fluorescence lifetimes for these species.

TABLE 9.9. Features of Pyrene Monomer and Excimer Fluorescence in Zeolites [a]

Zeolite	Conditions[b]	Intensity ratio (I_{ex}/I_{mon})	Peak ratio (III/I)	Monomer		Excimer	
				τ_1 (nsec)	τ_2 (nsec)	τ_1 (nsec)	τ_2 (nsec)
Na^+-X	H,R	2.32	1.65	10.6 (4%)	68 (96%)	17.7 (8%)	90.5 (92%)
Na^+-X	D,R	0.25	1.36	—	140 (100%)	*	*
Na^+-X	H,S	3.19	1.20	*	*	2.8 (2%)	50.6 (98%)
Na^+-X	D,S	0.06	1.17	12.5 (1%)	133 (99%)	*	*
Na^+-Y	H,R	0.55	1.61	8.9 (9%)	64.5 (91%)	3.9 (1%)	53.6 (99%)
Na^+-Y	D,R	1.31	1.32	6.3 (18%)	44.6 (82%)	5.8 (2%)	54.3 (98%)
Na^+-Y	H,S	0.28	1.51	7.6 (3%)	81.9 (97%)	4.8 (3%)	61.7 (97%)
Na^+-Y	D,S	0.38	1.30	8.8 (6%)	57.0 (94%)	4.9 (1%)	57.0 (99%)
Na^+-A	H,R	5.75	3.33	4.4 (6%)	43.9 (94%)	—	52.2 (100%)
Na^+-A	D,R	5.88	2.43	*	*	—	50.8 (100%)
Na^+-A	H,S	7.55	1.83	*	*	—	50.7 (100%)
Na^+-A	D,S	6.17	2.09	*	*	—	56.4 (100%)

[a] From Suib and Kostapapas.[72]

[b] Conditions: H = hydrated, R = rotary evaporated, D = dehydrated, and S = stirred; * indicates cases where the emission is too weak to detect.

The protective environment inside the cavities also significantly influences the triplet lifetimes of guest molecules. For example, it has been found that the triplet lifetime of β-phenylpropionone at room temperature is enhanced by approximately five orders of magnitude on inclusion in the channels of the hydrophobic zeolite silicalite[73] ($\tau_T = 0.18$ msec as compared to 1 nsec in toluene at room temperature). Silicalite is a type of framework structure zeolite mentioned earlier with a different type of channel system. It has near circular zig-zag channels, cross-linked by elliptical straight channels. The same probe β-phenylpropionone, however, did not show any phosphorescence in the large-pore zeolite Mordenite nor on an activated silica gel. The results illustrate the importance of the critical matching of the channel dimensions in the supporting matrix with the probe size and shape.

Earlier discussions on the photoreactions of solutes in organized molecular assemblies such as micelles showed that solubilization in the micellar cage even for very short periods of time does influence largely the nature and stereochemical course of organic photochemical reactions (cf. Chapter 4). Dibenzylketone photolysis was one such case that illustrated such micellar cage effects. Inclusion of molecules in the hydrophobic cavities of zeolites can lead to similar effects. Dramatic differences in the product distribution occur in the photolysis of 3-(4-methylphenyl)-1-phenyl-2-propanone (4-MeDBK) in several zeolites.[74] The percentage cage effect observed at 2% coverage follows the order sodalite > Na^+-X > Na^+-Mordenite > Na^+-Y > Na^+-A:

$$CH_3-\!\!\left\langle\bigcirc\right\rangle\!\!-CH_2-CO-CH_2-\!\!\left\langle\bigcirc\right\rangle\xrightarrow[-CO]{h\nu}$$

(A–X–B)

$$\left[CH_3-\!\!\left\langle\bigcirc\right\rangle\!\!-CH_2\cdot \quad \cdot CH_2-\!\!\left\langle\bigcirc\right\rangle\right] \qquad (9.13)$$

[A.   .B]

$$\longrightarrow \quad AA + AB + BB \qquad (9.14)$$

$$\text{percentage cage effect} = \left[\frac{AB - (AA + BB)}{AB + (AA + BB)}\right] \qquad (9.15)$$

The order merely reflects the relative degree of mobility of the photogenerated benzyl radicals in the zeolite. Benzyl radicals are least able to escape the geminate radical pair recombination in silicalite, while on Na^+-Y they escape completely to give hardly any cage effect as in homogeneous solutions. The order of ranking for the ^{13}C-isotope enrichment and percentage yields in the isomeric product 1-(4-methylphenyl)acetophenone was the same as that observed for the cage effect.

9.4   Photoprocesses in Crown Ether and Cryptate Complexes

9.4.1   General Features of Crown Ethers, Cryptands, and Their Complexes[75–84]

A third class of compounds that associate metal ions in their cavities (alkali metal ions in particular) are crown ethers and cryptands. Due to their large size and low charge density, alkali and alkaline earth metal ions generally do not form complexes. Until recently the only known complexes of these ions were those with β-diketones where the bonding is largely ionic. A major breakthrough has been the synthesis of several macrocyclic polyethers and their nitrogen–oxygen derivatives (called *crown ethers* and *cryptands*, respectively). Figure 9.8 presents some of the common crown ether and cryptand molecules. (Metal ion complexes of cryptands are called cryptates.)

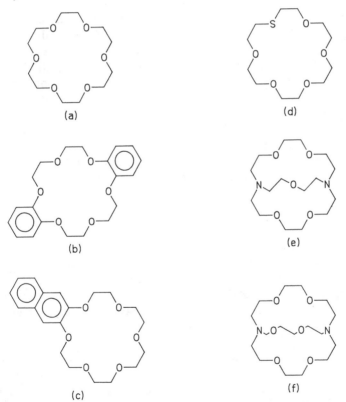

FIG. 9.8. Some common crown ethers, coronand, and cryptand molecules. (a) 18-crown-6, (b) dibenzo-18-crown-6, (c) 2,3-naphtho-21-crown-7, (d) 1-thio,4,7,10,13,16-pentaoxycyclo-octadecane (a coronand), (e) [2,2,1] cryptand, and (f) [2,2,2] cryptand.

Depending on the size of the ring opening, a crown ether composed of four oxygen atoms (herein abbreviated as crown-4) is selective for Li^+ (size of cavity, $Li^+ = 1.20$ Å), those with five oxygens for Na^+ (cavity 1.72–1.84 Å versus $Na^+ = 1.90$ Å) and crown-6 selective for K^+ (cavity 2.67–2.86 Å versus $K^+ = 2.66$ Å) crown-7 to Cs^+ (cavity = 3.4–3.7 Å versus $Cs^+ = 3.38$ Å) and so on. The name crown comes from the puckered arrangement of the carbon–oxygen organic linkages with the oxygen atoms arranged in a nearly planar fashion about the metal ion at the center of the ring. Special interest in these systems in on their ability to complex and transport alkali metal ions in an organic–hydrophobic environment, a process reminiscent of the ion transport in biological systems. Generally anions that coordinate to the metal ions forming ion pairs, tend to pull the latter out from the cavities.

Macropolycyclic ligands such as cryptands contain intramolecular cavities of the three-dimensional type (crypts). The three dimensional encapsulation of alkali and alkaline earth ions increases considerably the stability of the complexes. The K^+ complex of [2,2,2] cryptand, for example, is more stable than the corresponding diaza-18-crown-6 complex by a factor of 10^5. The nomenclature of cryptands uses a simple procedure of counting the number of bridges and the oxygen donor atoms in each bridge. Thus [1,1,1] cryptand refers to a structure with three bridges each equipped with an oxygen atom respectively. More general names of coronands have also been suggested for monocyclic crown ether type compounds containing any heteroatom. Our interests in this section are to examine how the complexation of metal ions with the crown ethers and cryptands influences the inherent photophysics and photochemistry of these ions and their ability to participate as redox quenchers in various photoredox reactions involving inorganic and organic dye molecules.

9.4.2 Luminescence of Metal Ions in Crown Ethers and Cryptates

Eu^{2+} ions generally do not luminesce in aqueous solutions at room temperature due to the rapid deactivation of the excited states by fast radiationless processes and by efficient photoreduction of water to yield Eu^{3+} and H_2. Balzani and co-workers have shown that the photophysical properties of Eu^{2+} ions are altered dramatically on their encapsulation in the cavities of cryptands such as [2.2.1] and [2.2.2]. As shown in Fig. 9.9 the cryptate complexes $[Eu^{2+} \subset 2.2.1]$ and $[Eu^{2+} \subset 2.2.2]$ both exhibit very strong luminescence at 77 K and even a weak emission in aqueous solutions at room temperature.[85] Table 9.10 summarizes some of the data on the absorption and emission properties of Eu^{2+} ions and their cryptate complexes in aqueous solution. The luminescence quantum yield of the ion is much higher when it is in the complex. Obviously, encapsulation partially shields Eu^{2+} from deactivation processes involving water molecules. The emission decay

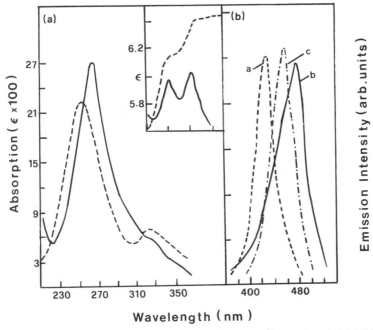

FIG. 9.9. (a) Absorption spectra of Eu^{2+} [2,2,2] (——) and Eu^{2+} ions(– – –) in 0.1 M NaClO$_4$ aqueous solutions. (b) Emission spectra of Eu^{2+} [2,2,2] in 0.1 M NaClO$_4$ aqueous solutions at room temperature (curve b) and at 77 K (curve a) and Eu^{2+} ions at 77 K (curve c). (From Sabbatini et al.[85] Copyright 1984 American Chemical Society.)

TABLE 9.10. Absorption and Emission Properties of Eu^{2+} Ion and its Cryptate Complexes in Aqueous Solution.[a]

Property	Temperature (K)	Eu^{2+}_{aq}	$[Eu^{2+} \subset 2.2.1]$	$[Eu^{2+} \subset 2.2.2]$
Absorption				
Maxima (cm^{-1}) and	293	40000(2230)	37680(1830)	39220(2700)
ε (mMole cm^{-1})		31250(650)	30300(650)	31450(600)
Emission				
Maxima (cm^{-1})	293		22220	21740
	77	22220	22730	23810
Lifetime (nsec)	293		1.5(2.1,7.0)[b,c]	3.0(14,41)[b,c]
	77	200(850)[b]	610(790)[b]	550(850)[b]
Quantum yield	293			10^{-3}
	77	0.13	1.0	1.0

[a] From Sabbatini et al.[85]
[b] Data in methanol : water mixtures (6 : 4 v/v).
[c] Double exponential fit values.

in aqueous solution is strictly first order but is more complex in MeOH–
H$_2$O mixtures.

9.4.3 Photoprocesses Involving Crown Ether and Cryptate Complexes

Encapsulation of the quencher metal ions in the crown ether or cryptate
voids in an elegant way of fixing one of the reactants at a predetermined
distance and orientation, especially in systems where the crown ether
framework is part of a larger molecule containing a chromophore. Studies of
such intramolecular effects on the photophysical and redox properties of the
adjacent chromophore are of increasing interest.

Sousa and Larson[86] have examined the effect of alkali metal cations (Na$^+$,
K$^+$, Rb$^+$, and Cs$^+$) on naphthalene emission in 2,3-naphtho-20-crown-6 and
in 1,8-naphtho-21-crown-6. Complexation of alkali metal chloride salts with
the former in 95% ethanol glass at 77 K causes a decrease in the fluorescence
quantum yield, an increase in phosphorescence yield, and a slight decrease in
phosphorescence lifetime (Fig. 9.10). However, complexation by the latter with
K, Rb, and CsCl salts causes a perceptible increase in ϕ_{fl}, a decrease in ϕ_{ph},
and a substantial decrease in τ_{phos}.

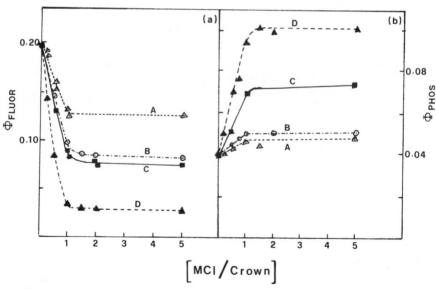

FIG. 9.10. Influence of alkali metal chloride salt complexes on the quantum yield of
(a) fluorescence and (b) phosphorescence of 1,8-naphtho-21-crown-6. A: NaCl, B: KCl, C: RbCl,
and D: CsCl. (From Sousa and Larson.[86] Copyright 1977 American Chemical Society.)

In order to determine the factors responsible for the fluorescence enhancement, Shizuka et al.[87] have measured the photophysical properties of dibenzo-18-crown-6 complexes with various alkali metal ions in methanol. Table 9.11 summarizes some of their results. Two main photophysical effects of M^+ on dibenzo-crown can be deduced: a decrease in the internal quenching rate (fluorescence enhancement) and an increase in intersystem crossing induced by the heavy-atom cation. With increasing ionic radii, Φ_{FM} and Φ_{fl} go through a maximum, suggesting that the metal ion effects are strongly linked to the stability of the crown–metal ion complex. The order of chelate strength is $K^+ > Na^+ > Rb^+ > Cs^+ > Li^+$.

Complexation of the benzo-15-crown-5 with Eu^{3+} and Tb^{3+} salts results in efficient quenching of the ligand (uncomplexed crown) fluorescence and in the case of Tb^{3+} salts accompanied by a sensitized luminescence from the metal ion.[88] Emission spectra demonstrating the energy transfer are presented in Fig. 9.11. The quantum yields for the ligand fluorescence and for sensitized emission in deaerated acetonitrile are: ligand fluorescence: 0.42 (free ligand), 0.31 ($Tb(NO_3)_3$), 0.29 ($Tb(SCN)_3$), and 0.37 ($Tb(ClO_4)_3$); Φ_{sens} ($\times 10^{-3}$): 0.4 (NO_3^-), 0.3 (SCN^-), and 2.2 (ClO_4^-), respectively:

$$M^{3+} = Eu^{3+}, \ Tb^{3+}$$
$$X^- = NO_3^-, \ ClO_4^-, \ SCN^-$$

An intramolecular energy transfer from the singlet of the ligand to the metal ion is suggested to explain the sensitized emission from the metal ions. The absorption and emission properties of the crown ether complexes are very sensitive to the nature of the anions suggesting that the processes involves an ion pair.

TABLE 9.11. Photophysical Properties of Dibenzo-18-Crown-6–Alkali Metal Ion Complexes in Methanol[87]

Metal ion	CH$_3$OH (300 K)		CH$_3$OH–EtOH (77 K)				
	Φ_{fl}	τ_{fl} (nsec)	Φ_{fl}	τ_{fl}^0	Φ_{phos} ($\times 10^2$)	Φ_{isc}	τ_T (sec)
—	0.15	2.5	0.35	5.7	3.5	0.65	0.67
Li^+	0.17	2.9	0.36	5.9	3.7	0.64	0.42
Na^+	0.23	3.6	0.40	6.2	4.1	0.60	0.28
K^+	0.24	3.8	0.41	6.3	6.0	0.59	0.22
Rb^+	0.20	3.2	0.34	5.6	3.5	0.66	0.075
Cs^+	0.16	2.8	0.28	4.7	1.6	0.72	0.055

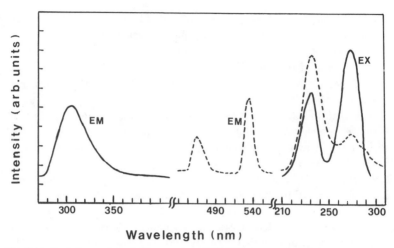

FIG. 9.11. Emission and excitation spectra of $Tb(ClO_4)_3$ complex of benzo-15-crown-5 in acetonitrile at 298 K. (——) Ligand emission (λ_{excit} = 286 nm) and excitation ($\lambda_{emission}$ = 305 nm). (·····) Sensitized Tb^{3+} complex emission (λ_{excit} = 286 nm) and excitation spectra ($\lambda_{emission}$ = 545 nm). (From de Costa et al.[88])

Crown-6-ether derivatives with pendant tryptophan or pyrene chromophores have been prepared and the quenching of the excited states by Tb^{3+} ions followed.[88a] The quenching is linear in the tryptophan complex up to a crown ether: Tb^{3+} mole ratio of 1:1, levelling off thereafter. The data implies quenching of the excited state by Tb^{3+} that are bound in the crown ether cavity. The derived binding constant is approximately $2.90 \times 10^4 \ M^{-1}$. The quenching is accompanied by the sensitized Tb^{3+*} ion emission. Studies using chiral compounds showed twice as efficient energy transfer in L-crown-L-Trypt/Tb^{3+} than that in D-Crown-L-Trypt/Tb^{3+}. Chiral recognition in crown ether is also inferred from the differences in the Tb^{3+} quenching of chiral pyreno-crown ether compounds.

Metalloporphyrin derivatives with benzo-crown ether groups attached to each of the meso positions have been synthesized and the photophysical properties of the metalloporphyrins examined on complexation of various metal ions with the crown ether.[89] Depending on the nature of the cation, various effects have been observed. As we would expect for a crown-5 complex, Na^+ ions form a stable complex. Large cations inserted into the cavity (K^+, Ba^{2+}, or NH_4^+) all cause dimerization of the metalloporphyrin (face-to-face) as evidenced by the absorption spectral changes and in the photophysical properties. Smaller ions such as Cu^{2+} or Ni^{2+} form weak complexes and quench metalloporphyrin excited states via nonredox quenching channels.

Insertion of Eu^{3+} ions in the crown ether cavity leads to novel effect

$$\boxed{Zn^{II}}\!\!-\!\!\left(Eu^{3+}\right) \underset{k_{ET}}{\overset{h\nu}{\rightleftharpoons}} \boxed{Zn^{II}}\!\!-\!\!\left(Eu^{2+}\right)$$

$$Eu^{2+} + \boxed{Zn}\!\!-\!\!\bigcirc \xleftarrow{k_2} \boxed{Zn^{II}}\!\!-\!\!\bigcirc + Eu^{2+}$$

$\boxed{Zn}$ = Znporphyrin $\qquad \bigcirc$ = crown ether

Intramolecular electron transfer quenching of the triplet state of the metalloporphyrin occurs. The photogenerated Eu^{2+} ions are too small to fit tightly into the crown ether cavity and are readily displaced. This displacement apparently occurs very rapidly in competition with the intramolecular back electron transfer. We observe only a bimolecular reverse electron transfer between the porphyrin π-radical cations and free Eu^{2+}. (It should be noted that intramolecular electron transfer reactions are first order, as has been demonstrated in micellar systems). The measured rate constants are $k_{ET} = 1560 \pm 160$ and 1300 ± 150 sec^{-1} for the benzo-crown-5 and benzo-6-crown, respectively. The k_2 values are $4 \pm 0.3 \times 10^9$ and 3.5×10^9 M^{-1} sec^{-1} for benzo-5-crown and benzo-6-crown, respectively. In the presence of electron donors such as EDTA or triethanolamine, irradiation of the metalloporphyrin$-Eu^{3+}-$crown ether complexes leads to formation of metalloporphyrin π-radical anions (via reductive quenching) followed by transfer of electrons to the Eu^{3+} ions housed in the crown ether cavity. Table 9.12 also includes data on the photophysical properties of these metalloporphyrin$-$crown ether complexes.

The redox quenching of $Ru(bpy)_3^{2+}$ or cyanine dye excited state by Ag^+ ions occurs reversibly with very low yields of products in homogeneous solvents.

$$Ru(bpy)_3^{2+} + Ag^+ \xrightarrow{h\nu} Ru(bpy)_3^{3+} + Ag^0 \qquad (9.16)$$

In the presence of crown ethers, the reaction is partially irreversible due to the stabilization of Ag^0 species in the voids of crown ether molecules.[90] Pyrene-functionalized crown ethers exhibit similar protection for the photogenerated Ag^0 species.[91]

Shinkai et al. have shown that photochemical and thermal isomerization reactions of various azo dyes chemically linked to crown ether moieties can be regulated to a significant extent by the introduction of alkali metal ions in the crown ether cavities.[92,93] For example, trans azo-benzophenone crown ether is photoisomerizable to the cis form using UV light, and the reverse (cis–trans) isomerization is achieved thermally or with visible light. As illustrated in the

TABLE 9.12. Influence of Metal Ion Complexation on the Photophysical Properties of Naphtho–Crown and Metalloporphyrin–Crown Ether Complexes [86,89]

Complex[a]	Cation	Temperature (K)	τ_{fl} (nsec)	τ_{trip} (sec)	Φ_{fl}	Remarks
2,3-NCE-6	—	77		2.9	0.2	$k_{isc} = 0.8 \times 10^7$
2,3-NCE-6	Na$^+$	77		3.4	0.13	$k_{isc} = 1.4 \times 10^7$
2,3-NCE-6	K$^+$	77		3.1	0.09	$k_{isc} = 2.3 \times 10^7$
2,3-NCE-6	Rb$^+$	77		2.7	0.085	$k_{isc} = 2.5 \times 10^7$
2,3-NCE-6	Cs$^+$	77		2.2	0.04	$k_{isc} = 6.7 \times 10^7$
1,8-NCE-6	—	77		2.45	0.08	$k_{isc} = 9.7 \times 10^6$
1,8-NCE-6	Na$^+$	77		2.3	0.09	$k_{isc} = 8.5 \times 10^6$
1,8-NCE-6	K$^+$	77		1.4	0.13	$k_{isc} = 6.3 \times 10^6$
1,8-NCE-6	Rb$^+$	77		1.3	0.14	$k_{isc} = 5.5 \times 10^6$
1,8-NCE-6	Cs$^+$	77		1.2	0.10	$k_{isc} = 7.4 \times 10^6$
ZnP-CE-5	—	298	1.7	0.00095	0.038	
ZnP-CE-5	Na$^+$	298	1.7	0.00096	0.038	
ZnP-CE-5	K$^+$	298	0.2	—	0.0008	porphyrin dimers
ZnP-CE-5	Eu^{3+}	298	1.3	0.00038	0.029	oxid. quenching

[a] 2,3-NCE-6 = 2,3-naphtho-20-crown-6; 1,8-NCE-6 = 1,8-naphtho-21-crown-6; ZnP-CE-5 = 5,10,15,20-tetra(benzo-15-crown-5)porphyrinatozinc.

scheme below, sodium ions complex strongly with the cis form and inhibit the thermal isomerization reaction:

References

1. M. L. Bender and M. Komiyama, "Cyclodextrin Chemistry," React. Struct. Concepts Org. Chem., Vol. 16. Springer-Verlag, Berlin and New York, 1978.
2. J. Szejtli, "Cyclodextrins and Their Inclusion Complexes." Académiai Kaidó, Budapest, 1982.
3. I. Tabushi and Y. Kuroda, Adv. Catal. 32, 417 (1983).
4. R. J. Bergeron, J. Chem. Educ. 54, 204 (1977).
5. W. Saenger, Angew. Chem., Int. Ed. Engl. 19, 344 (1980).
6. A. Villiers, C. R. Hebd. Seances Acad. Sci. 112, 536 (1891).
7. F. Schardinger, Wien. Klin. Wochensch. 17, 207 (1904).

8. F. Cramer, W. Saenger, and H. C. Spatz, *J. Am. Chem. Soc.* **89,** 14 (1967).
9. I. Tabushi, K. Shimokawa, N. Shimidzu, H. Shirawa, and K. Fujita, *J. Am. Chem. Soc.* **98,** 7855 (1976).
10. M. Hoshino, M. Imamura, and K. Ikehara, *J. Phys. Chem.* **85,** 1820 (1981).
11. N. J. Turro, T. Okubo, and C.-J. Chung, *J. Am. Chem. Soc.* **104,** 3954 (1982).
12. S. Scypinski and L. J. Cline Love, *Int. Lab.* **14,** 61 (1984).
13. S. Scypinski and J. M. Drake, *J. Phys. Chem.* **89,** 2432 (1985).
14. Y. Degani and I. Willner, *Chem. Phys. Lett.* **104,** 496 (1984).
15. H. E. Edwards and J. K. Thomas, *Carbohydr. Res.* **65,** 173 (1978).
16. A. Nakajima, *Bull. Chem. Soc. Jpn.* **57,** 1143 (1984).
17. A. DeKorte, R. Langlois, and C. R. Cantor, *Biopolymer* **19,** 1281 (1980).
18. S. Hamai, *Bull. Chem. Soc. Jpn.* **55,** 2721 (1982).
19. G. S. Cox, P. J. Hauptman, and N. J. Turro, *Photochem. Photobiol.* **39,** 597 (1984).
20. G. S. Cox and N. J. Turro, *Photochem. Photobiol.* **40,** 185 (1984).
21. T. Yorozu, M. Hoshino, M. Imamura, and H. Shizuka, *J. Phys. Chem.* **86,** 4422 (1982).
22. A. Ueno, K. Takahashi, U. Hino, and T. Osa, *J.C.S. Chem. Commun.* p. 194 (1981). (a) C. D. Tran and J. H. Fendler, *J. Phys. Chem.* **88,** 2167 (1984)
23. S. Kitamura, H. Yunokawa, and T. Huge, *Polym. J.* **14,** 85, 91 (1982).
24. A. Ueno, K. Takahashi, and T. Osa, *J.C.S. Chem. Commun.* p. 921 (1980).
25. T. Yorozu, M. Hoshino, and M. Imamura, *J. Phys. Chem.* **86,** 4426 (1982).
26. K. Kano, I. Takenoshita, and T. Ogawa, *Chem. Lett.* p. 321 (1982).
27. J. Emert, D. Kodali, and R. Catena, *J.C.S. Chem. Commun.* p. 758 (1981).
28. N. J. Turro, T. Okubo, and G. C. Weed, *Photochem. Photobiol.* **35,** 325 (1982).
29. M. Itoh and Y. Fujiwara, *Bull. Chem. Soc. Jpn.* **57,** 2261 (1984).
30. R. Arad-Yellin and D. F. Eaton, *J. Phys. Chem.* **87,** 5051 (1983).
31. K. Kano, I. Takenoshita, and T. Ogawa, *Chem. Lett.* p. 1035 (1980).
32. K. Kano, I. Takenoshita, and T. Ogawa, *J. Phys. Chem.* **86,** 1833 (1982).
33. H. Kobayashi, M. Takahashi, Y. Muramatsu, and T. Murata, *Bull. Chem. Soc. Jpn.* **54,** 2815 (1981).
34. G. S. Cox, N. J. Turro, N.-C. Yang, and M.-J. Chen, *J. Am. Chem. Soc.* **106,** 422 (1984).
35. N. J. Turro, J. D. Bolt, Y. Kuroda, and I. Tabushi, *Photochem. Photobiol.* **35,** 69 (1982).
36. N. J. Turro, T. Okubo, and C.-J. Chung, *J. Am. Chem. Soc.* **104,** 1789 (1982).
37. N. J. Turro, G. S. Cox, and X. Li, *Photochem. Photobiol.* **37,** 149 (1983).
38. S. Scypinski and L. J. Cline Love, *Anal. Chem.* **56,** 322 (1984).
39. S. Scypinski and L. J. Cline Love, *Anal. Chem.* **56,** 331 (1984).
40. S. Scypinski and L. J. Cline Love, *Int. Lab.* **14,** 61 (1984).
41. I. Tabushi, K. Fujita, and L. C. Yuan, *Tetrahedron Lett.* p. 2503 (1977).
42. F. J. DeLuccia and L. J. Cline Love, *Anal. Chem.* **56,** 2811 (1984).
43. M. Ohara and K. Watanabe, *Angew. Chem., Int. Ed. Engl.* **14,** 820 (1975).
44. R. Chenevert and R. Plante, *Can. J. Chem.* **61,** 1092 (1983).
45. T. Tamaki, *Chem. Lett.* p. 53 (1984).
46. Y. Hui, J. C. Russel, and D. G. Whitten, *J. Am. Chem. Soc.* **105,** 1374 (1983).
47. A. Ueno, K. Takahashi, and T. Osa, *J.C.S. Chem. Commun.* p. 837 (1980).
48. Y. Hui, J. R. Winkle, and D. G. Whitten, *J. Phys. Chem.* **87,** 23 (1983).
49. D. W. Breck, "Zeolite Molecular Sieves." Wiley, New York, 1974.
50. J. A. Rabo, ed., "Zeolite Chemistry and Catalysis," ACS Monograph Series, No. 171. Am. Chem. Soc., Washington, D.C., 1976.
51. G. D. Stucky and F. G. Dwyer, eds., "Intrazeolite Chemistry," ACS Symposium Series, No. 218. Am. Chem. Soc., Washington, D.C., 1983. (a) J. H. C. Van Hooff, *in* "Chemistry and Chemical Engineering of Catalytic Process" (R. Prins and G. C. A. Schuit, eds.), NATO Advanced Studies Institute Series E, No. E39, p. 161. Sijthoff & Noorolhoff, Alphen, aan den Rijn, Netherlands, 1980.

52. G. T. Pott and W. H. Stork, *Catal. Rev.* **12**, 163 (1975).

53. T. Arakawa, T. Takata, G.-Y. Adachi, and J. Shiokawa, *J.C.S. Chem. Commun.* p. 453 (1979).

54. T. Arakawa, T. Takata, G.-Y. Adachi, and J. Shiokawa, *J. Lumin.* **20**, 325 (1979).

55. T. Arakawa, T. Takata, M. Takakuwa, G.-Y. Adachi, and J. Shiokawa, *Mater. Res. Bull.* **17** 171 (1982).

56. T. Arakawa, M. Takakuwa, and J. Shiokawa, *Bull. Chem. Soc. Jpn.* **57**, 948 (1984).

57. P. A. Jacobs, J. B. Uytterhoeven, and H. K. Beyer, *J.C.S. Chem. Commun.* p. 128 (1977).

58. S. Leutwyler and E. Schumacher, *Chimia* **31**, 475 (1977).

59. S. M. Kuznicki and E. M. Eyring, *J. Am. Chem. Soc.* **100**, 6790 (1978).

60. R. P. Zerger, S. L. Suib, and G. D. Stucky, *Abstr., Am. Chem. Soc. Nat. Meet., 178th, Washington, D.C.*, 1979. p. 198.

61. R. P. Zerger, S. L. Suib, and G. D. Stucky, to be published. Cited in S. L. Suib, O. G. Bordeianu, K. C. McMohan, and D. Psaras, *in* "Inorganic Reactions in Organised Media" (S.L. Holt, ed.), ACS Symposium Series, No. 177. Am. Chem. Soc., Washington, D.C., 1981. p. 226, 238.

62. D. H. Strome and K. Klier, *J. Phys. Chem.* **84**, 981 (1980).

63. D. H. Strome and K. Klier, *in* "Adsorption and Ion Exchange in Synthetic Zeolites" (W. H. Frank, ed.), ACS Symposium Series, No. 135. Am. Chem. Soc., Washington, D.C., 1980. p. 155.

64. I. E. Maxwell and E. Drent, *J. Catal.* **41**, 412 (1976).

65. J. Texter, D. H. Strome, R. G. Herman, and K. Klier, *J. Phys. Chem.* **81**, 333 (1977).

66. S. L. Suib, O. G. Bordeianu, K. C. McMohan, and D. Psaras, *in* "Inorganic Reactions in Organised Media" (S. L. Holt, ed.), ACS Symposium Series, No. 177, p. 225. Am. Chem. Soc., Washington, D.C., 1981.

67. S. L. Suib, A. Kostapapas, and D. Psaras, *J. Am. Chem. Soc.* **105**, 1614 (1984). (a) S. L. Suib and K. A. Carrado, *Inorg. Chem.* **24**, 200 (1985).

68. W. H. Quayle, G. Peeters, and J. H. Lunsford, *J. Phys. Chem.* **84**, 2306 (1980).

69. L. R. Faulkner, S. L. Suib, C. L. Renschler, J. M. Green, and P. R. Bross, *in* "Chemistry in Energy Production" (R. G. Wymer and O. L. Keller, eds.), pp. 99–113. Am. Chem. Soc., Washington, D.C., 1982.

70. M. J. Camara and J. H. Lunsford, *Inorg. Chem.* **22**, 2498 (1983).

71. B. H. Baretz and N. J. Turro, *J. Photochem.* **24**, 201 (1984).

72. S. L. Suib and A. Kostapapas, *J. Am. Chem. Soc.* 106, 7705 (1984).

73. H. L. Casal and J. C. Scaiano, *Can. J. Chem.* **62**, 628 (1984).

74. N. J. Turro and P. Wan, *J. Am. Chem. Soc.* **107**, 678 (1985).

75. F. Vögtle, ed., "Host–Guest Complex Chemistry," Vols. I and II, Topics in Current Chemistry, Vols. 98 and 101. Springer-Verlag, Berlin and New York, 1981 and 1982.

76. M. Hiraoka, "Crown Compounds: Their Characteristics and Applications," Studies in Organic Chemistry, Vol. 12. Elsevier, New York, 1982.

77. G. A. Nelson, ed., "Coordination Chemistry of Macrocyclic Compounds." Plenum, New York, 1979.

78. R. M. Izatt and J. J. Christensen, eds., "Synthetic Multidentate Macrocyclic Compounds." Academic Press, New York, 1978.

79. J.-M. Lehn, *Struct. Bonding (Berlin)* **16**, 1 (1973); *Acc. Chem. Res.* **11**, 49 (1978).

80. J.-M. Lehn, *Pure Appl. Chem.* **52**, 2303, 2441 (1980).

81. D. J. Cram, *Acc. Chem. Res.* **11**, 8 (1978).

82. D. J. Cram, *in* "Applications of Biochemical Systems in Organic Chemistry" (J. B. Jones, C. J. Sih, and D. Pearlman, eds.), Part II, Chap. 5. Wiley, New York, 1976.

83. N. Poonia and A. V. Bajaj, *Chem. Rev.* **79**, 389 (1979).

84. F. DeJong and D. N. Reinholdt, *Adv. Phys. Org. Chem.* **17**, 279 (1980).

85. N. Sabbatini, M. Ciano, S. Dellonte, A. Bonazzi, F. Boletta, and V. Balzani, *J. Phys. Chem.* **88**, 1534 (1984); *Chem. Phys. Lett.* **90**, 265 (1982).
86. L. R. Sousa and J. M. Larson, *J. Am. Chem. Soc.* **99**, 307 (1977).
87. H. Shizuka, K. Takata, and T. Morita, *J. Phys. Chem.* **84**, 994 (1980).
88. S. M. B. de Costa, M. Manuela Queimado, and J. J. R. Frausto da Silva, *J. Photochem.* **12**, 31 (1980). (a) P. Tundo and J. H. Fendler, *J. Am. Chem. Soc.* **102**, 1760 (1980).
89. G. Blondeel, A. Harriman, G. Porter, and A. Wilowska, *J.C.S. Faraday II* **80**, 867 (1984).
90. R. Humphry-Baker, P. Tundo, and M. Grätzel, *Angew. Chem., Int. Ed. Engl.* **18**, 630 (1979).
91. T. Jao, G. S. Beddard, P. Tundo, and J. H. Fendler, *J. Phys. Chem.* **85**, 1963 (1981).
92. S. Shinkai, T. Minami, Y. Kusano, and O. Manabe, *J. Am. Chem. Soc.* **105**, 1851 (1983); **104**, 1967 (1982).
93. S. Shinkai, T. Ogawa, Y. Kusano, O. Manabe, K. Kikukawa, T. Ogawa, and T. Matsuda, *J. Am. Chem. Soc.* **104**, 1960 (1982).

Chapter 10

Photochemistry of Molecules in the Adsorbed State

10.1 Introduction

The studies reviewed so far all have been concerned with the influence of host assemblies, dynamical structures, or permanent systems on the photoprocesses of the guest molecules. There is a growing interest in extending this line of investigations to molecules in the adsorbed-state—either physisorbed or chemisorbed onto inorganic supports that are in the form of colloidal dispersions or as granules, powders, or gels (dry state). In this chapter, we attempt to outline some of the early results in these new areas. Common support materials are oxides such as silica, alumina, and clays, though a few studies also have been reported on materials such as cellulose and starch.

On a wide variety of oxide supports, the nature and extent of the surface reactive groups drastically affect the mode of distribution and the local organization of the adsorbates. Pretreatment and handling of the support material, solvent, and pH all affect the surface and the adsorption mode. These are often reflected in the observed photophysics and photochemistry of the adsorbates. There are two complementary goals: to observe how the surface interactions modify the behavior of the excited state and to use the photochemistry as a tool to probe the substrate surface. The field is very much in its infancy, but is well poised for a tremendous growth. It should be pointed out that there has also been extensive activity in the field of heterogeneous photocatalysis and synthesis with semiconductor oxide and nonoxide dispersions and colloids.[1-5] In these systems, the excitation light is absorbed by the support, producing electrons and holes, which upon arrival at the surface react with/on the adsorbates. Here, for lack of space, we exclude all discussions of such reactive surfaces.

10.2 Photochemistry in the Presence of Inorganic Colloids

10.2.1 Studies on Colloidal Silica

Colloidal silica (SiO_2) and clays (aluminosilicates) are two class of inorganic colloids that are receiving some scrutiny. Solutions of colloidal silica with particle radius in the range of 50–200 Å are commercially available today. The particles are spherical, possessing a negative surface, consisting of SiO^- groups and adsorbed OH^- ions. The counterions in most systems are Na^+, and they are readily replaced by other cations such as $Ru(bpy)_3^{2+}$. The silanol groups at the surface of SiO_2 particles are ionized at pH $\geqslant$ 5.5 and produce a diffuse double layer. The surface potentials are rather high (~ -170 mV).[6,7] As in anionic micellar systems encountered earlier (Chapters 2–4), in these systems, the reactions between two components can be localized on or near the surface. The negative surface charges then can aid quite significantly in the separation of photoredox products.

Wheeler and Thomas[8] have used two probes, $Ru(bpy)_3^{2+}$ and 4-(1-pyrenyl)-butyltrimethylammonium bromide (PN^+), to investigate the nature of the colloidal SiO_2 particles in water. The fluorescence spectra of these two probes show that the silica surface is very polar and similar to water. Inefficient quenching of emission of these probes by anionic quenchers (as compared to that in homogeneous solvents) confirms the presence of negative charges, though the charge is not as effective as that in sodium lauryl sulfate micelles. Studies with cationic quenchers show that the cations are strongly bound to the silica particles but do not move as readily around the surface as on anionic micelles.

The presence of a negatively charged interface in these SiO_2 particles should enable achievement of efficient charge separation. Calvin, Willner, and co-workers have demonstrated such processes using silica-bound sensitizers and viologen salts as acceptor relays.[9–13] With zwitterionic viologen, propylviologen sulfonate (PVS), on electron transfer, we produce two oppositely charged photoproducts, $Ru(bpy)_3^{3+}$ and PVS^-:

$$Ru(bpy)_3^{2+} + PVS \underset{k_2}{\overset{hv, k_q}{\rightleftharpoons}} Ru(bpy)_3^{3+} + PVS^- \qquad (10.1)$$

The former is strongly adsorbed to the silica particle but the reduced form of the electron acceptor is repelled by the negatively charged interface, hindering the recombination process as illustrated by

The forward quenching rate k_q is comparable in water and in SiO_2 colloids. However, the back electron transfer rate k_2 is retarded by more than 100-fold over that observed in the homogeneous phase (cf. Fig. 10.1). In the case of dicationic methyl viologen, MV^{2+}, the quenching rate is strongly enhanced on the silica surface (100-fold). The importance of the surface potential of SiO_2 colloids in retarding the back reaction has been demonstrated by varying the ionic strength and the pH of the medium. Similar results have been obtained with the tetracationic porphyrin sensitizer tetrakis(N-methylpyridyl)-porphyrinatozinc(II) (ZnTMPyP^{4+}).[9]

The photophysical behavior of uranyl ions (UO_2^+) is quite different when they are adsorbed onto colloidal SiO_2.[14] The emission is extremely long lived ($t_{1/2} \simeq 440$ μsec as compared to 11 μsec in water and yields about 2.8 times higher). The fluorescence spectrum is slightly blue shifted by 7 nm (occurring at 514 nm). The transient absorption spectrum has only one peak in silica (located at approximately 550 nm) but has two maxima in water (at approximately 485 and 570 nm). Quenching by anionic species such as I^-, Br^-, and SCN^- is inhibited by a factor of 10^3–10^4 due to the adsorption of the probe on the negatively charged silica particle.

By refluxing fumed silica particles with metasilicates (SiO_2^{2-}) in deionized water, Wheeler and Thomas[15] have prepared larger, porous colloidal SiO_2 particles of radius ~ 500 Å. These porous colloids, unlike the smaller SiO_2 particles apparently incorporate molecules in the interior, giving rise to certain features on the photochemistry of guest molecules in these systems. $Ru(bpy)_3^{2+}$, for example, associates strongly with these larger particles. Unlike the $Ru(bpy)_3^{2+}$ bound to smaller SiO_2 particles, here the dye cannot be extracted using SDS micelles. The emission shows two peaks ($\lambda_{max} = 573$ and 606 nm) in comparison to one ($\lambda_{max} = 613$ nm) in the SiO_2 particles. On polymerized silica particles, the lifetime increases to 2 μsec. Surprisingly, neutral molecules such as O_2 and $PhNO_2$ that normally quench $Ru(bpy)_3^{2+*}$ are unreactive in this system. Photoredox quenching of $Ru(bpy)_3^{2+*}$ by methyl viologen (MV^{2+}), however, has been observed. The quantum yield for the redox product (MV^+) is also quite high. [The relative yields are 15.6 (polymerized SiO_2), 9.6 (water), 3.5 (NALCO SiO_2), and 1.0 (SDS micelles), respectively.]

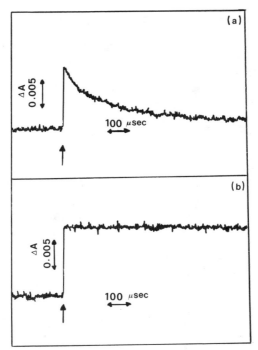

FIG. 10.1. Transient decay curves for viologen radical cation PVS$^{\bar{\ }}$ (at 602 nm) illustrating the influence of SiO_2 colloids on the reverse electron transfer reaction between $Ru(bpy)_3^{3+}$ and PVS$^-$. (a) Decay in homogeneous solutions (water) and (b) in aqueous SiO_2 colloids (0.2 M). (From Laane et al.[10])

10.2.2 Studies on Colloidal Clays

General Features of Clay Suspensions and Their Intercalation

Clays (or aluminosilicates) are a higher form of silica with unique structural features.[16–18] A wide variety of clay minerals occur in nature, having different chemical composition, morphology, colloidal, and surface properties. Most popular are the swelling clays, known as smectites. Unlike the three-dimensional network structures found in zeolites (cf. Section 9.3), these have layer lattice structures in which two-dimensional oxyanions are separated by layers of hydrated cations. Figure 10.2 presents schematically the structural framework of such layers. The oxygen atoms define the upper and lower sheets of tetrahedral sites (silica) and a central sheet of octahedral sites (alumina). The 2:1 relation between the tetrahedral and the octahedral sites within a layer allows smectite clays to be classified as 2:1 phyllosilicates. This structural designation differentiates the smectites from 1:1 clay minerals such as kaolinites. Here the layers are formed by coupling of one tetrahedral sheet to an octahedral sheet.

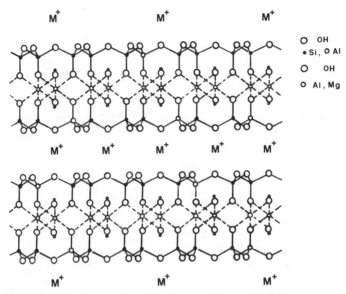

FIG. 10.2. Schematic representation of the structural framework of SiO_2 and Al_2O_3 layers in montmorillonite clays. (From A. Weiss.[18a])

The members of the smectite group of clays are distinguished by the type and location of the cations in the oxygen framework. In a unit cell formed from 20 oxygen atoms and 4 hydroxyl groups, there are 8 tetrahedral and 6 octahedral sites. When two-thirds of the octahedral sites are occupied by cations, the mineral is classified as a dioctahedral 2:1 phyllosilicate. A trioctahedral 2:1 phyllosilicate has all octahedral sites filled by cations. Minerals such as mica, pyrophillate, and talc (though not smectites) have the same 2:1 phyllosilicate oxygen framework found in smectites. In talc, all of the tetrahedral and octahedral sites in the oxygen framework are filled by Si^{4+} and Mg^{2+}, respectively, and the layers are electrically neutral.

In clays, isomorphic substitution within the lattice (Si^{4+} by Al^{3+}) leads to residual negative charges which are compensated by exchangeable cations on the surface of the layers. In addition to the exchangeable cations, which may be almost any kind of organic, inorganic, or organometalic cation, the interlayer space is able to accommodate water and/or a wide variety of organic species. This process of inclusion is known as intercalation. In nonswelling clays such as kaolinite, intercalation does not occur and the adsorbed molecules are confined to the external surface. Table 10.1 lists some of the common clay minerals along with some approximate indication of their surface area and cation exchange capacities. Clays such as montmorillonite carry significant amount of iron (mostly in the form of Fe^{3+}) in the lattice. This is

TABLE 10.1. Various Types of Clay Minerals, Their Classifications,
Surface Area and Cation Exchange Capacities [a]

Clay	Mineralogical group	Surface area[b] (m^2/g)	Cation exchange capacity[b] (meg/g)
Hectorite	Smectite, 2:1 trioctahedral	750	0.85
Laponite	Smectite, 2:1 trioctahedral	750	0.80
Montmorillonite	Smectite, 2:1 dioctahedral	750	0.95
Nontronite	Smectite, 2:1 dioctahedral	750	0.95
Sepiolite	Fibrous, 1:1 trioctahedral	250	0.15
Kaolinite	Kaolin, 1:1 dioctahedral	26.7	0.10

[a] From Van Olphen,[17] Habti et al.,[22] and Bergaya and Van Damme.[25]
[b] The surface areas and C.E.G. values quoted are representative ones. The values can vary from sources to sources.

a disturbing factor in photochemical studies, since Fe^{3+} and Fe^{2+} can act as quenchers of excited states.

Luminescence of Cations and Metal Complexes

Transition metal complexes such as $Ru(bpy)_3^{2+}$ are readily introduced onto smectite clays such as hectorite or montmorillonite and its photochemistry has been the subject of several investigations.[19-24] The absorption and luminescence of $Ru(bpy)_3^{2+}$ undergo significant perturbation in clays (cf. Fig. 10.3). The charge transfer absorption band shifts towards the red (maxima at 472, 462, and 465 nm in montmorillonite, hectorite, and kaolinite, respectively) with about 50% increase in the absorption cross section. The intensity of the $(\pi-\pi^*)$ band near 285 nm is dramatically reduced in intensity (and splits into two bands at 272 and 285 nm). The differences in the absorption spectra have been attributed to interaction between the $Ru(bpy)_3^{2+}$ complex and the clay surface. Possible explanations include distortion of the bipyridine ligands due to steric constraints and an enhancement in the quasi-metal-to-ligand charge transfer in the ground state.

In contrast to the differences in the absorption spectra, the luminescence profiles are very similar. Also the excitation spectra followed essentially the absorption spectral profiles. However, there are significant differences in the relative magnitude of the luminescence quantum yields and lifetimes. After correcting for the changes in the ground-state absorption coefficients, the emission yield is about 30% larger for $Ru(bpy)_3^{2+}$ in hectorite, whereas it is almost six times lower in montmorillonite. The greatly reduced yield in montmorillonite has been attributed to the quenching by Fe^{3+} present in the octahedral layers. (In Wyoming bentonite, a montmorillonite clay with high

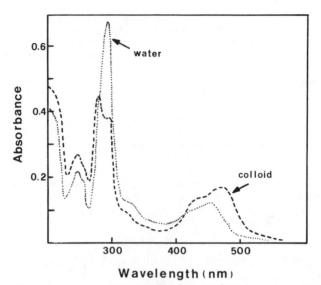

FIG. 10.3. Absorption spectra of $Ru(bpy)_3^{2+}$ (8.5×10^{-6} M) in water and adsorbed on montmorillonite clay colloids (1 g/liter in water). The latter spectrum was run with a solution of montmorillonite clay colloids (1 g/liter) in the reference compartment. (From DellaGuardia and Thomas.[21] Copyright 1983 American Chemical Society.)

iron content, the emission of $Ru(bpy)_3^{2+}$ is totally quenched!) The emission decay profiles in the presence of clays are complex, but they have been fitted to double exponential decays of the form:

$$I/I_0 = \alpha \exp(-k_1 t) + (1 - \alpha) \exp(-k_2 t) \qquad (10.2)$$

($\alpha = 0.6$, $\tau_1 = 62$ nsec, $(1 - \alpha) = 0.4$, and $\tau_2 = 350$ nsec). It has been suggested that the double exponential is due to the presence of two different adsorption sites on the clay particle: on the outer surface and intercalation between the phyllosilicate sheets. Intriguingly, the emission decay profiles were laser-intensity dependent. Based on an emission intensity study as a function of loading of the Ru complex, it has been proposed that clay interlayers segregate $Ru(bpy)_3^{2+}$ from exchangeable Na^+ ions, resulting in high local concentrations in the interlayers. $Ru(bpy)_3^{2+}$ was found to segregate MV^{2+} in the interlayers leading to poor efficiencies of quenching. $Ru(bpy)_3^{2+*}$ however is readily quenched by the neutral viologen PVS.

The luminescence of rare earth ions Eu^{3+} and Tb^{3+} that have been cation exchanged on montmorillonite and hectorite have been examined as a function of water coadsorbed on the clay.[25] On montmorillonite, the luminescence intensity was negligible, presumably due to the quenching by Fe^{3+} ions present in the clay lattice (via energy transfer?). The luminescence intensity on hectorite displayed complex behavior. The 618 nm band of Eu^{3+}

and the 544 nm band of Tb^{3+} were most sensitive to the extent of water content. Three distinct regions were identified. From approximately 30 to 6 or 8 H_2O/cation (region I), the luminescence intensity remains almost constant, at 6–8 H_2O/cation (region II), a sharp increase is observed and finally below 4–5 H_2O/cation, the emission intensity decreases sharply, as shown in Fig. 10.4. The results have been rationalized in terms of rare earth–clay interactions on one hand and rare earth–water interactions on the other.

It was pointed out earlier that, due to the inherent structural features, adsorbed molecules on 1 : 1 clays such as kaolinite are confined to the external surface. On swelling clays such as montmorillonite, molecules are intercalated

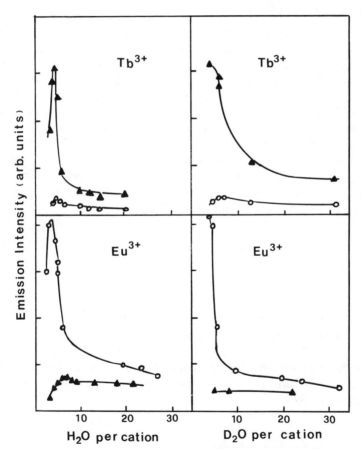

FIG. 10.4. Dependence of intensity of Eu^{3+} and Tb^{3+} emission bands (Eu^{3+} at 618 nm and Tb^{3+} at 544 nm) in hectorite film samples on the amount of co-adsorbed H_2O or D_2O per cation. For Eu^{3+} curves A and C represent excitation at 394 nm, and B and D represent excitation at 300 nm. For Tb^{3+} curves A and C represent excitation at 367 nm, and B and D represent excitation at 300 nm. (From Bergaya and Van Damme.[25])

between the layers and also adsorb on the external surface. Using 4-(1-pyrenyl)trimethylammonium bromide (PN^+) fluorescence as a probe, comparative studies have been made on these two classes of clays.[26,27] Studies with neutral quenchers such as CH_3NO_2 or $PhNO_2$ and with cations such as Tl^{3+} indicate that diffusion on the surface of montmorillonite is reduced below that observed in aqueous solutions. The apparent rate obtained with kaolin particles is increased. Introduction of co-adsorbates such as long-chain alcohols or surfactants affects significantly the mutual diffusion of the probe and the quencher molecules. Reactions of radical cations of tetramethylbenzidine (TMB^+) with colloidal montmorillonite have also been studied by laser photolysis techniques.[28]

Photochemical Water Cleavage

There have been a few attempts to utilize clay colloids as media for photochemical water cleavage experiments. Van Damme and co-workers have made some interesting studies on the mechanistic aspects of water cleavage using the clay suspensions. Cis and trans forms of $Ru(bpy)_2(H_2O)_2^{2+}$ were exchanged onto hectorite and sepiolite clays and their efficiencies for oxidation of water to molecular oxygen assessed in thermal and photochemical pathways. Only the cis form was found to be catalytically active.[29]

A common approach in studies of photochemical water cleavage has been to use some form of the organized assembly (to achieve efficient charge separation in photoredox reactions [Eq. 10.3)] of interest) and appropriate redox catalysts (to achieve water oxidation or reduction [Eqs. (10.4) and (10.5)], subsequently)

$$D + A \xrightleftharpoons{h\nu} D^+ + A^- \tag{10.3}$$

$$4D^+ + 2H_2O \xrightarrow[\text{cat}_1]{} 4D + 4H^+ + O_2 \tag{10.4}$$

$$2A^- + 2H_2O \xrightarrow[\text{cat}_2]{} 2A + 4OH^- + H_2 \tag{10.5}$$

A photocatalytic system has been constructed that employs two different colloids to anchor the crucial reactants and the redox catalysts.[30] The oxidizing system ($Ru(bpy)_3^{2+}$ and RuO_2 catalyst) was loaded onto an anionic fibrous clay sepiolite and the reducing system (Eu^{3+} ions and Pt catalysts) embedded on ill-defined alumina colloids. The following scheme outlines the reaction pathways:

Photochemical H_2 formation using the system $Ru(bpy)_3^{2+}$–methyl viologen–triethanolamine has been examined briefly on Wyoming bentanoite clays.[30a]

10.3 Photochemistry of Molecules on Oxide Surfaces

Except for some early investigations in the 1930s by de Boer et al.[31-33] and a few scattered reports thereafter, studies of photophysics and photochemistry of adsorbed molecules is a relatively new and underdeveloped area. The complex nature of the oxide surfaces and their varied interactions with adsorbates, coupled with many experimental difficulties have made the progress very slow. Elegant studies in the late 1960s by Leermakers[34-38] ended abruptly following his untimely death. In 1971, Nicholls and Leermakers reviewed[39] the early photophysical studies on organic molecules adsorbed on oxide supports (mostly ketones on silica gel). Also reviewed here are some photochemical reactions of adsorbates such as photoinduced cis–trans isomerization of stilbene, photochromism of spiropyran, and photocleavage of cyclohexadienones.

Studies have established that there are often considerable changes in the energy and the nature of the electronically excited states on adsorption. Consequently, the efficiencies of various photophysical and photochemical processes are altered. Interactions of adsorbate with the support can arise from both nonspecific interactions with the adsorbent as a whole and specific interactions with the active groups on the surface (hydroxyl groups on oxide surfaces for instance). The interactive forces responsible for the adsorption can be electrostatic, hydrogen bonding, or dispersion forces. The manner of binding of adsorbate to the surface is often derived from various physical studies such as examination of the infrared spectra, heats of adsorption, and adsorption isotherms. A good understanding of the surface structure and its modifications under various experimental conditions hardly needs any emphasis. Herein we consider photochemical studies on systems such as silica gel, porous vycor glass, and alumina.

10.3.1 Studies on Silica Gel and Porous Vycor Glass Surfaces

Silica gel probably is one of the most widely studied supports,[40] especially in the laboratories of de Mayo. Silica gel consists of a three-dimensional network of tetracoordinate oxysilicon tetrahedra and a partially hydrated surface consisting of both siloxane (—Si—O—Si—) and silanol (—Si—OH) groups. Some of the silanol groups exist as single species (more active) while others exist as hydrogen-bonded pairs (geminal or vicinal). These functions in the amorphous structure render the silica gel surface very inhomogeneous. Heat treatment of silica gel modifies the number and the nature of the surface silanol groups. During heating, there is loss of physisorbed water molecules followed by dehydroxylation of the silanol groups. The dehydroxylation process can be reversed by exposure to water vapor, up to 450°C. At higher temperatures, the regeneration process is very slow. Studies indicate that, with increasing temperature, all free water is removed up to 250°C. The number of silanol groups then decreases slowly. Starting with about 6 OH/100 $Å^2$, it is reduced to approximately 2 OH/100 $Å^2$ at 500°C and to about 1 OH/100 $Å^2$ at 800°C.

Porous vycor glasses (PVG) are derived from borosilicate glasses by chemical extraction of alkali and boric acid. The porous glass left behind consists mainly of silica gel (about 97%) and B_2O_3 (about 3%) with a very uniform pore size (average pore diameter 40 Å). Porous glass is, therefore, very similar to silica gel except for its amorphous structure and a larger surface area. Adsorption sites are similar and are derived from silanol groups.

Luminescence of Adsorbed Molecules

Tsubomura et al.[41-43] examined the absorption and luminescence features of several aromatic hydrocarbons adsorbed on PVG. At low concentrations, the absorption and fluorescence spectra are in good agreement with those in solution. At higher concentrations, new bands ascribable to excimer fluorescence are observed. For anthracene, the excimer fluorescence is in agreement with that observed in photolytic dissociation of dianthrane in a low-temperature matrix. In the latter the sandwich dimer of anthracene is formed by photolysis. Quite unexpectedly, the fluorescence of aromatic hydrocarbons was detectable even in the presence of liquid oxygen! (For naphthalene, even short-lived phosphorescence was observable.)

De Mayo and co-workers have made very extensive studies on the effect of surface modification on the photophysics of naphthalene and pyrene adsorbed on silica gel.[44-48] Heating of the silica gel in vacuum to 500–700°C causes the adsorbed hydrocarbons to have broad and structureless fluorescence spectra, low quantum efficiency, and substantially nonexponential decay. Introduction of co-adsorbant polar molecules such as water, methanol, hexanol, or decanol on the surface results in an increase in the quantum efficiency, higher resolution of vibronic structure, and a large increase in the decay time (approaches a single exponential). Furthermore, with increasing amounts of additives, the ratio of vibronic bands (I/III) decreases, suggesting a change in the polarity of the surface. Methanol as a co-adsorbing additive was more efficient than water in resolving the vibrational fine structure and longer lifetimes (383 nsec as compared to 370 nsec, and 430 nsec in neat methanol and cyclohexane, respectively). On a molar basis, 1-decanol was even more effective than methanol or water.

As shown in Fig. 10.5 at higher coverages ($\sim 1\%$) we also observe an excimer-like emission from pyrene in the silica gel.[45-49] The excitation spectrum of the broad-band emitter was red shifted with respect to the absorption responsible for the monomer fluorescence. These along with time-resolved fluorescence studies on the dynamics of excimer formation indicate that associated complexes of adsorbates in the ground state are responsible for some of the excimer-like fluorescence.

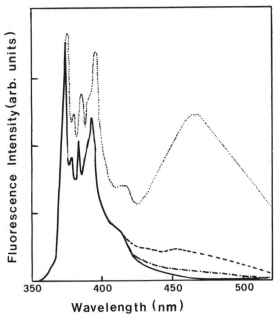

FIG. 10.5. Emission spectra of adsorbed pyrene on silica gel (SiO_2) as a function of surface coverage at 3% ($\cdots$) (λ_{excit} = 341 nm), at 1% (– – –) (λ_{excit} = 341 nm), at 0.2% (–·–·) (λ_{excit} = 341 nm) and at 0.2% (——) (λ_{excit} = 331 nm). (From Bauer et al.[47] Copyright 1982 American Chemical Society.)

The decays of neither the monomer nor the excimer exhibit single exponential behavior. If we take the relative intensity ratios (I_{460}/I_{390}) as a measure of the ground-state association, then the extent of the association increases with the extent of activation of the sample. The observation of associated complexes of pyrene on silica surfaces is a remarkable observation, since in solution, a ground state bimolecular complex is dissociative. Introduction of alcohols as co-adsorbates (glycerol or 1-decanol) affects the distribution of the fluorophore on the surface. With increasing coverage, the monomer fluorescence exhibits single exponential, and the excimer emission can be deconvoluted with a double exponential with a negative pre-exponential term indicating a true dynamic excimer formation. Adamantanol as a co-adsorbate did not produce this effect. The behavior of the covalently linked bichromophoric molecule 1,3-bis(1-pyrenyl)propane was similar.[45] In a related study, Levitz et al.[50] found that the presence of nonionic surfactants such as Triton X-100 also influences the distribution of adsorbed pyrene molecule on the silica gel. At very low (pyrene/surfactant) ratios in the adsorbed layer, pyrene monomer fluorescence exhibited single exponential (τ = 310 nsec) independent of the surface coverage.

The pyrene emission could also be quenched by halonaphthalenes when both the probe and quenchers are adsorbed on silica gel.[45] The dependence of the mean lifetime τ on the quencher concentration and the differing slopes in the Stern–Volmer plots for intensities and lifetimes suggest that both static and dynamic quenching processes operate on the silica surface:

$$\bar{\tau} = (A_1 \tau_1^2 + A_2 \tau_2^2)/(A_1 \tau_1 + A_2 \tau_2)^2 \qquad (10.6)$$

It has also been found that tumbling of the coated silica gel grains causes intra- and intergranular motion of the aromatic molecules as evidenced by the effects on the singlet and triplet excited-state lifetimes and on the quenching behavior.

Utilizing the monomer and excimer fluorescence as a probe, Lochmuller et al.[51] have examined the organization and distribution of molecules chemically bound to the silica surface. The study employed [3-(3-pyrenyl)propyl] dimethylchlorosilane (3PPS) chemically bonded to microparticulate silica gel. The results obtained, in general, are similar to those obtained for adsorbed pyrene by de Mayo and co-workers. Examination of the fluorescence intensity as a function of surface concentration showed that the chemically bonded molecules were not evenly distributed but are rather clustered into regions of high density. Time-resolved fluorescence studies[52] indicate that the organization and proximity are controlled primarily by the inhomogeneous distribution of reactive silanols on the surface. Studies in different solvents (acetonitrile, methanol, tetrahydrofuran, and hexane) reveal solvent-induced conformational changes resulting in approximately 20% change in the organization of the bound molecules on the surface.

Adsorption of molecules considerably reduces their mobility on the surface and enhances the possibility of observing room-temperature phosphorescence (RTP). Ford and Hurtubise[53] found intense RTP from benzo(f)quinoline and related molecules when they are adsorbed on several silica gel samples. Silica gel chromotoplates containing a polymeric binder with carboxyl groups were the best samples for inducing strong RTP. Adsorption of $[Ir(bpy)_3OH]^{2+}$ onto silica gel apparently does not affect the efficiency with which this transition metal complex photosensitizes the valence isomerization of norbornadiene.[54]

It is well known in catalysis that metal oxides such as V_2O_5, MoO_3, and CrO_3 in a highly dispersed state are efficient catalysts for numerous chemical conversions thermally as well as photochemically (phenol, ammonia synthesis, oxidation, reduction of CO and CO_2, etc.). In mechanistic studies of heterogeneous photocatalysis, the reactive excited states are readily identified via their luminescence if the oxides are dispersed on transparent supports such as porous vycor glass or silica gel. There are a growing number of studies that attempt to identify such luminescence of excited states in dispersed oxides (V_2O_5,[55–58] MoO_3[59–62] and correlate their energy and lifetime with the observed photocatalytic efficiency.

Anpo and co-workers, for example, find a correlation between the lifetime of the charge transfer triplet state of supported oxides (on porous vycor glass) with their efficiency for photoreduction of CO:[59]

photocatalytic activity: $V_2O_5-PVG > MoO_3-PVG > CrO_3-PVG$

(triplet), μsec: 218   63   2.9

Iwamoto et al.[58] similarly find that the relative intensity of phosphorescence of V_2O_5 supported on various materials shows the same order as the catalytic activity of these oxides for reactions such as hydroxylation of benzene:

$$V_2O_5-SiO_2 > V_2O_5-MgO > V_2O_5-\gamma-Al_2O_3 > V_2O_5-\eta-Al_2O_3 > V_2O_5-TiO_2$$

The implications are that vanadyl groups responsible for the phosphorescence are potential active sites for phenol synthesis.

$Ru(bpy)_3^{2+}$ binds ionically to the anionic silanol sites of porous vycor glass (PVG).[63] The absorption and emission spectra of the adsorbed complex are very similar to the spectra in aqueous and alcoholic solutions [absorption maxima 452 nm in PVG and in water; emission maxima 620 nm (PVG) and 600 nm (H_2O); emission lifetimes 740 nsec (PVG, exponential) and 600 nsec (H_2O, exponential)]. When the dye-adsorbed PVG samples are excited in the presence of O_2, N_2O_2, and SO_2 the MLCT excited state is quenched and the quenching parallels adsorption of the gases onto porous vycor glass.[64] Quenching is limited to the adsorbed species and occurs by static and dynamic processes depending on the strength of adsorption of the quencher. The triplet–triplet absorption spectrum of pyrene adsorbed onto PVG has been found to be red-shifted (~ 100 nm) compared to that in homogeneous solutions.[65]

Dynamics of Radical Pairs on Silica Surfaces

In earlier chapters, we have encountered several cases where the dynamic solubilization of molecules in host aggregates produces, in effect, a cage effect on the bimolecular reactions of molecular excited states and on the dynamics of photogenerated species such as redox products and radical pairs. Constrained to move around in a finite volume, the probabilities of certain encounters are increased. This results in major deviations from reaction pathways commonly observed in homogeneous solutions. Adsorption of molecules on surfaces also restricts their mobility and rotational and translational diffusion. (This is also true for photogenerated products on the surface such as radical pairs). Consequently we can anticipate similar effects in photoreactions occurring on surfaces.[66-75] Concurrent to the studies on micelles, De Mayo and co-workers have observed independently, several such novel effects which we shall review here. (The notion of surface mobility of photogenerated radical pairs and their influence on product distribution can be found in the work of Leermakers *et al.* in the late 1960s, though some of their results are complicated by hydrolysis reactions.)

A class of organic photoreactions that is ideally suited for the study of dynamics of photogenerated radical pairs are the photofragmentation reactions of the type:

$$R_1 \text{---} X \text{---} R_2 \xrightarrow{h\nu} \underset{\text{geminate pair}}{[R_1 \text{---} \overset{\cdot}{X} \quad R_2]} \xrightarrow{-X} [R_1 \cdot \ \cdot R_2] \qquad (10.7)$$

In a solution solvent cage, a spin-correlated geminate radical pair, either singlet or triplet, can undergo a number of processes: (1) it may recombine to regenerate the starting materials $R_1 \text{---} X \text{---} R_2$; (2) it may either recombine in a different manner $(R_1' \text{---} X \text{---} R_2)$ or it may disproportionate to form new products $(R_1 H, R_2' H, \text{etc.})$; (3) it may be transformed into a new radical pair $[R_1 \cdot \ \cdot R_2]$, or (4) it may diffuse apart to form free radicals $(R_1 \cdot$ and $R_2 \cdot)$.

The rate of processes (1) and (4) depend on the radical pair spin multiplicity since they occur only with singlet radical pairs. Because intersystem crossing must precede the reactions of a triplet radical pair, spin independent processes (3) and (4) compete more effectively with geminate recombination than for singlet pairs. Changing either the rate of intersystem crossing of the radical pair or its rate of diffusional separation will affect the relative contribution of processes (1)–(4) to the decay of the particular radical pair. For example, increasing the viscosity of the medium or generating the radical pair in an environment in which its movement is restricted will affect the diffusional behavior of the radical pair. Such effects are readily evaluated by measuring the changes in the amount of geminate radical recombination under various conditions.

The dynamics of radical pair mobility have been examined on silica gel surfaces for several photogenerated radical pairs. The earliest system investigated by Leermakers *et al.* was cyanopropyl radicals generated via photolysis of azobis(isobutyronitrile):

$$\underset{(A \text{---} X \text{---} A)}{(CH_3)_2 C(CN) \text{---} N = N \text{---} C(CN)(CH_3)_2} \xrightarrow[-N_2]{h\nu} \underset{\quad A \qquad\qquad A}{[(CH_3)_2(CN)C \cdot \ \cdot C(CN)(CH_3)_2]}$$

$$\longrightarrow \underset{(A \text{---} A)}{(CH_3)_2(CN)C \text{---} C(CN)(CH_3)_2} + \underset{(A \text{---} A')}{(CH_3)_2(CN)C \text{---} N = C = C(CH_3)_2} \qquad (10.8)$$

$$+ \underset{\text{(Hydrolysis product of } A \text{---} A')}{(CH_3)_2(CN)C \text{---} NH \text{---} CO \text{---} CH(CH_3)_2}$$

For photolysis in silica gel–benzene slurries, Leermakers reported the formation of recombination product (A—A) only and that the cyanopropyl radicals (A) were not free to rotate on the surface to produce the unsymmetrical coupling product ketenimine (A—A'). Later investigations by Johnston *et al.*[66] found that ketenimine is very susceptible to hydrolysis and that the ketenimine and its hydrolysis products are indeed formed in high yields both for silica gel slurries in benzene and on dry silica gel. Presumably, both rotational and translational motion do occur in this system to yield effects similar to that observed in solvent cages. Leffler *et al.* have studied azocumene[67] and diacylperoxides[68] photolysis on silica gel and found substantial mobility for the photogenerated radicals on the silica surface:

azocumene photolysis:

$$Ph(CH_3)_2C—N{=}N—C(CH_3)_2Ph \xrightarrow[-N_2]{h\nu} [Ph(CH_3)_2C\cdot \quad \cdot C(CH_3)_2Ph] \qquad (10.9)$$
$$\underset{(A—X—A)}{} \qquad\qquad\qquad\qquad \underset{A \qquad\quad A}{}$$

$$\longrightarrow \quad \underset{(A—A)}{dicumyl} + \underset{(A—A')}{HT\ cumyl\ dimer} + \underset{(disproportionation\ products)}{\alpha\text{-Methyl styrene} + cumene}$$

diacylperoxide photolysis:

$$(R_1{=}Ph\ CH_2CH_2—)$$

$$R_1COO—OOCR_1 \xrightarrow{h\nu} [R_1COO\cdot\cdot OOCR_1] \xrightarrow[-CO_2]{} [R_1\cdot\ \cdot R_1]$$

$$\longrightarrow \quad R_1COOH + R_1COOR_1 + R_1H + R_1R_1 + dibenzyl \quad (10.10)$$

Photo-Fries rearrangements of anilide[69,70] yielding ortho and para migration and elimination products show small differences in the product distribution when carried out on dry silica gel, implying limited translational motion on the surface. The para isomer yields slightly higher than that observed in methanol:

photo-Fries rearrangement:

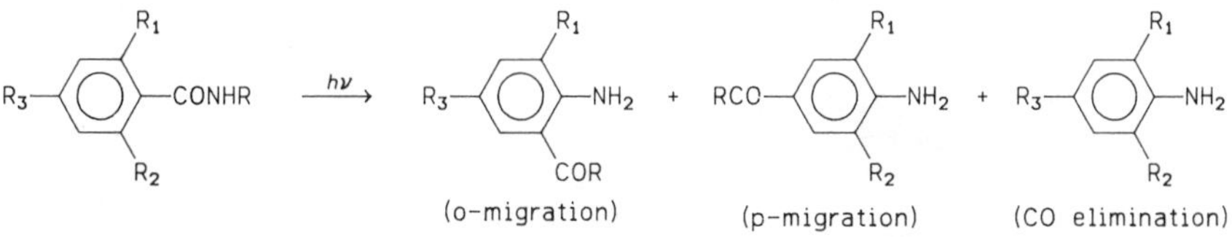

$$(10.11)$$

Depending on their spin multiplicity and the photolysis temperature, benzyl radical pairs generated via photolysis of dibenzylketones, dibenzyl-sulfones, or benzylphenylacetates show significant differences in their mobility on the silica surfaces:[71,72]

benzylic compounds photolysis:

$$A—X—B \xrightarrow[-X]{h\nu} A—A + A—B + B—B \qquad (10.12)$$

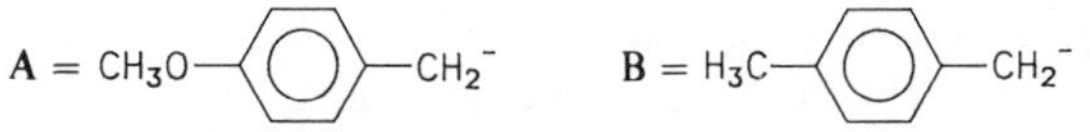

X = CO (dibenzylketone, DBK), CO$_2$ (benzylphenylacetate, BPA) and SO$_2$ (dibenylsulfone, DBS)

Table 10.2 presents some representative results. Results indicate that the translation motion is insensitive to the state of hydration of the silica gel and the presence of co-adsorbates. The yields of unsymmetrical coupling product (A—B) increase dramatically on lowering of the temperature.

Baretz and Turro[73,74] investigated the homolytic cleavage of adsorbed benzylic ketones (2,4-diphenylpentan-3-one, DPP; 1,3,4-triphenylbutan-2-one, TPB; and 1,3-diphenylbutan-2-one, DPB) when the ketones were adsorbed onto porous silica and porous glass samples. In neat solvents such as benzene or pentane, the main course of reaction in DPP photolysis is decarbonylation followed by coupling ($\sim 93\%$) or disproportionation ($\sim 3\%$) of $PhCHCH_3 \cdot$ radicals. In the adsorbed state, the extent of disproportion and

TABLE 10.2. Influence of Adsorption of Substrate on Silica Surfaces in Photofragmentation Reactions [a]

Substrate/medium[b]	Product distribution				
	Recombination products			Disproportionation	
DPP photolysis	AA(DPP)	AB'(PPA)	AXA(DPP)	A'(ETB)	A''(STY)
Pentane (27°C)	93%	0.05%	0.5%	3%	3%
Porous glass (27°C)	76	2	8	8	6
Porous silica (27°C)	83	1	8	3	5
TLC SiO$_2$ (27°C)	78	2	12	3	5
Pentane (−77°C)	90	0.2	0.5	4	5
Porous glass (−77°C)	15	15	45	5	20
TLC SiO$_2$ (−77°C)	31	12	40	2	15
DBK photolysis	AA	AB	BB	(relative yields)	
Dioxan (20°C)	1.0	2.5	1.0		
Silica gel (1%)(20°C)	1.2	3.7	1.0		
Silica gel (10%)(20°C)	1.0	3.2	1.0		
Silica gel (50%)(20°C)	1.0	3.1	1.0		
Silica gel (10%)(−55°C)	1.2	9.0	1.0		
Silica gel (50%)(−55°C)	1.0	6.2	1.0		
Silica gel (50%)(					
Silica gel (50%)(−165°C)		95%			
BPA photolysis	AA	AB	BB		
Silica gel (1%)(20°C)	1.0	6.1	1.0		
Silica gel (10%)(20°C)	1.4	5.5	1.0		
Silica gel (50%)(20°C)	1.0	4.1	1.0		
Silica gel (50%)(−55°C)		96%			

[a] From Baretz and Turro,[73] Turro et al.,[74] and Johnston and Wong.[75]

[b] DPP = 2,4-diphenylpentan-3-one; DBK = dibenzylketone; and BPA = benzylphenyl-acetate.

diastereoisomeric interconversion increases significantly, especially at low temperatures. Application of magnetic fields also influences the product distribution:

diphenylpentanone photolysis:

$$Ph(CH_3)CH—CO—CH(CH_3)Ph \xrightarrow[-CO]{h\nu}$$
$$\text{meso A—X—A (DPP)}$$

$$PhCH_2CH_3 + PhCH=CH_2 + Ph(CH_3)CH \cdot CO \cdot CH(CH_3)Ph \qquad (10.13)$$
$$\text{A' (ETB)} \qquad \text{A'' (STY)} \qquad \text{A—A (DPP) } d,l$$

$$+ Ph(CH_3)CH—CH(CH_3)Ph + Ph(CH_3)CH—CHO$$
$$\text{(A—A)(DPB)} \qquad \text{(A—B')(PPA)}$$

Similar to the behavior observed in micelles, photolysis of benzylketones on silica surfaces also shows substantial ^{13}C isotope enrichments.[75]

Other Photoreactions on Silica Surfaces

Photolysis of benzoin ethers demonstrates nicely how the variation in the reaction course can vary on changing of the reaction medium from homogeneous solvents to silica surfaces.[76] In solution, the reaction proceeds via singlet and triplet radical pairs (type I cleavage) producing pinacol ether and benzil:

$$Ph—CO—CH—Ph \xrightarrow[\text{type I}]{h\nu} [Ph—CO \cdot \cdot CH—Ph] \longrightarrow$$
$$\qquad\qquad \overset{|}{OCHR_2} \qquad\qquad\qquad \overset{|}{OCHR_2}$$
$$\text{(A—X—B)} \qquad\qquad\qquad \text{[A—X} \cdot \cdot \text{B]}$$

$$PhCOCOP + PhCH——CH——Ph \qquad (10.14)$$
$$\qquad\qquad\qquad\quad \overset{|}{OCHR_2} \overset{|}{OCHR_2}$$
$$\qquad\quad \text{(A—A)} \qquad\qquad \text{(B—B)}$$
$$\qquad\quad \text{benzil} \qquad\qquad \text{pinacol}$$

Type II pathway contributes only in a minor way. On a silica gel type II products are formed in substantial amounts, presumably hydrogen bonding effects facilitate the formation of critical cyclic intermediates:

$$Ph—CO—CH—Ph \xrightarrow[\text{type II}]{h\nu} Ph—\overset{OH}{\underset{R_2C—O}{\overset{|}{C}—\overset{\vdots}{C}H}} \longrightarrow PhCOCH_2Ph + Ph—\overset{OH}{\underset{R_2C—O}{\overset{|}{|}—CHPh}}$$
$$\quad\; \overset{|}{OCHR_2}$$

$$(10.15)$$

Zawadski and Ellis[77] have explored the effect of adsorption of organometallic compounds on silica gel on their photoreactivity. For example, several (arene) $(CO)_3Cr^0$ complexes have been adsorbed onto wet silica gel and suspended in cyclohexane solutions. Photolysis in the Cr−arene charge transfer bands have provided evidences for the operation of steric effects in the slurry photoreactivity. In comparative studies in normal and large pore size silica gels, with identical electronic spectra, the large pore size gel gave higher quantum yields for the disappearance of the complex. Obviously, the recombination of the photodissociated CO is less competitive in large sized gels.

The photocycloaddition of alkenes or allenes to steroidal enones[77a] has been shown to occur in the adsorbed state on silica gel and alumina surfaces. This reaction, normally occurring from the less hindered side has been found to be directed towards the more hindered side on adsorption of the steroidal enone on a dry silica gel or alumina surface:

	(4α5α)	(4α5β)	(4β5β)
in methanol (25°C)	57%	32%	10%
on SiO$_2$ (11°C)	49%	trace	51%
on SiO$_2$ (−70°C)	42%	trace	58%

The adsorption of the steroid on the surface disfavors the required conformational inversion in the intermediate 1,4-biradical for the trans addition. Such control of regiospecificity on surfaces can have potential synthetic utility.

Photodimerization of acenaphthylene[45,47] is another reaction ideally suited to study the effect of substrate adsorption on the photoreaction pathways, stereochemistry in particular. Photodimerization can yield both cis and trans dimers:

acenaphthylene   cis dimer   trans dimer

The short-lived singlet state is thought to generate only the cis dimer. The long-lived triplets that allow diffusion produce both cis and trans products. In homogeneous solvents, the reaction shows some polarity dependence, favoring the cis dimer in polar solvents. If the substrate adsorption on silica surfaces is inhomogeneous and/or the surface mobility restricted, then we can anticipate effects on the observed (cis/trans) photodimer yields. Indeed, on silica and alumina surfaces, the cis/trans ratio has been found to be very sensitive to the extent of surface coverage by acenaphthylene and to the

TABLE 10.3. Influence of Adsorption on Silica Gel Surfaces and
on Coadsorbates in the Photodimerization
of Acenaphthylene [a]

Co-adsorbate	Medium and coverage	(Cis/trans) dimer yield		% dimer from triplet (direct)
		Direct	Sensitized	
None	CH_3OH	0.18		
None	C_6H_6		0.54	
None	SGel (13%)	8.76	0.76	14–18
None	SGel (3%)		1.02	
1-Adamantanol	SGel (13%)	2.28	0.34	39–41
1-Heptanol	SGel (13%)	4.47	0.61	27–30
1-Decanol	SGel (13%)	1.96	0.55	51–53
1-Hexadecanol	SGel (13%)	1.59	0.56	59–60
Acenaphthene	SGel (13%)	2.83	0.54	38–40
Acenaphthalene	SGel (3%)	4.04	0.52	

[a] Data from Bauer et al.[45,47]

dilution of the surface concentration with inert additives such as alkanols, adamantanol, and acenaphthene. Table 10.3 provides representative data on the (cis/trans) product yields ratios for direct and sensitized dimerization in the presence of silica gel. The dimerization can be quenched by triplet quencher molecules such as ferrocene, suggesting *inter alia* intragranular diffusion of adsorbed molecules. Dimerization of 9-cyanoanthrene on silica gel surfaces has also received some scrutiny.[47]

10.4 Studies on Alumina Surfaces

Closely related to the studies on silica gel are similar studies carried out on alumina (Al_2O_3). Surface structures of alumina are rather complex. Various phases are known including γ- and η-aluminas which form the so-called catalytic aluminas.[78] Most of the photophysical studies have been on γ-alumina; Kessler and Wilkinson have cited various reasons for this choice: (1) its surface and dehydration procedure is well established; (2) in the absence of O_2, chemisorbed aromatic hydrocarbons are relatively stable; (3) controlled variation in the pretreatment procedures of the surface causes decisive changes in the properties of the excited states of aromatic compounds allowing studies to be made as a function of surface activity, and (4) several adsorbed hydrocarbons exhibit intense luminescence even at room temperatures.

Photophysical properties of numerous aromatic hydrocarbons have been studied on alumina surfaces,[79–87] mostly in the laboratories of Oelkrug et al. On activated γ-alumina, the fluorescence properties depend on the activation

References

1. M. Formenti and S. J. Teichner, *Catalysis (London)* **2**, 87 (1978).
2. A. J. Bard, *Science* **207**, 139 (1980).
3. R. I. Bickley, *Catalysis (London)* **5**, 308 (1983).
4. K. Kalyanasundaram, *in* "Energy Resources by Photochemistry and Catalysis" (M. Grätzel, ed.), p. 217. Academic Press, New York, 1983.
5. K. Kalyanasundaram and M. Grätzel, *in* "Chemistry and Physics of Solid Surfaces" (R. Vanselov and R. Howe, eds.), Vol. 5, p. 111. Springer-Verlag, Berlin and New York, 1984.
6. R. K. Iler, "The Chemistry of Silica," 2nd Ed. Wiley, New York, 1979. (a) R. K. Iler, "The Colloid Chemistry of Silica and Silicates." Cornell Univ. Press, Ithaca, New York, 1955.
7. D. W. Fuerstenau, *in* "The Chemistry of Biosurfaces" (M. L. Hair, ed.), Vol. 1, p. 143. Dekker, New York, 1971.
8. J. Wheeler and J. K. Thomas, *in* "Inorganic Reactions In Organised Media" (S. L. Holt, ed.), ACS Symposium Series, No. 177, p. 97. *Am. Chem. Soc.*, Washington, D.C., 1981.
9. I. Willner, J. W. Otvos, and M. Calvin, *J. Am. Chem. Soc.* **103**, 3203 (1981).
10. C. Laane, I. Willner, J. W. Otvos, and M. Calvin, *Proc. Natl. Acad. Sci. U.S.A.* **78**, 5928 (1981).
11. I. Willner, J.-M. Yang, C. Laane, J. W. Otvos, and M. Calvin, *J. Phys. Chem.* **85**, 3277 (1981).
12. I. Willner and Y. Degani, *J.C.S. Chem. Commun.* p. 761 (1982); *Isr. J. Chem.* **22**, 163 (1982).
13. Y. Degani and I. Willner, *J.C.S. Chem. Commun.* p. 710 (1983); *J. Am. Chem. Soc.* **105**, 6228 (1983).
14. J. Wheeler and J. K. Thomas, *J. Phys. Chem.* **88**, 750 (1984).
15. J. Wheeler and J. K. Thomas, *J. Phys. Chem.* **86**, 4540 (1983).
16. B. K. G. Theng, "The Chemistry of Clay Organic Reactions." Wiley, New York, 1974.
17. H. Van Olphen, "An Introduction to Clay Colloid Chemistry," 2nd Ed. Wiley, New York, 1977.
18. T. J. Pinnavaia, *Science* **220**, 365 (1983). (a) A. Weiss, *Angrew. Chem. Int. Ed. Engl.* **20**, 850 (1981).
19. D. Krenske, S. Abdo, H. Van Damme, M. Cruz, and J. J. Fripiat, *J. Phys. Chem.* **84**, 2447 (1983).
20. S. Abdo, P. Canesson, M. Cruz, J. J. Fripiat, and H. Van Damme, *J. Phys. Chem.* **85**, 797 (1981).
21. R. A. DellaGuardia and J. K. Thomas, *J. Phys. Chem.* **87**, 990 (1983).
22. A. Habti, D. Keravis, P. Levitz, and H. Van Damme, *J.C.S. Faraday II* **80**, 67 (1984).
23. P. K. Ghosh and A. J. Bard, *J. Phys. Chem.* **88**, 5519 (1984).
24. R. A. Schoonheydt, P. DePauw, D. Vliers, and F. C. DeSchryver, *J. Phys. Chem.* **88**, 5113 (1984).
25. F. Bergaya and H. Van Damme, *J.C.S. Faraday II* **79**, 505 (1983).
26. R. A. DellaGuardia and J. K. Thomas, *J. Phys. Chem.* **87**, 3550 (1983).
27. R. A. DellaGuardia and J. K. Thomas, *J. Phys. Chem.* **88**, 964 (1984).
28. L. Kovar, R. DellaGuardia, and J. K. Thomas, *J. Phys. Chem.* **88**, 3595 (1984).
29. H. Nijs, M. I. Cruz, J. J. Fripiat, and H. Van Damme, *Nouv. J. Chim.* **6**, 551 (1982); *J.C.S. Chem. Commun.* p. 1026 (1981).
30. H. Nijs, J. J. Fripiat, and H. Van Damme, *J. Phys. Chem.* **87**, 1279 (1983). (a) C. Detellier and G. Villemure, *Inorg. Chim. Acta* **86**, L19 (1984).
31. J. H. de Boer, *Z. Phys. Chem., Abt. B* **14**, 163 (1931); **15**, 281 (1932); **16**, 397 (1932); **17**, 161 (1932); **18**, 49 (1932).
32. J. H. de Boer and C. J. Dippel, *Z. Phys. Chem., Abt. B* **25**, 399, 408 (1933).
33. J. H. de Boer and J. F. H. Custers, *Z. Phys. Chem., Abt. B* **21**, 208, 217 (1933); **25**, 225 (1933).
34. P. A. Leermakers, L. D. Weis, and H. T. Thomas, *J. Am. Chem. Soc.* **87**, 4603 (1965).
35. L. D. Weis, B. W. Bowen, and P. A. Leermakers, *J. Am. Chem. Soc.* **88**, 3177 (1966).

36. P. A. Leermakers, H. T. Thomas, L. D. Weis, and F. C. James, *J. Am. Chem. Soc.* **88,** 5075 (1966).

37. T. R. Evans, A. F. Toth, and P. A. Leermakers, *J. Am. Chem. Soc.* **89,** 5060 (1967).

38. L. D. Weis, T. R. Evans, and P. A. Leermakers, *J. Am. Chem. Soc.* **90,** 6109 (1968).

39. C. H. Nicholls and P. A. Leermakers, *Adv. Photochem.* **8,** 315 (1971).

40. A. V. Kisielew, "Structure and Properties of Porous Materials." Academic Press, New York, 1968.

41. H. Ishida, H. Takahashi, and H. Tsubomura, *Bull. Chem. Soc. Jpn.* **43,** 3130 (1970).

42. H. Ishida, H. Takahashi, and H. Tsubomura, *J. Am. Chem. Soc.* **92,** 275 (1970).

43. H. Ishida and H. Tsubomura, *J. Photochem.* **2,** 285 (1973–1974).

44. R. K. Bauer, P. de Mayo, L. V. Natarajan, and W. R. Ware, *Can. J. Chem.* **62,** 1279 (1984).

45. R. K. Bauer, P. de Mayo, K. Okada, W. R. Ware, and K. C. Wu, *J. Phys. Chem.* **87,** 460 (1983).

46. R. K. Bauer, P. de Mayo, W. R. Ware, and K. C. Wu, *J. Phys. Chem.* **86,** 3781 (1982).

47. R. K. Bauer, R. Borenstein, P. de Mayo, K. Okada, M. Rafalska, W. R. Ware, and K. C. Wu, *J. Am. Chem. Soc.* **104,** 4635 (1982).

48. K. Hara, P. de Mayo, W. R. Ware, A. C. Weedon, G. S. K. Wong, and K. C. Wu, *Chem. Phys. Lett.* **69,** 105 (1980).

49. C. Francis, J. Liu, and L. A. Singer, *Chem. Phys. Lett.* **94,** 162 (1983).

50. P. Levitz, H. Van Damme, and D. Keravis, *J. Phys. Chem.* **88,** 2228 (1984).

51. C. H. Lochmuller, A. S. Colborn, M. L. Hunnicutt, and J. M. Harris, *Anal. Chem.* **55,** 1344 (1983).

52. C. H. Lochmuller, A. S. Colborn, M. L. Hunnicutt, and J. M. Harris, *J. Am. Chem. Soc.* **106,** 4077 (1984).

53. C. D. Ford and R. J. Hurtubise, *Anal. Chem.* **51,** 659 (1979); **52,** 656 (1980).

54. P. A. Grutsch and C. Kutal, *J.C.S. Chem. Commun.* p. 893 (1982).

55. A. M. Gritscov, V. A. Shvets, and V. B. Kazanski, *Kinet. Catal. (Engl. Transl.)* **15,** 1257 (1974).

56. A. M. Gritscov, V. A. Shvets, and V. B. Kazanski, *Chem. Phys. Lett.* **35,** 511 (1975).

57. M. Anpo, I. Tanakashi, and Y. Kubokawa, *J. Phys. Chem.* **84,** 3440 (1980).

58. M. Iwamoto, A. Furukawa, K. Matsukami, T. Takenaka, and S. Kagawa, *J. Am. Chem. Soc.,* **105,** 3719 (1983).

59. M. Anpo, I. Tanahashi, and Y. Kubokawa, *J. Phys. Chem.* **86,** 1 (1982).

60. Y. Iwasawa and S. Ogasawa, *Bull. Chem. Soc. Jpn.* **53,** 3709 (1980).

61. M. Anpo, T. Suzuki, Y. Kubokawa, F. Tanaka, and S. Yamashita, *J. Phys. Chem.* **88,** 5778 (1984).

62. M. Anpo, I. Tanakashi, and Y. Kubokawa, *J.C.S. Faraday I* **78,** 2121 (1982).

63. A. Basu, H. D. Gafney, D. J. Perettie, and J. B. Clark, *J. Phys. Chem.* **87,** 4532 (1983).

64. S. Wolfgang and H. D. Gafney, *J. Phys. Chem.* **87,** 5395 (1983).

65. P. L. Picuolo and J. W. Sutherland, *J. Am. Chem. Soc.* **101,** 3123 (1979).

66. L. J. Johnston, P. de Mayo, and S. K. Wong, *J.C.S. Chem. Commun.* p. 1106 (1982); *J. Org. Chem.* **49,** 20 (1984).

67. J. E. Leffler and J. J. Zapanicic, *J. Am. Chem. Soc.* **102,** 259 (1980).

68. J. E. Leffler and J. T. Barbas, *J. Am. Chem. Soc.* **103,** 7768 (1981).

69. D. Avnir, P. de Mayo, and I. Ono, *J.C.S. Chem. Commun.* p. 1109 (1978).

70. M. M. Abdel-Malik and P. de Mayo, *Can. J. Chem.* **62,** 1275 (1984).

71. D. Avnir, L. J. Johnston, P. de Mayo, and S. K. Wong, *J.C.S. Chem. Commun.* p. 958 (1981).

72. B. Frederick, L. J. Johnston, P. de Mayo, and S. K. Wong, *Can. J. Chem.* **62,** 403 (1984).

73. B. H. Baretz and N. J. Turro, *J. Am. Chem. Soc.* **105,** 1309 (1983).

74. N. J. Turro, C.-C. Cheng, and W. Mahler, *J. Am. Chem. Soc.* **106,** 5022 (1984).

75. L. J. Johnston and S. K. Wong, *Can. J. Chem.* **62,** 1999 (1983).

76. P. de Mayo, A. Nakamura, P. W. K. Tsang, and S. K. Wong, *J. Am. Chem. Soc.* **104,** 6824 (1982).

77. M. E. Zawadski and A. B. Ellis, *J. Org. Chem.* **48,** 3156 (1983); *Organometallics* **3,** 192 (1984). (a) R. Farwaha, P. de Mayo, and Y. C. Toong, *J. Chem. Soc. Chem. Commun.* p. 739 (1983).
78. H. Knozinger and P. Ratnasamy, *Catal. Rev. Sci. Eng.* **17,** 31 (1978).
79. D. Oelkrug, M. Radjaipour, and H. Erbsse, *Z. Phys. Chem. (Wiesbaden)* **88,** 23 (1974).
80. D. Oelkrug, H. Erbse, and M. Plauschinat, *Z. Phys. Chem. (Wiesbaden)* **96,** 275 (1975).
81. D. Oelkrug and M. Radjaipour, *Z. Phys. Chem. (Wiesbaden)* **123,** 169 (1980).
82. D. Oelkrug, M. Plauschinat, and R. W. Kessler, *J. Lumin.* **18/19,** 434 (1979).
83. R. W. Kessler, S. Uhl, W. Honnen, and D. Oelkrug, *J. Lumin.* **24/25,** 551 (1981).
84. M. Paluschinat, W. Honnen, G. Krabichler, S. Uhl, and D. Oelkrug, *J. Mol. Struct.* **115,** 351 (1984).
85. W. Honnen, G. Krabichler, S. Uhl, and D. Oelkrug, *J. Phys. Chem.* **87,** 4872 (1983).
86. V. A. Fenin, V. A. Shvets, and V. B. Kazanski, *Kinet. Catal. (Engl. Transl.)* **19,** 1289 (1978).
87. G. N. Asmdov and O. V. Krybov, *Kinet. Catal. (Engl. Transl.)* **19,** 1004, 1208 (1978).
88. R. W. Kessler and F. Wilkinson, *J.C.S. Faraday I* **77,** 309 (1981).
89. R. W. Kessler, G. Krabichler, S. Uhl, D. Oelkrug, W. P. Hogen, J. Hyslop, and F. Wilkinson, *Opt. Acta* **30,** 1099 (1983).
90. G. Beck and J. K. Thomas, *Chem. Phys. Lett.* **94,** 553 (1983).
91. H. G. Hecht and R. L. Crackel, *J. Photochem.* **15,** 263 (1981).
92. H. G. Hecht and J. L. Jansen, *J. Photochem.* **9,** 33 (1978).

Appendix

Studies of photophysics and photochemistry of molecules in various forms of microheterogeneous systems is currently an extremely popular and important area of research. Progress is very rapid and research is highly competitive. Coverage of the literature in this volume is restricted to end of 1984. The purpose of this appendix is to supply an updated bibliography of material published since the completion of the work presented in this volume. The lengthiness of this list (all 1985 publications) is evidence of the intense level of research activity in the topic of this volume. The list is classified according to chapter subject.

Chapters 2–4—Micelles

1. E. Blatt and W. H. Sawyer, *Biochim. Biophys. Acta* **822,** 43 (1985). Depth-dependent fluorescence quenching in micelles and membranes.
2. K. P. Ananthapadmanaban, E. D. Goddard, N. J. Turro, and P. L. Kuo, *Langmuir* **1,** 352 (1985). Fluorescence probes for critical micelle concentration.
3. R. Malliaris, J. Le Moigne, J. Strumm, and R. Zana, *J. Phys. Chem.* **89,** 2709 (1985). Temperature-dependence of micelle aggregation number and the rate of intramicellar excimer formation in aqueous surfactant solutions.
4. S. Hautecloque, D. Grand, and A. Bernas, *J. Phys. Chem.* **89,** 2705 (1985). Salt effects on the photoionization yields in micellar systems.
5. F. Steinmuller and H. Rau, *J. Photochem.* **28,** 297 (1985). Association phenomena and enhancement of light induced charge separation in Ru-$[(C_{13}H_{22})_2bpy]_3^{2+}$—$MV^{2+}$ system in large SDS micelles.
6. B. Lelerbours, Y. Chevalier, and M. P. Pileni, *Chem. Phys. Lett.* **117,** 89 (1985). Micellar effects on charge separation.
7. H. Shioyama, A. Takami, and N. Mataga, *Bull. Chem. Soc. Jpn.* **58,** 1029 (1985). Comparative studies on anionic molecular assemblies as media for electron transfer quenching of $Ru(bpy)_3^{2+}$ by cationic quenchers.

8. M. J. Politi, O. Brandt, and J. H. Fendler, *J. Phys. Chem.* **89**, 2345 (1985). Ground- and excited-state proton transfers in reverse micelles. Polarity restrictions and isotope effects.

9. W. D. Turley and H. W. Offen, *J. Phys. Chem.* **89**, 2833 (1985). Micellar microfluidities at high pressures.

10. S. Hashimoto and J. K. Thomas, *J. Phys. Chem.* **89**, 2771 (1985). Laser photolysis studies of electron transfer reactions in micellar–metal ion systems.

11. N. J. Turro, M. A. Paczkowski, and P. Wan, *J. Org. Chem.* **50**, 1399 (1985). Magnetic isotope effects in photochemical reactions: observation of carbonyl hyperfine coupling to phenylacetyl and benzoyl radicals.

12. K. Kano, Y. Ueno, and S. Hashimoto, *J. Phys. Chem.* **89**, 3161 (1985). Fluorescence studies on characterization and solubilizing abilities of SDS, CTAB, and Triton X-100.

13. R. Arce and L. Kevan, *J. Chem. Soc., Faraday Trans. I*, **81**, 1669 (1985). Photochemical behavior of TMB and its protonoted form in SDS micelles under 337 nm laser irradiation.

14. R. Arce and L. Kevan, *J. Chem. Soc., Faraday Trans. I* **81**, 1025 (1985). Photoionization of TMB in CTAC miceles under 337 nm laser irradiation.

15. I. R. Gould, M. B. Zimmt, N. J. Turro, B. H. Baretz and G. F. Lehr, *J. Am. Chem. Soc.* **107**, 4607 (1985). Dynamics of radical pair reactions in micelles.

16. M. Mitsuzuka, K. Kikuchi, H. Kokubun, and Y. Usui, *J. Photochem.* **29**, 363 (1985). Primary processes of the photosensitization of methylene blue in aqueous SDS micellar solutions.

17. S. J. Formoshinho, K. Migrid, and G. Mode, *J. Chem. Soc., Faraday Trans. I* **81**, 1891 (1985). Photophysics of uranyl ions in aqueous solution: 5: fluorescence quenching in Triton X-100 micellar solutions.

18. Y. Kaizu, H. Ohta, K. Kobayashi, K. Takuma, and T. Matsuo, *J. Photochem.* **30**, 93 (1985). Lifetimes of lowest excited states of $Ru(bpy)_3^{2+}$ and its amphiphilic derivatives in micellar systems.

19. S. W. Snyder, D. E. Rainer, P. T. Rieger, J. N. Demas, and B. A. DeGraff, *Langmuir* **1**, 548 (1985). Luminescence quenching of Ru(II) photosensitizers by Cu^{2+} in Triton surfactant media.

20. E. Roelants, E. Gelade, J. Smid, and F. C. DeSchryver, *J. Colloid Interface Sci.* **107**, 337 (1985). A study of temperature dependence of mean aggregation number and kinetic parameters of quenching in CTAC, TTAC micelles.

21. G. Gabor and N. J. Turro, *Photochem. Photobiol.* **42**, 447 (1985). Variations of Stern–Volmer quenching constants in micellar solutions as a function of aqueous or micellar quencher.

22. N. J. Turro, X. Lei, I. R. Gould, and M. B. Zimmt, *Chem. Phys. Lett.* **120**, 397 (1985). External magnetic field dependent influence of lanthanide ions on the chemistry of radical pairs in micelles.

23. G. G. Warr and F. Grieser, *Chem. Phys. Lett.* **116**, 505 (1985). The effect of long-chain fluorescence probes on the size of the SDS micelles.

24. K. Ohkubo, M. Chiba, and K. Yamashita, *J. Mol. Catal.* **32**, 1 (1985). Micellar acceleration of electron transfer from photoexcited 1-benzylnicotinamide to a hydrophilic metal complex.

25. A. R. Lehemy, R. Rossetti, and L. E. Brus, *J. Phys. Chem.* **89**, 4091 (1985). TTF photoionization in micellar solution: a time resolved raman scattering study of interfacial solvation.

26. M. Okazaki, S. Sakta, R. Konaka, and T. Shiga, *J. Am. Chem. Soc.* **107**, 7214 (1985). Application of spin trapping to probe radical pair model in magnetic field dependent photoreduction of naphthoquinone in SDS micelles.

27. I. Okura, T. Kita, and S. Aono, *J. Mol. Catal.* **32**, 361 (1985). Retardation of back electron transfer in photoredox processes by addition of surfactant micelles.

28. N. Ramnath, V. Ramesh, and V. Ramamurthy, *J. Photochem.* **31**, 75 (1985). Micellar structure and micellar control of photochemical reactions.

29. G. Jones, W. R. Jackson, S. Kanottanoporn, and W. R. Bergman, *Photochem. Photobiol.* **42,** 477 (L985). Photophysical and photochemical properties of coumarin dyes in amphiphilic media.

30. M. V. Encinas and E. A. Lissi, *Photochem. Photobiol.* **42,** 491 (1985). Deactivation of pyrene by indole in micellar solutions.

31. K. Ohkubo and Y. Arikawa, *J. Mol. Catal.* **33,** 65 (1985). Micellar promoted stereoselective photoreduction of $Co(III)(en)(ac)_4$ by a chiral long chain Ru complex.

32. L. B. A. Johnson, T. Vallmark, and G. Lindblom, *J. Chem. Soc., Faraday Trans., I,* **81,** 1389 (1985). Polarized light spectroscopic study of indocarbocyanine dyes solubilized in amphiphilic aggregates.

33. Y. Degani and I. Willner, *J. Phys. Chem.* **89,** 5685 (1985). Complex formation between anthraquinone-2,6-disulphonate, and a neutral zinc porphyrin: effect of CTAB micelles on complex stability and photoinduced electron transfer.

34. M. Nakagaki, H. Komatsu, and T. Honda, *Bull. Chem. Soc. Jpn.* **58,** 3197 (1985). Effect of alkyl chain of thiacarbocyanine dyes as the sensitizers for the photoreduction of MV^{2+} in the micellar phase.

35. D. LeRoux, M. Takakubo, and J.-C. Mialocq, *J. Chim. Phys.* **82,** 739 (1985). Photosensitized charge transfer and recombination of ionic products in micellar solutions.

36. K. Takagi, K. Aoshima, Y. Sawaki, and H. Iwamura, *J. Am. Chem. Soc.* **107,** 47 (1985). Electron relay chain mechanism in the sensitized photo-isomerization of stilbazole in aqueous anionic micelles.

37. T. J. Burkey and D. Griller, *J. Am. Chem. Soc.* **107,** 246 (1985). Micellar systems as devices for enhancing the lifetimes and concentration of free radicals.

38. K. S. Schanze, M. S. Dong, and D. G. Whitten, *J. Am. Chem. Soc.* **107,** 507 (1985). Photochemical reactions in organized assemblies. 43. Micelle and vesicle solubilization sites. Determination of micropolarity and microviscosity using photophysics of a dipolar olefin.

39. M. H. Abdel-Kader, A. M. Braun, and N. Paillous, *J. Chem. Soc., Faraday Trans. I* **81,** 245 (1985). Investigations of the dynamic behavior of counterions of anionic micellar system by fluorescence quenching experiments.

40. S. N. Guha, P. N. Moorthy, and K. N. Rao, *J. Photochem.* **28,** 37 (1985). Transient spectra and decay kinetics of semithinine radicals in SDS micellar media.

41. T. Miyashita, T. Murakata, Y. Yamaguchi, and M. Matsuda, *J. Phys. Chem.* **89,** 497 (1985). Kinetic studies of emission quenching of photoexcited Ru(II) complexes by univalent and divalent pyridinium cations in SDS micellar solution.

42. K. Kasatani, M. Kawaski, H. Sato, and N. Nakashima, *J. Phys. Chem.* **89,** 542 (1985). Micellized sites of dyes in SDS micelles as revealed by time resolved energy transfer studies.

43. E. Blatt, K. P. Ghiggino, and W. H. Sawyer, *Chem. Phys. Lett.* **114,** 47 (1985). Fluorescence anisotropy studies of n-(9-anthroyloxy) fatty acids in Triton X-100 micelles.

44. T. Nagamura, N. Takeyama, and T. Matsuo, *J. Colloid Interface Sci.* **103,** 202 (1985). Self-assembling and photochemical properties of ammonium amphiphiles containing naphthyloxy group.

45. N. J. Turro and P. L. Kuo, *Langmuir* **1,** 170 (1985). Fluorescence probes for aqueous solutions of nonionic micelles.

46. E. A. Lissi and E. Abuin, *J. Colloid Interface Sci.* **105,** 1 (1985). Aggregation numbers of SDS micelles formed on poly(ethylene oxide) and poly(vinylpyrrolidone) chains.

47. F. Grieser, M. Lay, and P. J. Thistlethwaite, *J. Phys. Chem.* **89,** 2065 (1985). Excited-state torsional relaxation in 1,1'-dihexyl-3,3,3',3'-tetramethylindocarbocyanine iodide: application to the probing of micelle structure.

48. J.-P. Chauvet, M. Bazin and R. Santus, *Photochem. Photobiol.* **41,** 83 (1985). On the triplet–triplet energy transfer from chlorophyll to carotene in Triton X-100 micelles.

49. J. Kuczynski and J. K. Thomas, *Langmuir* **1**, 158 (1985). Photochemical behavior of CTAB-stabilized CdS. Effect of surface charge on electron transfer across the colloid–water interface.
50. A. Plonka and L. Kevan, *J. Phys. Chem.* **89**, 2087 (1985). Effect of micellar interface modifications on the reactivity of embedded photoproduced cations.
51. A. Plonka and L. Kevan, *J. Chem. Phys.* **82**, 4322 (1985). Temperature dependence of nonexponential decay of photoproduced $NNN'N'$-tetramethylbenzidine cation radicals in SDS micelles: determination of activation energy distributions and relation to dynamic cooperativity.

Chapter 5—Inverse Micelles and Microemulsions

1. R. A. MacKay and M. Grätzel, *Ber. Bunsenges. Phys. Chem.* **89**, 526 (1985). Photoreduction of thionine and surfactant thionine by Fe(II) ions in anionic micelles and microemulsions.
2. P. Brochette and M.-P. Pileni, *Nouv. J. Chim.* **9**, 551 (1985). Photoelectron transfer in reversed micelles: 3: influence of sensitizer location on reactivity.
3. E. Bardez, E. Monnier, and B. Valeur, *J. Phys. Chem.* **89**, 5031 (1985). Dynamics of excited-state reactions in reversed micelles: 2: proton transfer reaction involving various fluorescence probes according to their sites of solubilization.
4. R. Zana, P. Lianos, and J. Lang, *J. Phys. Chem.* **89**, 41 (1985). Fluorescence probe studies of interactions between poly(oxyethylene) and surfactant micelles and microemulsion droplets in aqueous solutions.
5. M. Sanchez-Rubio, M. Santos-Vidals, D. S. Rushforth and J. E. Puig, *J. Phys. Chem.* **89**, 411 (1985). Conductance and fluorescent probe studies of sodium dodecyl sulfate/n-butyl alcohol/toluene/brine microemulsions.
6. S. M. de B. Costa, M. R. Aires de Barros, and J. P. Conde, *J. Photochem.* **28**, 153 (1985). Porphyrin-quinone excited-state interactions in reversed micelles.
7. I. Willner, Z. Goren, D. Mandler, D. Maidan, and Y. Degani, *J. Photochem.* **28**, 215 (1985). Transformation of single electron transfer photoproducts into multielectron charge relays: The functions of water–oil two-phase systems and enzyme catalysis.
8. M. P. Pileni, B. Lerebours, P. Brochette, and Y. Chevalier, *J. Photochem.* **28**, 273 (1985). Comparison of charge separation in direct and reverse micelles.
9. C. K. Grätzel, M. Jirousek, and M. Grätzel, *Colloids Surf.* **13**, 221 (1985). Photoinduced oxidation of *N*-vinylcarbazole in a functionalized oil-in-water microemulsion.
10. M. J. Politi, O. Brandt, and J. H. Fendler, *J. Phys. Chem.* **89**, 2345 (1985). Ground- and excited-state proton transfers in reversed micelles. Polarity restrictions and isotope effects.
11. S. Atik, J. Kuczynski, B. H. Milosavijevic, K. Chandrasekaran, and J. K. Thomas, *ACS Symp. Ser.* **272**, 303 (1985). Photochemical reactions in microemulsions and allied systems.

Chapter 6—Lipid, Surfactant Vesicles and Liposomes

1. T. Matsuo, *J. Photochem.* **29**, 41 (1985). Role of organized molecular assemblies in artificial photosynthesis.
2. I. Tabushi, M. Kinnaid, and I. Kugiyama, *J. Photochem.* **29**, 217 (1985). Artificial photosynthesis: liposome charge separation aided by phase transfer of electron carrier.
3. C. D. Stubbs, S. R. Meech, A. G. Lee, and D. Phillips, *Biochim. Biophys. Acta* **815**, 351 (1985). Solvent relaxation in lipid bilayers with dansyl probes.
4. T. Kunitake, M. Shiomomura, Y. Hashiguchi, and T. Kawanaka, *J. Chem. Soc. Chem. Commun.* **833** (1985). Carbazole containing bilayer membranes and efficient energy migration.

5. N. Ohta and L. Kevan, *J. Phys. Chem.* **89**, 2415 (1985). Photoionization thresholds of chlorophyll a and TMB in vesicle and micelle frozen solution.

6. J. H. Fendler, *Isr. J. Chem.* **253**, (1985). Polymerized surfactant vesicles.

7. D. Daems, M. Van den Zegel, N. Boens, and F. C. DeSchryver, *Eur. J. Biophys.* **12**, 97 (1985). Fluorescence decay of pyrene in small and large unilamellar LDPC vesicles below and above T_c.

8. D. A. Barrow and B. R. Lentz, *Biophys. J.* **48**, 221 (1985). Membrane structure domains: resolution limits using DPH fluorescence decay.

9. L. Davenport, R. E. Dale, R. H. Bisby, and R. B. Cundall, *Biochemistry* **24**, 4097 (1985). Tranverse location of DPH in model lipid bilayer membrane system by resonance excitation energy transfer.

10. V. E. Maier and V. Ya. Shafirovich, *J. Chem. Soc., Chem. Commun.*, 1063 (1985). Light induced H_2 evolution in oxidizing media promoted by cationic sites encapsulated by phospholipid membranes.

11. Y. M. Tricot, D. N. Furlong, A. W. H. Mau, and W. H. F. Sasse, *Aust. J. Chem.* **38**, 527 (1985). DHP vesicles: differences in the spectral properties of $Ru(bpy)_3^{2+}$ adsorbed on inner and outer surfaces.

12. C. Lambert, M. A. J. Rodgers, G. Jori, E. Reddi, and J. D. Spikes, *NATO Adv. Study Inst. Ser.* **A 85**, 325 (1985). Photophysics of some liposome bound porphyrins.

13. B. R. Suddaby, P. E. Brown, J. C. Russell, and D. G. Whitten, *J. Am. Chem. Soc.* **107**, 5609 (1985). Surfactant and hydrophobic derivatives of *t*-stilbene as probes for vesicle, micelle solubilization sites: studies using fluorescence and photoisomerization as probes.

14. W. D. Turley and H. W. Offen, *J. Phys. Chem.* **89**, 3962 (1985). Fluorescence detection of gel–gel phase transition in DMPC vesicles at high pressure.

15. B. S. Packard and D. E. Wolf, *Biochemistry* **24**, 5176 (L985). Fluorescence lifetimes of carbocyanine lipid analogs in phospholipid bilayers.

16. J. K. Yamamoto and R. F. Borch, *Biochemistry* **24**, 3338 (1985). Photoconversion of 7-dehydrocholesterol to vitamin D_3 in synthetic phospholipid bilayers.

17. T. Kunitake, M. Shimomura, Y. Hashiguchi, and T. Kawanaka, *J. Chem. Soc., Chem. Commun.* 833 (1985). Carbazole-containing bilayer membranes and efficient energy migration.

18. N. Ohta and L. Kevan, *J. Phys. Chem.* **89**, 2415 (1985). Photoionization thresholds of chlorophyll a and *NNN'N'*-tetramethylbenzidine in vesicle and micelle frozen solutions.

19. H. H. Paradies, *Colloids Surf.* **13**, 263 (1985). Excited-state proton transfer in dimethyl-*n*-dioctadecylammoniumchloride and dihexadecylphosphate vesicles containing the phyto-hormone indolyl-3-acetic acid.

20. S. Hidaka, E. Matsumoto, and F. Toda, *Bull. Chem. Soc. Jpn.* **58**, 207 (1985). Photoinduced reduction of methylviologen by ascorbate using chlorophyllin in liposome system.

21. P. L. G. Chong and T. E. Thompson, Biophys. J., **47**, 613 (1985). Oxygen quenching of py-renelipid fluorescence in phosphotidylcholine vesicles. A probe for membrane organization.

22. A. M. Kleinfeld and M. F. Lukacovic, *Biochemistry* **24**, 1883 (1985). Energy transfer study of cytochrome b_5 using the anthroyloxy fatty acid membrane probes.

23. R. Homan and M. Eisenberg, *Biochim. Biophys. Acta* **812**, 485 (1985). A fluorescence quenching technique for the measurement of paramagnetic ion concentrations at the membrane–water interface. Intrinsic and X537A-mediated cobalt fluxes across lipid bilayer membranes.

24. R. Rafaeloff, Y.-M. Tricot, F. Nome, and J. H. Fendler, *J. Phys. Chem.* **89**, 533 (1985). Colloidal catalyst-coated semiconductors in surfactant vesicles. In situ generation of Rh-coated CdS particles in DODAC surfactant vesicles and their utilization in photo-sensitized charge separation and hydrogen generation.

Chapter 7—Monolayers, Bilayers, and Liquid Crystalline Solvents

1. R. Subramanian and L. K. Patterson, *J. Am. Chem. Soc.* **107**, 5820 (1985). Effect of molecular organization on the photophysical behavior: excimer kinetics and diffusion of 1-pyrene-decanoic acid in lipid monolayers at N_2-water interface.
2. W. J. Leigh, *J. Am. Chem. Soc.* **107**, 6114 (1985). Organic reactions in liquid crystalline solvents: 3: substituted β-propiophenone as photochemical probes of solute-solvent interaction in liquid crystals.
3. M. L. Agarwal, J. P. Chauvet, and L. K. Patterson, *J. Phys. Chem.* **89**, 2979 (1985). Effects of molecular organization on photophysical behavior: Lifetime and fluorescence of chlorophyll singlets in monolayers of dioleoyl phosphotidylcholine at N_2-water interface.
4. T. M. Liu and D. Mauzerall, *Biophys. J.* **481** (1985). Distributed kinetics of decay of photovoltages at lipid bilayer-water interface.
5. R. Subramanian, L. K. Patterson, and H. Levanon, *Photochem. Photobiol.* **41**, 511 (1985). Photophysics of liquid crystals: fluorescence behavior of chlorophyll a in dodecyl-cyanobiphenyl.
6. H. Hediger and R. Steiger, *J. Colloid Interface Sci.* **103**, 343 (1985). Study of organized monolayer assemblies by photoacoustic spectroscopy.
7. R. T. Arieta, I. C. Arieta, P. M. Pachori, A. E. Popp, and J. S. Huebner, *Photochem. Photobiol.* **42**, 1 (1985). Photoelectric effects from chlorophyll a in bilayer membranes.
8. S. Vaidhyanathan, L. K. Patterson, D. Möbius, and H.-R. Gruniger, *J. Phys. Chem.* **89**, 491 (1985). Molecular architecture in cyanine dye aggregates at the air-water interface. Effect of monolayer composition and organization on fluorescent behavior.
9. R. Subramanian and L. K. Patterson, *J. Phys. Chem.* **89**, 1202 (1985). Effects of molecular organization on photophysical and photochemical behavior. Interactions involving 1-pyrenedodecanoic acid and 4-dodecylaniline at N_2-water interface.
10. A. Ilani, T. M. Liu, and D. Mauzerall, *Biophys. J.* **47**, 679 (1985). The effect of oxygen on the amplitude of photodriven electron transfer across the lipid bilayer-water interface.

Chapter 8—Polymers and Polyelectrolytes

1. J. Rabani and R. E. Sasson, *J. Photochem.* **29**, 7 (1985). Separation of photoredox products by local potential fields.
2. S. Tazuke, N. Kitamura, and Y. Kawanishi, *J. Photochem.* **29**, 123 (1985). Problems of back electron transfer in electron transfer sensitization.
3. T. Nakahira and M. Grätzel, *Makromol. Chem., Rapid Commun.* **6**, 341 (1985). Visible light sensitization of $Pt-TiO_2$ photocatalysied by surface adsorbed poly(4-vinylpyridine) derivatized with $Ru(bpy)_3^{2+}$ complex.
4. A. Altamore, C. Carlini, F. Ciardelli, M. Panatonni, and J. L. Howben, *Macromolecules* **18**, 729 (1985). Optically active polymers containing side chain *t*-stilbene chromophores directly bound to backbone: photochemistry, photophysics of copolymers of *t*-4-vinylstilbene and methacrylate.
5. F. Wilkinson, C. J. Willsher, and R. B. Pritchard, *Eur. Polym. J.* **21**, 333 (1985). Laser flash photolysis of dyed fabrics and polymers: 1: rose bengal as a photosensitizing dye.
6. J. Paczkowski and D. C. Neckers, *Macromolecules* **18**, 1245 (1985). Polymer bound sensitizers for the formation of singlet oxygen: new studies of polymeric derivatives of rose bengal.
7. M. Shini, Y. Hanatani, and M. Tanaka, *J. Macromol. Sci. Chem.* **A 22**, 279 (1985). Interactions between dyes and polyelectrolytes: XV: effect of mixed dimer formation on excitation energy transfer between bound dyes.

8. J. F. Pratte, S. E. Webber, and F. C. DeSchrvyer, *Macromolecules* **18**, 1284 (1985). Intersystem crossing and triplet-state properties of dinaphthyl compounds.

9. R. D. Burkhardt, O. Lee, S. Boileau, and S. Boivin, *Macromolecules* **18**, 1277 (1985). Spectral and triplet photophysical properties of poly (N-vinyl-carbonyl) carbozole and its monomeric analog.

10. J. P. C. Bootsma, G. Challa, A. J. W. C. Visser, and F. Muller, *Polymer* **26**, 951 (1985). Polymer bound flavins: 5: characterization by static and time-resolved fluorescence techniques.

11. J. E. Guillett, W. K. McInnis, and A. E. Redpath, *Can. J. Chem.* **63**, 1333 (1985). Prospects for solar synthesis: II: study of photocyclization of α-(o-to-lyl) acetophenone in solution and in cross linked ethylene–vinylacetate beads.

12. M. A. Winnik, A. M. Sinclair, and G. Beinert, *Can. J. Chem.* **63**, 1300 (1985). An exciplex probe for end-to-end cyclization of polystyrene.

13. S. N. Semerek and C. W. Frank, *Can. J. Chem.* **63**, 1328 (1985). Energy migration in aromatic vinyl polymers: 5: poly (2-vinylnaphthalene) and polystyrene.

14. J. S. Hargreaves and S. E. Webber, *Can. J. Chem.* **63**, 1320 (1985). Photon harvesting polymers: intracoil energy transfer in anthryl and flourescein-tagged polystyrene.

15. F. R. F. Fan, A. Mau, and A. J. Bard, *Chem. Phys. Lett.* **116**, 400 (1985). A chemiluminescent polymer based on poly (9,10-diphenylanthracene).

16. F. W. Wong and R. E. Lowry, *Polymer* **26**, 1046 (1985). Excimer fluorescence technique for the study of polymer segmental mobility: application to pyrene labelled PMMA and PMA in solution.

17. M. A. Winnik, A. M. Sinclair, and G. Beinert, *Macromolecules* **18**, 1517 (1985). Cyclization dynamics of polymers: 17: probe effects on the detection of polymer end-to-end cyclization.

18. A. Mar and M. A. Winnik, *J. Am. Chem. Soc.* **107**, 5376 (1985). End-to-end cyclization dynamics of hydrocarbon chains: temperature effects on an intramolecular phosphorescence quenching reaction in solution.

19. J. E. Guillett, Y. Takahashi, A. R. McIntosh, and J. R. Bolton, *Macromolecules* **18**, 1788 (1985). A new polymer model for active sites in artificial photosynthesis.

20. R. E. Sasson and J. Rabani, *J. Phys. Chem.* **89**, 5500 (1985). A study of the effect of polyelectrolytes on the photochemical system containing a Ru(bpy)$_3^{2+}$ derivative, methyl viologen, and ferrocyanide.

21. Y. Morishima, *Photochim. Photobiol.* **42**, 457 (1985). Synthesis and photoredox properties of polyelectrolytes functionalized with 5-deazaflavin.

22. J. W. Park and Y. H. Paik, *Bull. Korean Chem. Soc.* **6**, 287 (1985). Novel effects of polyelectrolytes on the fluorescence quenching of Ru(bpy)$_3^{2+}$ by methyl viologen and Cu^{2+}.

23. K. Sumi, M. Furue, and S. Nozakura, *J. Polym. Sci., Polym. Chem.* **23**, 3059 (1985). Efficiency of polymer sensitizer in photosensitized reactions of Ru(bpy)$_3^{2+}$-containing polymers.

24. R. Ramaraj, R. Tamilarasan, and P. Natarajan, *J. Chem. Soc., Faraday Trans, I* **81**, 2763 (1985). Luminescence quenching and flash photolysis studies of arylamidomethylthionine copolymers.

25. J. S. Hargreaves and S. E. Webber, *Macromolecules* **18**, 734 (1985). Water-soluble photon-harvesting polymers: Intracoil energy transfer in anthryl- and fluorescein-tagged poly (vinylpyrrolidinone).

26. R. Harrop, P. A. Williams, and J. K. Thomas, *J. Chem. Soc., Chem. Commun.*, 280 (1985). Photophysical studies of a copolymer of acrylic acid and 1-pyreneacrylic acid adsorbed on calcium carbonate.

27. M. A. Winnik, O. Pekcan, and M. D. Croucher, *Can. J. Chem.* **63**, 129 (1985). Phosphorescence of naphthalene labelled colloidal polymer particles. The α-methyl relaxation of one microphase in a multicomponent material.

28. M. N. Szentirmay, N. E. Prieto, and C. R. Martin, *J. Phys. Chem.* **89**, 3017 (1985). Luminescence probe studies of ionomers. 1. Steady-state measurements from Nafion membrane.

29. Y. Moroshima, T. Kobayashi, and S. Nozakura, *J. Phys. Chem.* **89**, 4081 (1985). Kinetic analysis of anamolous behavior in the fluorescence quenching of phenanthryl groups covalently linked to polyelectrolytes.

30. D. Y. Chu and J. K. Thomas, *J. Phys. Chem.* **89**, 4065 (1985). Effect of conformation of poly (methacrylic acid) on the photophysical and photochemical processes of $Ru(bpy)_3^{2+}$.

31. F. R. F. Fan, H.-Y. Liu, and A. J. Bard, *J. Phys. Chem.* **89**, 4418 (1985). Integrated chemical systems. Photocatalysis at TiO_2 incorporated into Nafion and clay.

32. P. C Lee and D. Meisel, *Photochim. Photobiol.* **41**, 21 (1985). Photophysical studies of pyrene incorporated in Nafion membranes.

33. N. Kakuta, J. M. White, A. Campion, A. J. Bard, M. A. Fox, and S. E. Webber, *J. Phys. Chem.* **89**, 397 (1985). Surface analysis of semiconductor-incorporated polymer systems. 1. Nafion and CdS-Nafion.

34. T. Nishikubo, J. Uchida, E. Takahashi, and T. Iizawa, *Makromol. Chem.* **186**, 1555 (1985). Study of photopolymers. 25. Photochemical reaction of dilute solutions of polymers containing pendant cinnamic ester moieties and photosensitizing groups.

35. A. M. Sinclair and M. A. Winnik, *J. Am. Chem. Soc.* **107**, 5798 (1985). Cyclization dynamics of polymers. 18. Capture radius effects in the end-to-end cyclization rate of polymers. Excimers versus exciplexes.

36. M. N. Szentirmay, N. E. Prieto, and C. R. Martin, *Talanta* **32**, 745 (1985). Luminescence probe studies of ionomers. II. Steady-state measurements from sulphonated polyethylene and teflon membranes.

37. I. F. Pierola, N. J. Turro, and P.-L. Kuo, *Macromolecules* **18**, 508 (1985). Proton transfer reactions of poly (2-vinylpyridine) in the first singlet excited state.

38. F. C. DeSchryver, J. Vandendriessche, K. Demeyer, P. Collart, and N. Boens, *Polym. Photochem.* **6**, 215 (1985). Polymer photophysics. A model approach.

Chapter 9—Crown Ethers, Cyclodextrins, and Zeolites

1. S. L. Suib and K. A. Carrado, *Inorg. Chem.* **24**, 200 (1985). Zeolite photochemistry: energy transfer between rare earth and actinide ions in zeolites.

2. G. Y. Adachi, K. Sorita, K. Kawata, K. Tomokiyo, and J. Shiokawa, *Inorg. Chim. Acta* **109**, 117 (1985). Luminescence of Eu(II) complexes with 18-crown-6 derivatives.

3. S. Scypinski and J. M. Drake, *J. Phys. Chem.* **89**, 2432 (1985). Photophysics of coumarin inclusion complex with cyclodextrin: evidence for normal and inverted complex formation.

4. H. Shizuka, M. Kameta, and T. Shionaki, *J. Am. Chem. Soc.* **107**, 3956 (1985). Excited-state proton transfer reaction of naphthylammonium ions-18-crown-6 complexes.

5. I. R. Gould, P. L. Kuo, and N. J. Turro, *J. Phys. Chem.* **89**, 3030 (1985). Quenching of pyrene fluorescence by Cs ions in micellar system: protection by surface active crown ethers.

6. J. C. Scaiano, H. L. Casal, and J. C. Netto-Ferraira, *ACS Symp. Ser.* **278** (1985). Intrazeolite photochemistry: use of β-phenylpropiophenone and its derivatives as probes of cavity dimensions and mobility.

7. S. Hashimoto and J. K. Thomas, *J. Am. Chem. Soc.* **107**, 4655 (1985). Fluorescence study of pyrene and naphthalene in cyclodextrin–amphiphile complex systems.

8. B. R. Suddaby, R. N. Dominey, Y. Hui, and D. G. Whitten, *Can. J. Chem.* **63**, 1315 (1985). Photochemical reactions in organized assemblies: 40: amylose and carboxymethylamylose inclusion complexes with photoreactive molecules.

9. H. L. Casal and J. C. Scaiano, *Can. J. Chem.* **63**, 1308 (1985). Intrazeolite chemistry: II: evidence for site inhomogeneity from studies of aromatic ketone phosphorescence.

10. M. C. Gonzalez and A. C. Weedon, *Can. J. Chem.* **63**, 602 (1985). Preparation and properties of a linked porphyrin–cyclodextrin.

11. N. J. Turro, C. C. Cheng, and X. G. Lei, *J. Am. Chem. Soc.* **107**, 3739 (1985). Size and selectivity in zeolite chemistry: a remarkable effect of additive on products produced in the photolysis of ketones.

12. J. H. Liu and R. G. Weiss, *J. Photochem.* **30**, 303 (1985). Photohydroxylation, photocyanation, and fluorescence quenching of 2-fluoroanisole complexed with cyclodextrins: a comparison with 4-fluroanisole-CD complexes.

13. K. Kano, H. Matsumoto, S. Hashimoto, M. Sisido, and Y. Imanishi, *J. Am. Chem. Soc.* **107**, 6117 (1985). Chiral pyrene excimer in γ-cyclodextrin cavity.

14. N. J. Turro, X. Lei, C. C. Cheng, D. R. Corbin, and L. Abrams, *J. Am. Chem. Soc.* **107**, 5824 (1985). Photolysis of dibenzylketones in the presence of pentasil zeolites: examples of size, shape selectivity, and molecular diffusional traffic control.

15. T. Yomoto, K. Hayakawa, K. Kawase, H. Yamakita, and H. Toda, *Chem. Lett.* 1021 (1985). Photoisomerization of 7-substituted norbornadiene-cyclodextrin inclusion complex.

16. H. Shizuka, M. Fukushima, T. Fuji, T. Kobayashi, H. Ohtani, and M. Hoshino, *Bull. Chem. Soc. Jpn.* **58**, 2107 (1985). Proton induced quenching of methoxynaphthalene studied by laser flash photolysis and inclusion effect of β-cyclodextrin on the quenching.

17. S. Shinkai, S. Nakamura, M. Nakashima, and O. Manabe, *Bull. Chem. Soc. Jpn.* **58**, 2340 (1985). Photoresponsive crown ethers.

18. J. H. Liu and R. G. Weiss, *Isr. J. Chem.* **25**, 228 (1985). Protection of 4-fluoroanisole from aromatic nucleophilic photosubstitution by cyclodextrins.

19. R. A. Femia and L. J. Cline Love, *J. Colloid Interface Sci.* **108**, 271 (1985). Mixed organized media: effect of micellar-cyclodextrin solutions on the phosphorescence of phenanthrene.

20. N. Sabbatini and V. Balzani, *J. Less Common Met.* **112**, 381 (1985). Photoinduced electron transfer processes involving Eu-cryptates.

21. T. Arakawa, M. Takakuwa, and J. Shiokawa, *Inorg. Chem.* **24**, 3807 (1985). Luminescence study of adsorption of ammonia and other simple molecules on an activated Eu-exchanged mordenite.

Chapter 10—Adsorbates on Silica, Silica Gel, Porous Vycor Glass, and Clays

1. M. Anpo, N. Aikawa, Y. Kubokawa, and M. Che, *J. Phys. Chem.* **89**, 5017 (1985). Photoluminescence and photocatalytic activity of highly dispersed TiO_2 anchored on porous vycor glass.

2. T. Nakamura and J. K. Thomas, *Langmuir* **1**, 568 (1985). Photochemistry of molecules adsorbed in clay systems. Effect of the nature of adsorption on the kinetic description of the reactions.

3. N. J. Turro, I. R. Gould, M. B. Zimmt, and C. C. Cheng, *Chem. Phys. Lett.* **119**, 484 (1985). Ketone photochemistry on solid silica: a diffuse reflectance laser flash photolysis study.

4. V. I. Zemskii, S. V. Libov, I. K. Meshkovskii, and A. V. Sechkarev, *Russ. J. Phys. Chem.* **59**, 92 (1985). Fluorescence spectra and surface relaxation of organic molecules adsorbed on porous glass.

5. R. C. Simon, H. D. Gafney, and D. L. Morse, *Inorg. Chem.* **24**, 2565 (1985). Reactions of $W(CO)_5$ adsorbed onto porous vycor glass with various ligands.

6. P. de Mayo, L. V. Natarajan, and W. R. Ware, *J. Phys. Chem.* **89**, 3526 (1985). Surface

photochemistry: the effect of temperature on the singlet quenching of pyrene adsorbed on silica gel by 2-bromonaphthalene.

7. D. Avnir, R. Busse, M. Ottolenghi, E. Wellner, and K. A. Zachariasse, *J. Phys. Chem.* **89**, 3521 (1985). Fluorescence probes for silica and reversed phase silica surfaces: 1,3-dipyrenyl-propanes and pyrene.

8. Y. Liang and A. M. Ponte Goncalves, *J. Phys. Chem.* **89**, 3290 (1985). Time-resolved measurements of fluorescence of rhodamine B on semiconductor and glass surfaces.

9. T. Vo-Dinh, G. W. Suter, A. J. Kallir, and U. P. Wild, *J. Phys. Chem.* **89**, 3026 (1985). Phosphorescence line narrowing of coronene and phenanthrene adsorbed on cellulose substrates.

10. T. Kennelly, H. D. Gafney, and M. Braun, *J. Am. Chem. Soc.* **107**, 4431 (1985). Photoinduced disproportionation of $Ru(bpy)_3^{2+}$ on porous vycor glass.

11. G. Villemure, H. Kodama, and C. Detellier, *Can. J. Chem.* **63**, 1139 (1985). Photoreduction of water by visible light in the presence of montmorillonite.

12. J. Kuczynski and J. K. Thomas, *J. Phys. Chem.* **89**, 2720 (1985). Photophysical properties of CdS deposited in porous vycor glass.

13. D. N. Furlong, J. W. Loder, A. W. H. Mau, and W. H. F. Sasse, *Aust. J. Chem.* **38**, 363 (1985). Interfacial charge separation in the photoreduction of water.

14. J. K. Thomas and J. Wheeler, *J. Photochem.* **28**, 285 (1985). Photochemical reactions carried out within a silica gel matrix.

15. J. Stahlberg and M. Almgren, *Anal. Chem.* **57**, 817 (1985). Polarity of chemically modified silica surfaces and its dependence on mobile phase composition by fluorescence spectrometry.

16. N. J. Turro, C. C. Cheng, P. Wan, C. Chung, and W. Mahler, *J. Phys. Chem.* **89**, 1567 (1985). Magnetic isotope effects in the photolysis of dibenzylketones on porous silica.

17. S. L. Suib and K. A. Carrado, *Inorg. Chem.* **24**, 863 (1985). Uranyl clay photocatalysts.

18. B. H. Milosavijevic and J. K. Thomas, *J. Chem. Soc., Faraday Trans I*, **81**, 735 (1985). Photochemistry of compounds adsorbed on cellulose: 4: diffusion controlled mechanism of $Ru(bpy)_3^{2+}$ luminescence quenching by copper (II).

19. S. Wei, S. Wolfgang, T. C. Strekas, and H. D. Gafney, *J. Phys. Chem.* **89**, 974 (1985). Spectral and photophysical properties of $Ru(bpy)_3^{2+}$ adsorbed onto porous vycor glass.

20. B. H. Milosavijevic and J. K. Thomas, *Chem. Phys. Lett.* **114**, 133 (1985). Photochemistry of compounds adsorbed into cellulose: effect of environment on photoinduced electron transfer in constringent media.

21. D. Avnir, V. R. Kaufman, and R. Reisfeld, *J. Non-Cryst. Solids* **74**, 395 (1985). Organic fluorescence dyes trapped in silica and SiO_2/TiO_2 thin films: photophysical, film, and cage properties.

22. J. D. Barnes, D. A. Oduwole, M. A. Trivedi, and B. Wiseall, *Appl. Surf. Sci.* **20**, 249 (1985). On the catalytic reactivity of solid surfaces. III. Stabilization of photolytically produced abnormal methyl radicals on silica surfaces.

Index

371